AAPG REPRINT SERIES NO. 19

# *Ancient Deltas*



Selected Papers Reprinted from

AAPG BULLETIN

Published by The American Association of Petroleum Geologists
Tulsa, Oklahoma, U.S.A., 1976

## CONTENTS



 Published July 1976. Second printing December 1980. Library of Congress Catalog Card No. 76-21972. ISBN 0-89181-006-4.

## Ancient Deltas: Preface

Reservoirs of deltaic origin are very common throughout the geologic column and they contain more hydrocarbons than any other types of sandstone reservoirs. AAPG Reprint Series No. 18 includes six papers on several modern deltas which provide criteria for the recognition of ancient deltaic sediments and also concepts of deltaic sedimentation. The articles in this volume illustrate the application of criteria derived from the study of modern deltas to the interpretation of ancient deltas which range in age from Miocene to Devonian.

The twelve papers included herein vary widely in scope and can be divided into three groups. The first group includes four papers which are concerned with local oil field and outcrop studies. These articles describe the character and distribution of an Oligocene alluvial-deltaic sandstone reservoir in South Texas, a Mississippian distributary-channel sandstone which occurs on the east flank of the Illinois basin in western Indiana, the entrapment of hydrocarbons in Pennsylvanian deltaic sandstones of West Texas, and an example of oil production from a Miocene distributary-channel sandstone in offshore Louisiana.

The second group of papers is based on studies which are more regional in scope; they describe deltaic facies of entire basins or portions of basins. Included in this group are nine papers on the following subjects: the Cretaceous "McNairy Delta" of the Upper Mississippi embayment, the Mississippian Michigan River Delta of the Illinois basin, the Tertiary Niger Delta system of Nigeria, large Eocene deltas of the Texas Gulf Coast, Cretaceous deltaic tar sands of Canada, the Devonian deltaic facies of the central Appalachians, and Pennsylvanian deltaic reservoirs of eastern Oklahoma.

The last paper in this volume presents a detailed stratigraphic analysis of several Pennsylvanian deltas of Oklahoma and also outlines an approach to the problem of prospecting for hydrocarbons in deltaic deposits.

The original suggestion for a compilation of papers on deltas was made by AAPG Managing Editor Gary D. Howell and Editor John W. Shelton. I am very pleased to have been selected to prepare this compilation of papers. The papers were assembled by E. M. Tidwell of the AAPG Editorial Staff. Composition was by Nancy Greeson.

**RUFUS J. LE BLANC**
**Shell Development Company**
**Houston, Texas**
**February 1976**

**Reprinted from:**
BULLETIN OF THE AMERICAN ASSOCIATION OF PETROLEUM GEOLOGISTS
VOL. 38, NO. 1 (JANUARY, 1954), PP. 96-117, 21 FIGS.

# GENESIS OF OLIGOCENE SANDSTONE RESERVOIR, SEELIGSON FIELD, JIM WELLS AND KLEBERG COUNTIES, TEXAS[1]

ROBERT H. NANZ, JR.[2]
Houston, Texas

## ABSTRACT

The character, distribution, and probable genesis of a typical lenticular oil-bearing sand body of the Gulf Coast Tertiary, namely, Zone 19*B*, an Oligocene sand of the Seeligson field, Texas, were determined for the purpose of developing and defining criteria by which such sand bodies can be readily identified and delimited in the subsurface.

The stratigraphic framework was specified by cross sections, fence diagrams, isopach maps, and facies maps, prepared from electric logs and core descriptions, and the sediments were characterized by size, shape, thin-section, X-ray, differential thermal, and chemical analyses.

Principal results are: (1) the isopachal trend of the sand body is nearly normal to the regional depositional strike; (2) the sand grades from medium grain at the base to fine at the top and can be divided into three zones on maximum grain size; (3) a typical specimen is a mixture of (*a*) well sorted sand, the framework size fraction, (*b*) interstitial clay, the tail of the size-distribution curve, and (*c*) interstitial calcite cement; (4) the mineral composition is similar to that of a Recent river sand, but dissimilar to that of beaches, of the same geological province.

The sand body is probably an Oligocene alluvial-plain deposit as indicated by its similarity to such Recent deposits in external form, vertical textural zonation, colors of associated shales, mineralogy, presence of mud fragments, and absence of indigenous fauna.

The often stated opinion that oil-field trends of the Gulf Coast, such as the Frio-Vicksburg, are "shore-line" trends has led many geologists to infer that the reservoir sands of these oil fields are strand-line deposits. The conclusions of the present study emphasize the risk of such deductions. The trends may be indirectly related to shore lines by reason of marine source beds, but many, or possibly most, of the reservoir sands may be of continental origin. The trapping in this zone arises from updip normal faulting and gentle doming of a lenticular alluvial sand body which is oriented in the direction of regional dip.

The greatest value of these findings lies in the extrapolation and extension of this knowledge of local genesis to other parts of the section. In such a fashion a regional depositional and tectonic history can be outlined, centers of deposition for various geologic ages can be delimited, and areas favorable for the presence of permeable beds and oil accumulation can be approached.

## INTRODUCTION

Extensive petroleum accumulations have been found only in or adjacent to sedimentary deposits. Consequently, this rock group is of primary importance to petroleum geologists. Of the total known volume of sedimentary rocks, sandstone and conglomerate compose about one-third; they are exceeded in abundance only by shale, and are the most common type of petroleum reservoir rock. In comparison with their economic and academic importance, however, knowledge of their nature and genesis is surprisingly meager.[3] This is attested by a review

[1] Read before the S.E.P.M. at the Joint Annual Meeting of A.A.P.G., S.E.P.M., and S.E.G. in Houston, Texas, March 23, 1953. Manuscript received, September 8, 1953. Publication No. 39, Exploration and Production Research Division, Shell Development Company.

[2] Shell Development Company. The writer is particularly indebted to R. J. Le Blanc, W. S. Adkins, and F. J. Pettijohn for their continued assistance in the formulation and execution of the investigation. The advice and guidance given by L. R. Newfarmer and S. W. Lowman in the early stages of the endeavor are also greatly appreciated. The enthusiastic cooperation of R. K. Guthrie and J. W. Amyx of the Seeligson Engineering Committee contributed greatly to the success of the project. Grateful acknowledgment is due J. Laverde, B. Gray, J. Welsh, J. Lawson, R. A. Rowland, H. Steinfink, Mrs. D. E. Groetsch, L. Hinton, and Mrs. R. Blackwood for their aid in the analysis of the sediments and presentation of the results. The original manuscript has been improved by the valuable constructive criticism of F. E. Lozo, R. J. Le Blanc, and Mrs. J. Breeding.

[3] A noteworthy exception is the quantitative evaluation by P. D. Krynine (1940).

of the thirty-eight contributions to *Stratigraphic Type Oil Fields*,[4] in which nearly all the authors recognize and stress the need for more reliable basic information on the processes and products of sedimentation.

### OBJECTIVES OF PRESENT INVESTIGATION

Lens- or stringer-type bodies are a common geometrical class of sandstone deposits, and it is the express purpose of this study to determine the character and probable genesis of a representative of this group. The study is aimed at an understanding of the distribution of permeable beds, one of the several major factors bearing on oil accumulation. Specific objectives are to determine: (1) the external form and geographic trend of a specific sand body,[5] (2) the detailed character and spatial variation of rock properties within the sand deposit, and (3) the probable genesis of the sand body, through a comparison with modern deposits of the same geologic province.

This investigation is part of a coordinated effort to characterize typical sediments and their related genetic processes through a two-pronged method of attack based on the premise that studies of Recent and ancient sediments are complementary. Studies of Recent sediments provide a knowledge of the patterns of deposition and the nature of the original material, whereas studies of the ancient sediments indicate the character and geometry of the preserved rock record and furnish data on the nature and rate of post-depositional changes—structural, textural, and compositional. The program is directed toward the development and establishment of criteria by which ancient sand bodies can be readily identified and delimited in the subsurface. In petroleum exploration and exploitation, such criteria are of value in predicting the location of oil-bearing sands in a sedimentary province, and in determining the subsurface trend and probable extent of reservoir sands already encountered by drilling.

### LOCALITY SELECTED FOR STUDY

Sand 19*B* of the Seeligson field in Jim Wells and Kleberg counties, Texas, a lens-type reservoir, was selected as the site for this investigation for the following reasons: (1) it is a geologically young deposit (Oligocene) and has probably retained some indication of its environmental origin; (2) it is so situated geographically that depositional conditions at the time of its formation are probably represented in Recent environments of the province at the present time; (3) it has experienced only mild structural deformation; (4) it is relatively well represented by core samples; (5) the oil field in which it is located is well developed and unitized so that information and samples are available from the various operating companies.

*Geographical and geological setting.*—Sand 19*B* is a part of the Oligocene-Frio strata of Seeligson field, a large oil and gas field in the central part of the Rio

[4] *Amer. Assoc. Petrol. Geol.* Tulsa, 1941.

[5] The term sand is used in this paper for sand and sandstone.

Grande embayment in the Frio-Vicksburg trend of oil fields. The general relationship of this sand to the Cretaceous and Tertiary deposits of this embayment is depicted in Figure 1.

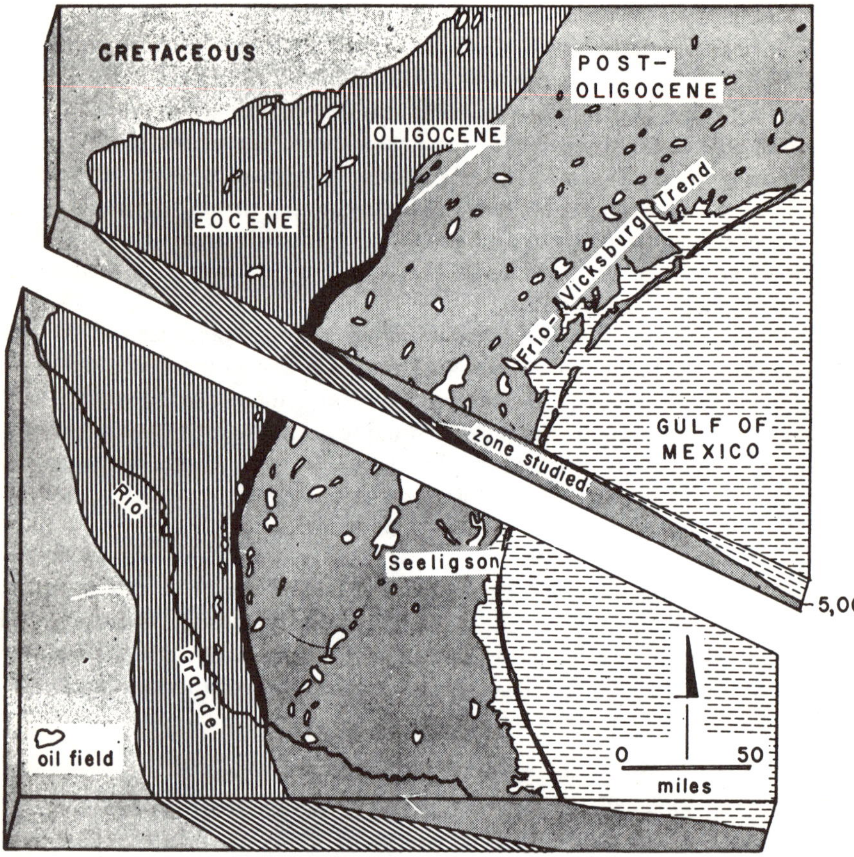


FIG. 1.—Geographical and geological setting of Seeligson field.

*Structural and stratigraphic framework.*—Three domal structures, one major and several minor normal faults, combined with an erratic sand section of more than 40 separate sands, give rise to more than 140 individual reservoirs in the field. The structural configuration of the top of sand 19*B* is shown in Figure 2. The general nature of the associated strata and the position of sand 19*B* are illustrated in Figure 3, a south-north cross section along line *XX′* of Figure 2.

Sand 19*B* is an irregular belt of sand approximately 3 miles wide (north-south), 5 miles long[6] (east-west), and 40 feet in average thickness. It represents about one-ninth of a cubic mile of sediment.

[6] Actual length unknown, but 5 miles lie within area studied.

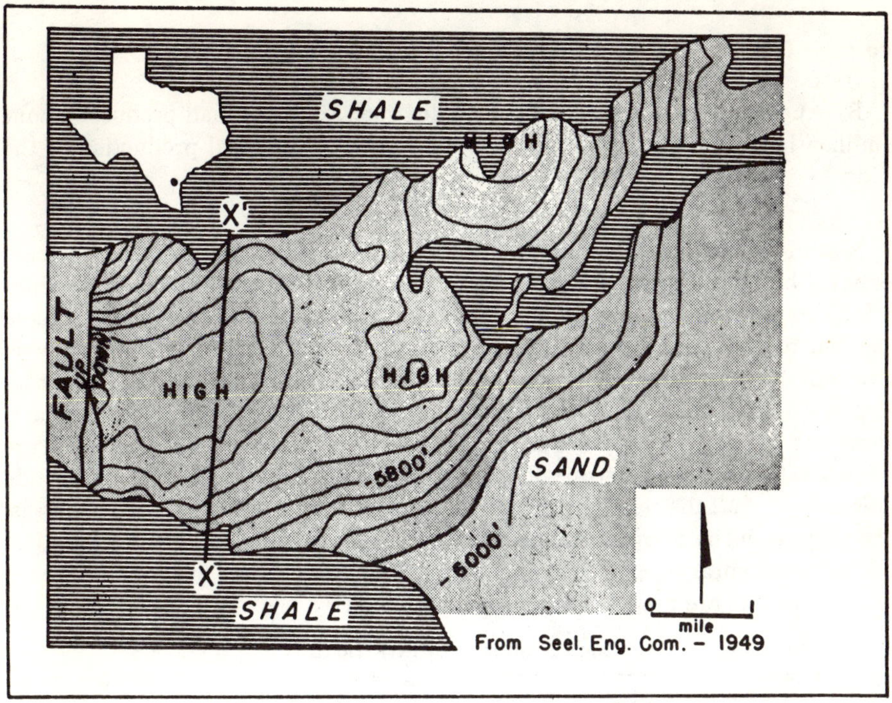


FIG. 2.—Structural configuration at top of sand 19*B*.

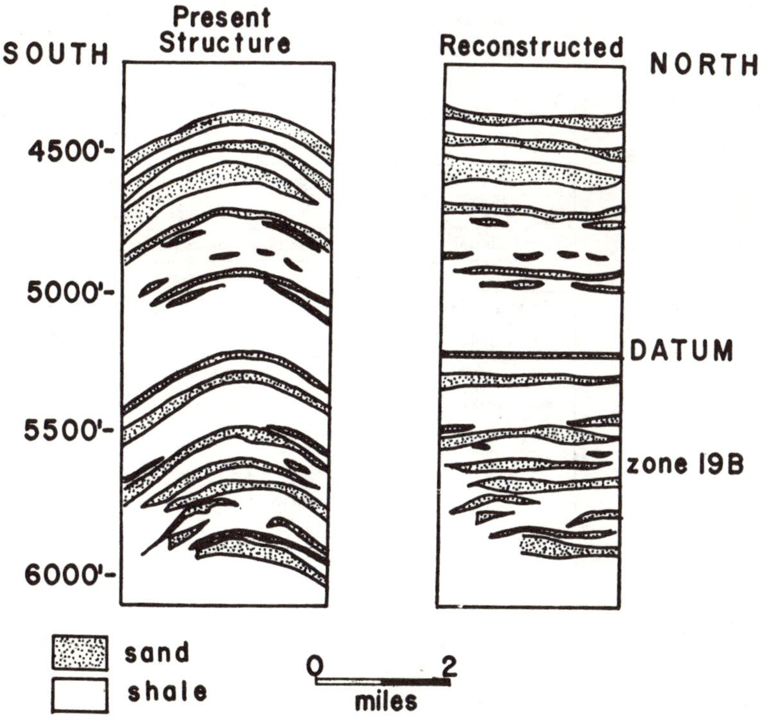


FIG. 3.—Nature of Frio strata in Seeligson field.

By August, 1949, after 6 years of production, Zone 19*B* had produced about 8 million barrels of oil, approximately one-eighth of the total production of the field.

## METHOD OF INVESTIGATION

A sand deposit has two major attributes, *external form* and *sediment character*. As noted by others, sediment character is determined by two fundamental properties, composition and texture. In other words, sand grains of various kinds have been put together in a certain systematic way. Mass properties such as porosity, permeability, and color are determined by, or derived from, the two fundamental properties. This philosophy and the method of attack are outlined in Table I, where the measured properties are classified and the analytical techniques are indicated. Obviously, the scope of this paper can not be broad enough to include a discussion of all the relationships which arise from results of the listed analyses. The purpose of the complete tabulation in Table I is merely to show the practical field of sedimentary petrology, and only the results of the analyses indicated by check-marks are used in the present discussion.

TABLE I. METHOD OF INVESTIGATION

| | *Fundamental Property* | *Components of Property* | *Method of Evaluation* | *Present Study* |
|---|---|---|---|---|
| External form of sand body | Size | Areal extent | Facies map | √ |
| | | | Cross section | √ |
| | | Thickness | Ispoach map | √ |
| | Shape | | Fence diagram | √ |
| Character of sediments | Texture | Grain-size distribution | Mechanical-size analysis | √ |
| | | Grain form | Microscopical analysis | |
| | | Grain orientation | Two-dimensional fabric analysis | |
| | | Cementation | Thin-section microscopical analysis | √ |
| | Composition | Total mineral | Thin-section microscopical analysis | √ |
| | | Clay mineral | X-ray, thermal analyses | √ |
| | | Heavy mineral | Loose-mount microscopical analysis | |
| | | Bulk chemical | Wet chemical analysis | √ |
| | | Trace elements | Spectrochemical analysis | √ |
| | Derived mass properties | Pore-size distribution | Capillary-pressure method | |
| | | Permeability | Air permeability | |
| | | Porosity | Grain density and saturation | |
| | | Color | Core description | √ |

All analyses were made on 2-inch wire-line cores; no cuttings were used.

## RESULTS

The results of the study fall into three groups: (1) external form of the sand body, (2) texture of the sediments, and (3) composition of the sediments. The method of determination and the indicated geological significance of each of these are considered.

EXTERNAL FORM AND TREND OF SAND BODY

*Determination of form and trend.*—Electric logs from more than 500 wells through Zone 19*B* were available for the study, and because of the paucity of diagnostic fossils in this Frio section of sand and shale, lithologic character as inferred from electric logs served as the basis for correlation and three-dimensional mapping. No driller's logs were used.

In order to establish a stratigraphic framework for the later, more detailed considerations of sand 19*B*, numerous cross sections were prepared embracing the strata above and below this sand body. On the premise that shale intervals are more likely to be constant stratigraphic levels than sand intervals, these sections were constructed by comparing and correlating electric-log characteristics in the shale intervals between sand zones. These comparisons were made in detail with the aid of a light table. A typical section prepared in this fashion and based on intrashale datum *A* is shown in Figure 4*a*. Although electric-log cross sections of this type furnish a satisfactory correlation medium, the detail shape of any single sand is not apparent. Fisk (1944) and others have conclusively demonstrated the necessity and desirability of great vertical exaggeration in deciphering the intricate complexities of Recent sediments, and, in view of the objectives of this report, a rapid means was sought to make comparably exaggerated cross sections of the subsurface sands from the electric-log data. If regular logs are used, vertical exaggeration in cross sections is limited by the scales of the individual curves, for if they overlap, the sections become confusing. To overcome this difficulty, positions of self-potential deflections on each electric log were ruled along a single vertical base line to give an abbreviated log (Fig. 4*b*). These logs were then used for construction of vertically exaggerated cross sections such as the section of Figure 4*c*. The advantages of the exaggerated cross section for evaluation of the genetic character of the sands are evident from a comparison of Figures 4*a* and 4*c*.

After establishment of a stratal framework, which specified the relationship of sand 19*B* to surrounding sands, the three-dimensional configuration of the sand was evaluated by the preparation of a fence diagram based on intrashale datum *B* which is directly above the sand (Fig. 5).

Another means of three-dimensional portrayal is an isopach map of total sand thickness (Fig. 6). This map is of greatest value in defining the trend of the deposit, whereas the fence diagram (Fig. 5) is useful for evaluating the effect of shale interbeds, changes in stratigraphic level of the sand, and the nature of associated shales.

*Geological significance of form and trend.*—Of great genetic significance is the fact that the trend of the sand deposit is essentially at right angles to the regional depositional strike. According to present knowledge, Recent sand deposits most

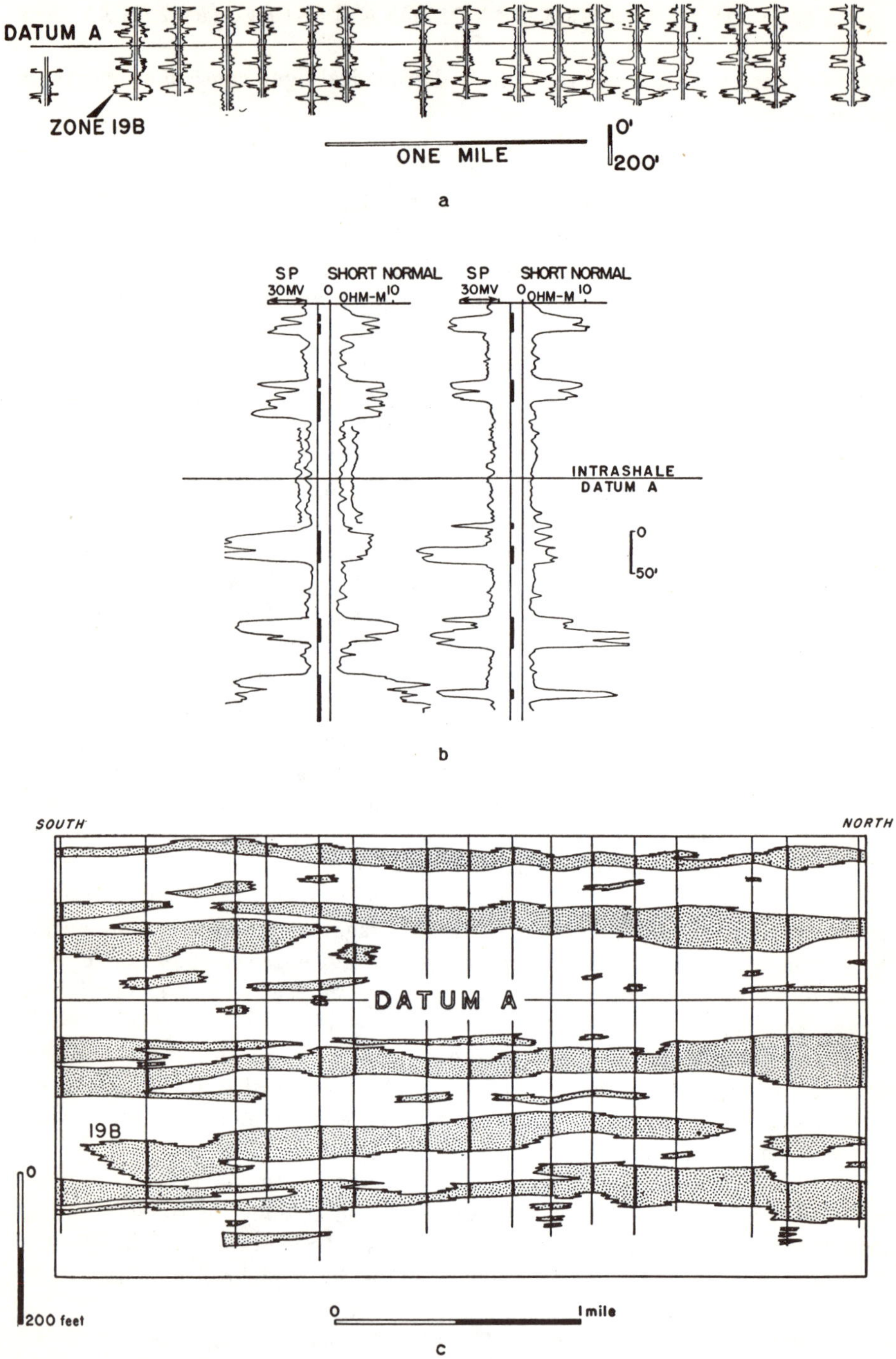


FIG. 4.—Preparation of vertically exaggerated cross sections by use of correlations in shale intervals.

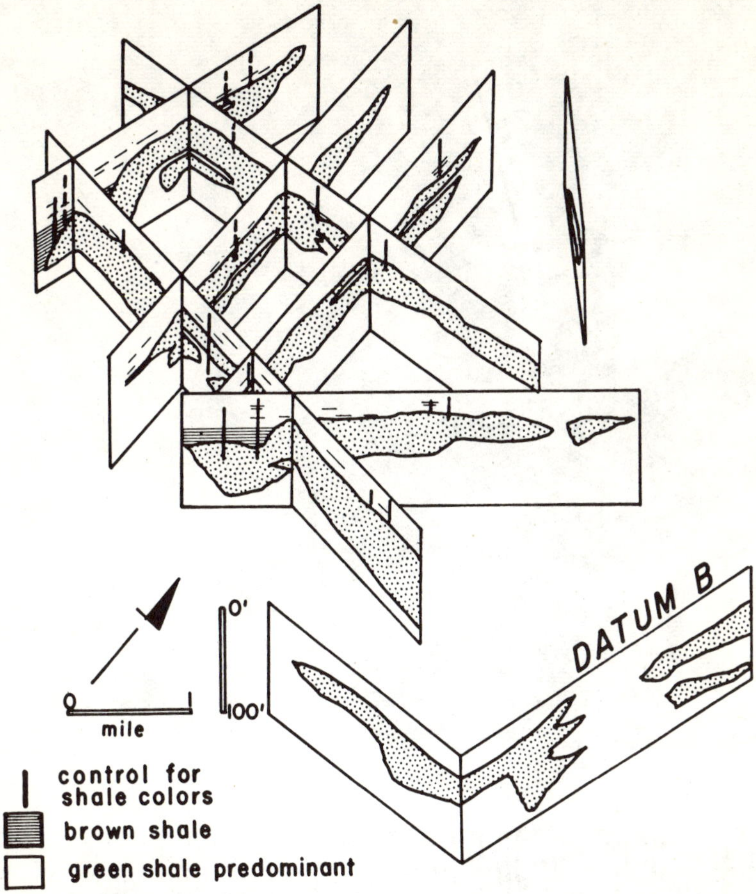


FIG. 5.—Three-dimensional shape of sand 19*B* and nature of associated shales.

similar in distribution to this subsurface sand are deltaic-plain deposits such as those of the Rio Grande in the vicinity of Brownsville, Texas (Fig. 7), especially in the framed area of Figure 7, about 20 miles northwest of Brownsville.

The configuration of the base of the sand was contoured in an attempt to obtain a better understanding of the depositional history of the deposit. This is shown in Figure 8, where arrow I indicates the lowest part of the sand and possibly an early position of the Oligocene river. Arrow II may be a later distributary channel which broke out of the old river system near point *A*. The remainder of the sand could have been formed through the meandering of the river, and the horizontal sand limits now recorded in the subsurface probably represent the boundaries of the meander belt. Fisk (1947) and others have presented abundant factual data to demonstrate the general nature of alluvial-valley and deltaic-plain sedimentation. From their work it is evident that a meandering stream

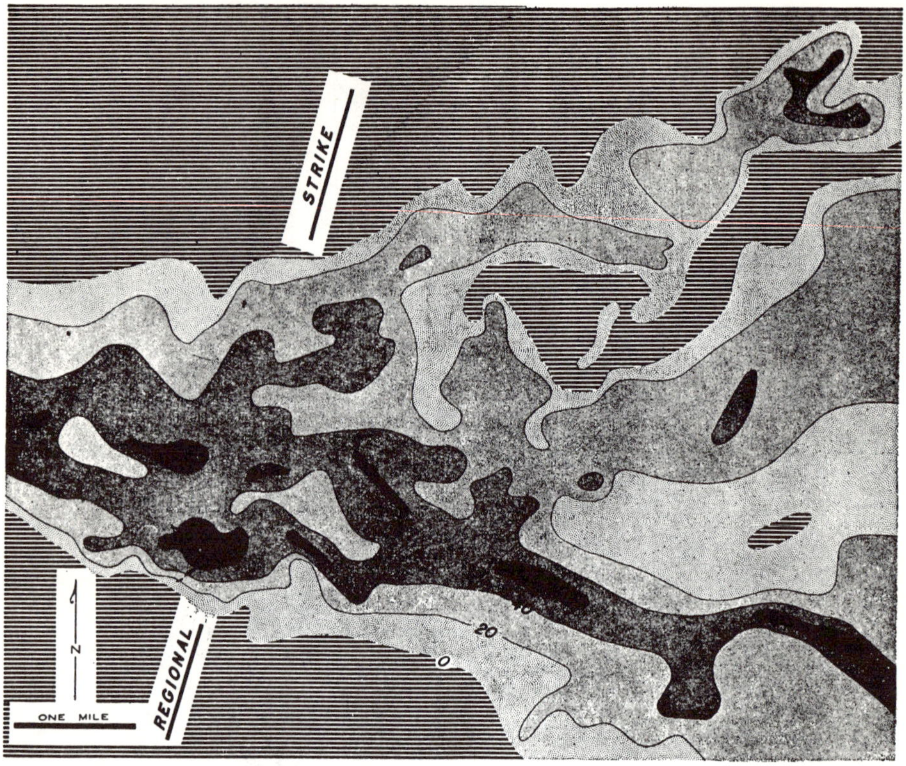


FIG. 6.—Isopach map of total sand thickness.

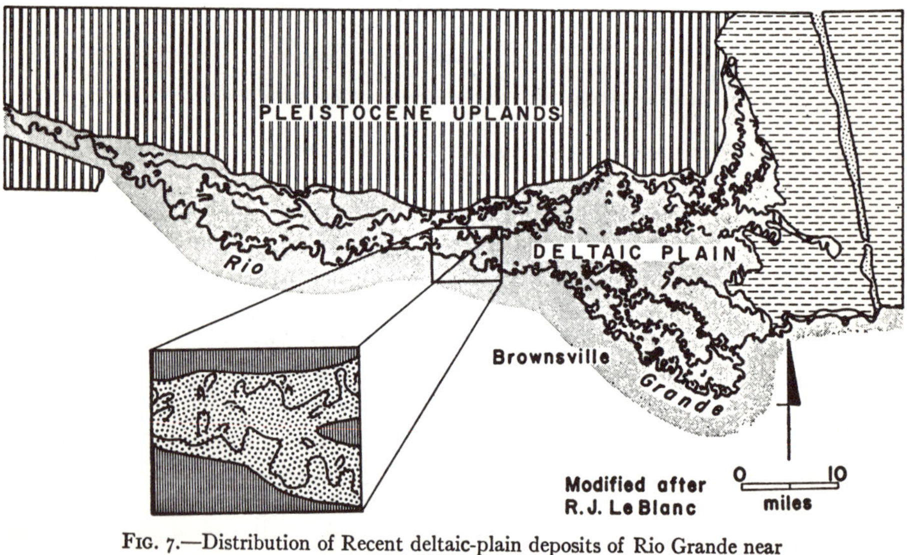


FIG. 7.—Distribution of Recent deltaic-plain deposits of Rio Grande near Brownsville, Texas.

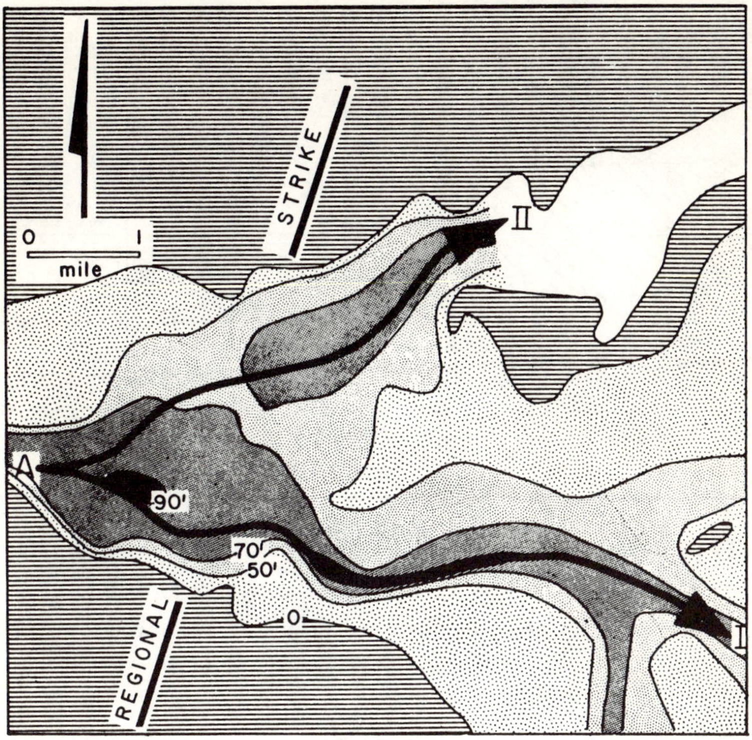


FIG. 8.—Configuration of base of sand body and positions of hypothetical channels.

widens its meander belt by reworking material on the outer side of the channel and depositing sand on the inner or bar side of a meander bend. The result of this mechanism, by which a meandering stream deposits a layer of continental sand adjacent to marine clay which could be the source of hydrocarbons, is shown in an actual example of river migration in Recent time in Figure 9. Continuation of these depositional processes in various Recent alluvial valleys through shifting meander, meander cut-offs, and new distributary channels has given rise to sand distribution patterns which are strikingly similar in form to sand 19*B*.

*Possible significance of shale colors.*—The colors of shales associated with the sand body are indicated in Figure 5. These data were obtained from core descriptions made by several companies. No shale samples were available. In a general sense the shales below the highest stratigraphic level of the sand are either brown, or brown and green, whereas shales above the sand are predominantly green. Recent silts of floodplain origin, such as the deposits between meander belts in Figure 7, are notably oxidized and commonly brown, whereas most lagoonal or marine shales are not brown. The differences in color of the shales indicate a break in the sedimentation near the top of the sand, and this could possibly be due to shifting of the Oligocene river to a new locality with the deposition of lagoonal

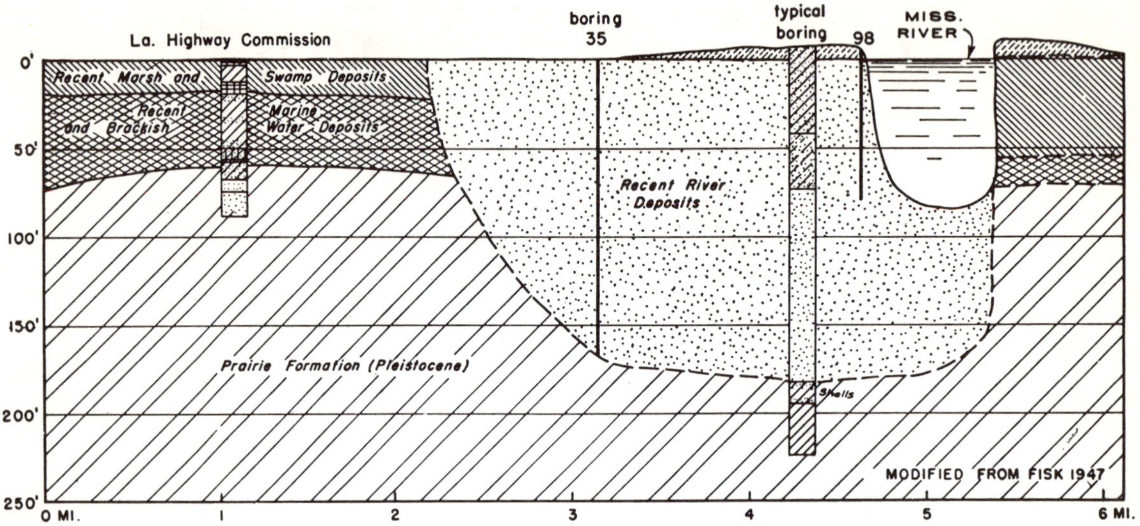


FIG. 9.—Recent deltaic-plain deposits of Mississippi River adjacent to Recent marine sediments in vicinity of New Orleans, Louisiana.

or marine clays over the abandoned river deposits. Strata of the Calcasieu River trench, Cameron Parish, Louisiana, grade upward from alluvial to marine and show similar shale color differences.

## TEXTURE OF SEDIMENTS

The texture of a sedimentary rock is usefully characterized by knowledge of the size, shape, and manner of aggregation of the rock constituents. Of these attributes, the present paper is confined solely to a consideration of the determination and geological significance of grain size.

*Determination of texture.*—Grain-size distributions were determined by conventional sieve and pipette analyses, and the results of the analysis of 53 specimens from sand 19*B* are given in Figure 10. The specimens fall into four textural groups: conglomeratic sand, sand, shaly sand, and silty shale. As seen in Figure 10, most of the samples are well sorted, fine- to medium-grained sand with 5–20 per cent of interstitial silt and clay material. The well sorted sand is the *framework* of the rock and it can be defined by statistical measures of size and uniformity of size. Within this framework are various amounts of *interstitial detritus* (silt and clay). The two end-members of this continuous series, namely, clean sand and sand whose pores are completely filled with silt and clay, are shown in Figure 11. In addition to clay, secondary calcite may occupy part of the original pore space. The generalized texture of the sand is illustrated in Figure 12. The importance of the recognition of these basic textural factors to the geological interpretation of size distributions and to the evaluation of textural and derived mass properties such as porosity and permeability can not be overemphasized. The sediment

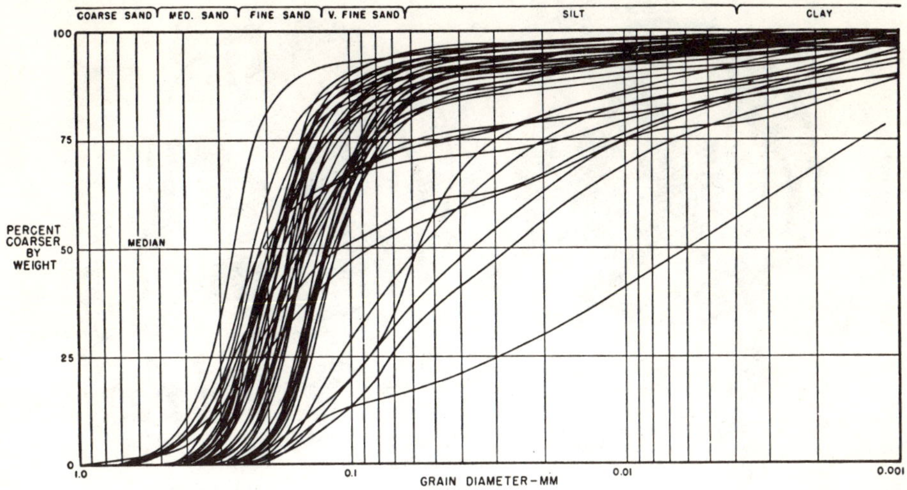


FIG. 10.—Grain-size distributions for 53 samples from sand 19*B*.

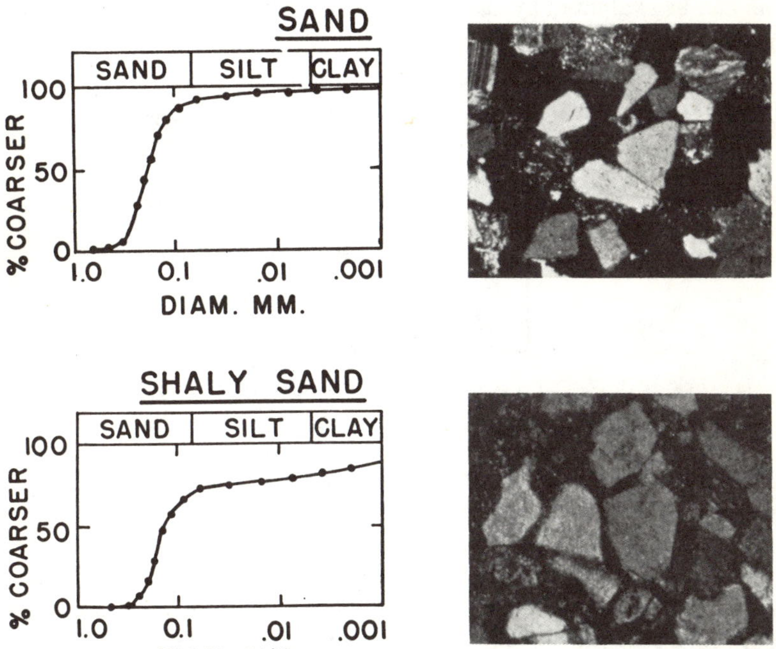


FIG. 11.—Basic textural types.

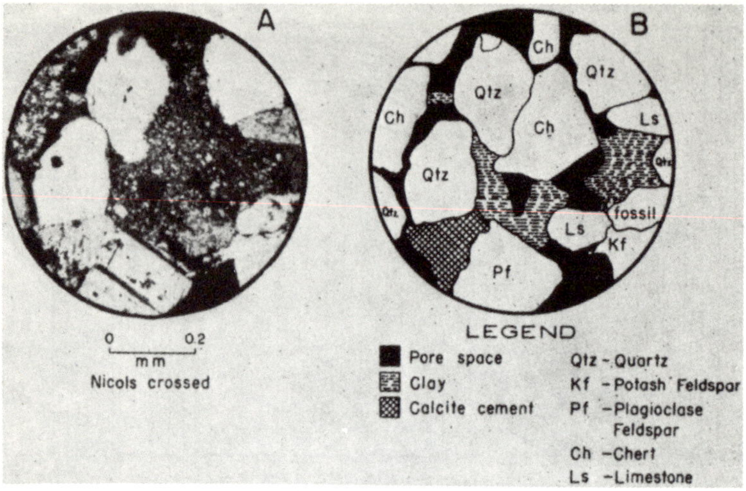


FIG. 12.—Generalized texture of sand deposit.

is not a simple mixture of materials of all sizes in a unimodal size distribution and can not be adequately represented by a median. Instead, it is a mixture of two kinds of materials, *framework fraction* and *interstitial detritus*. The framework fraction can be evaluated by the use of modal or maximum grain sizes as shown in Figure 13.

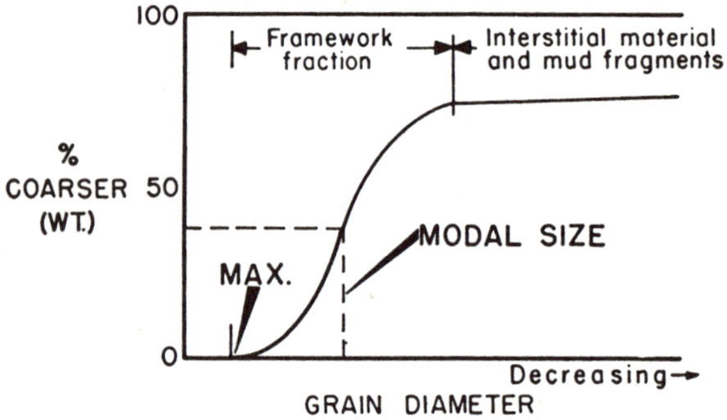


FIG. 13.—Statistical parameters used in present study.

*Geological significance of texture.*—The primary purpose in making the size analyses was to evaluate vertical and lateral grain-size variations in the sand body. A measure of the degree of vertical variation was obtained by analysis of 22 samples distributed throughout the entire thickness (60 feet) of sand 19*B* in a single well. Modal and maximum grain sizes for each of the samples are given in Figure 14. From the modal size it is apparent that the sand is coarsest at the base and in general grades upward to fine sand. This is not graded bedding in the classical sense but is merely a large-scale gradation in grain size. On the basis of maximum size, a quality of the size curve which has received little attention, a

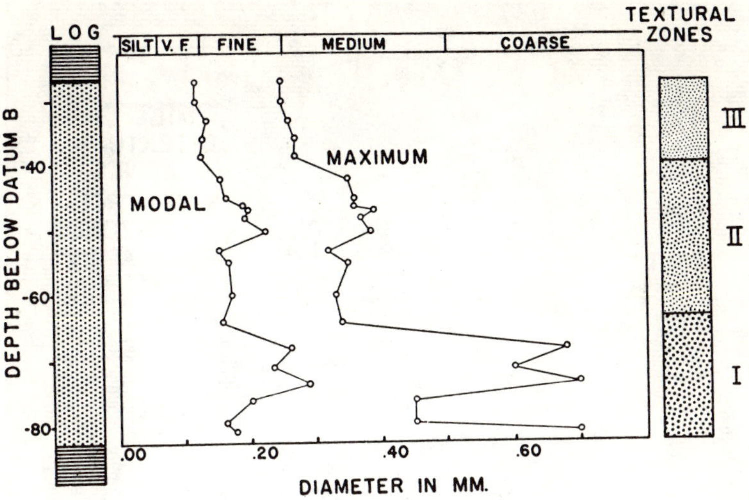


Fig. 14.—Variation of grain size with depth in single well.

three-fold zonation of the sand is suggested. In order to evaluate the areal validity of this zonation, the maximum grain sizes of samples from various wells in the field were compared by using intrashale datum *B*. These data are given in Figure 15. The zonation is apparently valid. No systematic lateral variation in grain size was recognized, but this could be due to inadequately spaced samples.

The relationship of the vertical textural zonation to the form of the sand body is illustrated in Figure 16. The salient features of this diagram are: (1) the trend of the sand body is nearly normal to the regional strike; (2) the sand body can be extended westward across the major normal fault; (3) both the top and base of the sand are irregular; and (4) the coarsest textural subzone (*I*) occupies only the lower part of the sand.

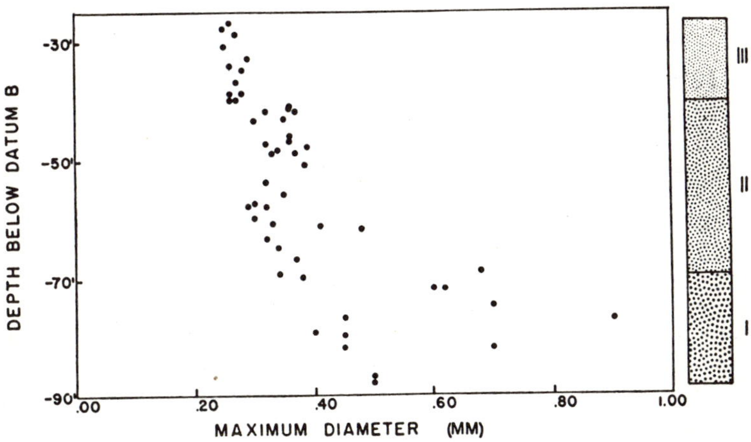


Fig. 15.—Vertical zonation of sand body on basis of maximum size.

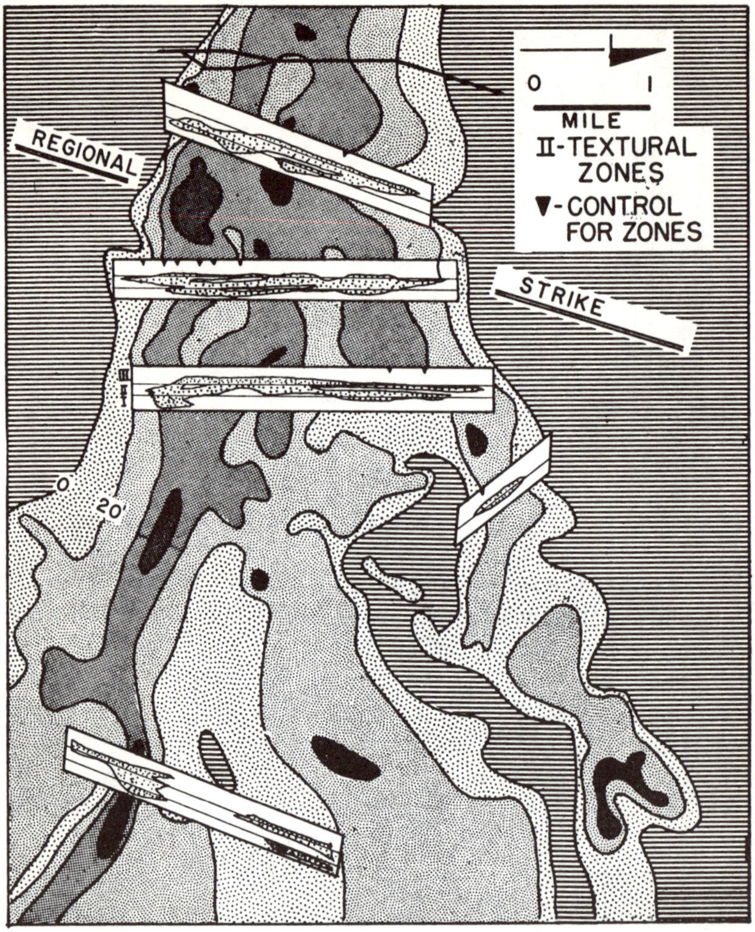


FIG. 16.—Relationship of textural zones to form of sand body.

The relationships displayed in Figure 16 are represented in the deposits of many Recent deltaic plains and alluvial valleys, for example, the Rio Grande as noted previously (Fig. 7), and the alluvial valley of the Sabine River (Fig. 17). A general gradation from coarsest at the base to finest at the top, and an apparent textural zonation, similar to the subsurface case, are strongly suggested in Figure 17. Representative results of investigations by R. J. Le Blanc and the writer of grain-size variations in typical modern sand bodies are given in Figure 18. The relationships depicted, namely, an increase in grain size with depth in the deltaic-plain sands and a relatively constant size for barrier-island sands, have numerous exceptions, but in general the concept appears to be valid. These observations notably support the hypothesis of deltaic-plain origin expressed in the first part of this paper.

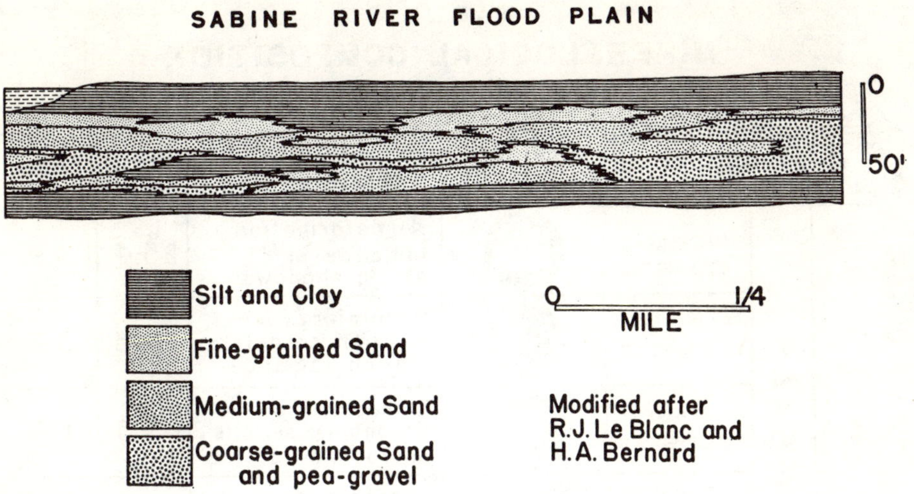


FIG. 17.—Textural zonation in Recent alluvial deposits of Sabine River, Louisiana.

COMPOSITION OF SEDIMENTS

The complexity of sandstone composition has not been generally recognized, and the false notion that most sandstones are simple aggregates of quartz grains is still prevalent. Pettijohn (1949) has recently summarized the factual data on sandstone composition, and, although some aspects of the classification of sandstones are not yet satisfactory, the significant point is that sandstones have diverse compositions which are of value in determining sediment source, tectonic environment, and local depositional environment. The complexity of sandstone composition arises from the fact that sandstones can contain fragments of *all*

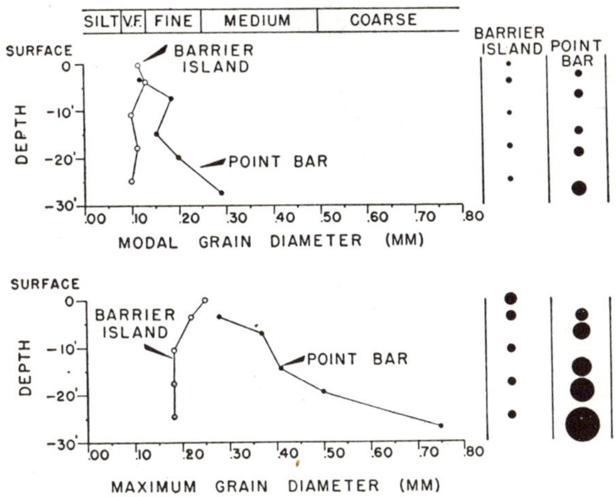


FIG. 18.—Variation of grain size with depth in selected typical Recent sand deposits.

TABLE II.
## MINERALOGICAL COMPOSITION OF THE OLIGOCENE SAND

| Components | | | Typical Specimens | Brief Descriptions | Average of 22 % | Standard Deviation |
|---|---|---|---|---|---|---|
| DETRITALS | MINERALS | Normal Quartz | 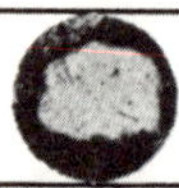 | No deformation lamellae or strain shadows | 25.4 | 4.9 |
| | | Strained Quartz |  | Undulatory extinction or deformation lamellae | 5 5 | 3.3 |
| | | Potash Feldspar | 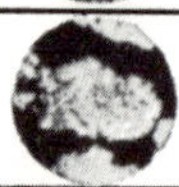 | Orthoclase with secondary sericite. Microcline rare. | 5.3 | 1.8 |
| | | Plagioclase Feldspar | 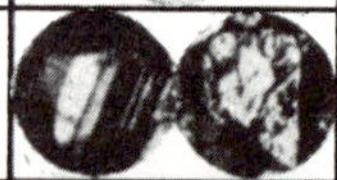 | Labradorite or andesine, alteration common. | 4.7 | 1.9 |
| | | Micas |  | Brown biotite and green glauconite. Muscovite rare. | 0.5 | 0.4 |
| | | Opaques | 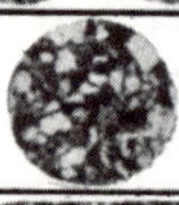 | Authigenic pyrite common, Detrital Magnetite rare. | 0 5 | 0.5 |
| | ROCK FRAGMENTS | Shale |  | Mud fragments com. True shale and silt rare | 2.8 | 5 4 |
| | | Chert | 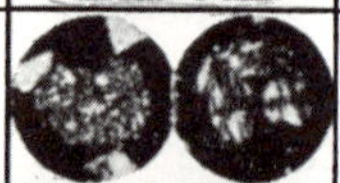 | Aggregate-type common, fibrous rare. | 14.5 | 2 6 |
| | | Limestone and Fossils | 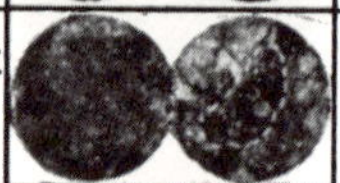 | Limestone, pure or with fossils and glauconite, common. | 6.3 | 2 1 |
| | | Quartzite |  | Sandstone fragments rare. Metamorphic quartzite, rare. | 0.6 | 0.6 |
| | | Volcanic Glass |  | Plain and spherulitic, index less than 1.537, rare. | 0.8 | 0.3 |
| | | Rhyolite or Trachyte |  | Felt-like albite laths, common Skeletal feldspar in glass, rare. | 8.3 | 3.1 |
| SECONDARY OR INTERSTITIAL | | Calcite | 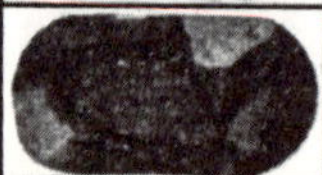 | As overgrowths or pore filling, common. Replacing qtz., rare. | 19.2 | 5.6 |
| | | Clay |  | Illite predominant. Associated with calcite. | 5.5 | 3.4 |

rock types, igneous, metamorphic, and older sedimentary, in addition to organic components, animal shells, and secondary chemical precipitates.

*Determination of composition by thin-section analysis.*—Sandstones have two major components: detrital fragments (including fossils) and secondary cement or interstitial materials. The detrital fraction can be divided into mineral and rock fragments, and the latter can be further divided into rock types. The scheme is illustrated in the first two columns of Table II. Photomicrographs and brief descriptions of the mineral constituents are given in Table II.

Quantitative mineral analyses were made from thin sections of the rocks utilizing the Chayes point-counting mechanical stage (Chayes, 1948). This stage differs from the ordinary in that it employs a notched wheel on the horizontal pinion which allows a continuous spot-sampling of the mineral population along successive traverses. The average mineral composition of 22 samples is given in Table II. Standard deviation values which indicate degrees of variability for each component are also given in Table II. In a normal distribution of values about the average, approximately two-thirds of the values lie within one standard deviation of the average. The high-standard deviation for shale (including mud fragments) is the result of a few very high values.

*Geological significance of mineral composition.*—Unfortunately, analytical data as comprehensive as the figures of Table II are not available for any specific Recent sand body. This is partly because of the fact that, in the past, unconsolidated near-surface sands have not been studied in thin section. Although the studies are incomplete at this time, considerable progress has been made in the characterization of Recent sands of the Gulf Coast. Undisturbed samples of modern sands are obtained by impregnation *in situ* with quick-setting plastic and are then studied in thin section. For comparison with the analytical data on sand 19*B*, the mineral compositions of typical Recent sand deposits of the Rio Grande embayment are given in Figure 19. The similarity of the subsurface sand to the Recent deltaic-plain sand, and its dissimilarity to the Recent beach and dune sands of the area are readily apparent. This comparison supports the hypothesis that sand 19*B* is a deltaic-plain deposit.

The concept of mineralogical maturity, recently stressed by Pettijohn (1949) and Krynine (1941), is strongly supported by these data on Recent sands (Fig. 19). The beach and dune sands are more mature because of their origin through reworking of alluvial deposits and concentration of the more resistant minerals, especially quartz. If it is recognized that a separate relative standard of maturity may be characteristic of each sedimentary basin, and if the effect of sediment contributions from outside the province can be evaluated, the concept of mineralogical maturity is of great value in locating and delineating lenticular sands.

Possible concern over the presence of glauconite, fossils, and limestone fragments in the Oligocene deltaic-plain deposits is dispelled by the knowledge that these constituents are present in the Recent Rio Grande sample attesting their detrital origin from older rocks. The limestone grain shown in Table II, which

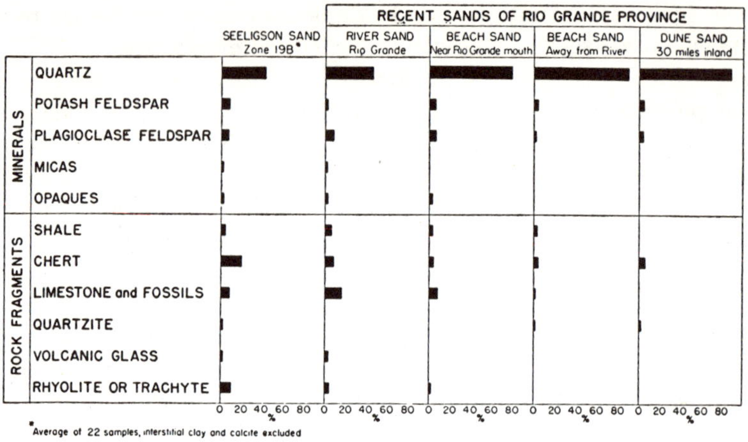


FIG. 19.—Comparison of mineral character of Oligocene sand with Recent sands of same geologic province.

contains a fossil and glauconite, is further evidence of the transported character of these constituents.

*Chemical composition.*—The average chemical composition of ten sand samples is given in Table III. For the objectives of the present paper, these results have no particular interpretational value. They are presented merely to further characterize the sand.

*Clay-mineral composition.*—The combined results of investigations of the mineralogical composition of the clay-size fractions of the sands are given in Table

TABLE III. CHEMICAL COMPOSITION OF ZONE 19*B* SANDS

| *Unit Determined* | *Average of Ten Analyses (Percentage by Weight)* |
|---|---|
| $SiO_2$ | 65.0 |
| $Al_2O_3$ | 9.57 |
| $Fe_2O_3$ | 1.59 |
| FeO | 1.08 |
| MgO | 0.4 |
| CaO | 10.1 |
| $Na_2O$ | 2.14 |
| $K_2O$ | 1.43 |
| $H_2O$ (>105° C.) | 0.82 |
| $H_2O$ (<105° C.) | 0.23 |
| $CO_2$ | 6.9 |
| C | 0.06 |
| $SO_3$ | 0.04 |
| S | 0.16 |
| Cu | 0.002 |
| V | 0.017 |
| Zn | <0.03 |
| Cr | 0.003 |
| Total | 99.572 |

TABLE IV. AVERAGE COMPOSITION OF CLAY-SIZE MATERIALS

| | *Interstitial Detritus* | *Mud Fragment* |
|---|---|---|
| Illite | Abundant | Abundant |
| Montmorillonite | Common (poorly organized) | Common (poorly organized) |
| Kaolinite | Very rare | Very rare |
| Chlorite | Rare | Rare |
| Quartz | Common | Common |
| Feldspar | Common | Common |
| Organic Matter | Common | Common |
| Carbonates | Abundant | Abundant |

IV. The predominant clay mineral is illite but no specific interpretation of this fact can be made from the present knowledge of the provenance of clay materials.

*Geological significance of mud fragments.*—Fragments of soft shale (or mud) are common in Zone 19*B*. They are also common on modern sand bars of the Rio Grande (Fig. 20). Because of their similarity to the Recent mud fragments, fragments from sand 19*B* are interpreted as suspension material which was temporarily deposited in quiet stretches of the river on or near sand bars, was partially lithified, and was then incorporated into the sand deposit. Additional support for this hypothesis is the fact that the mud fragments and the interstitial clay have the same clay-mineral composition (Table IV).

*Source of sediments.*—The nature of the source areas which contributed sediments to sand 19*B* is shown in Figure 21. The figures indicate relative abundances in per cent. Grains derived from older rocks, together with mud fragments derived from suspension material of this hypothetical river, formed the bed load of the stream. To this sand fraction were later added interstitial clay, probably

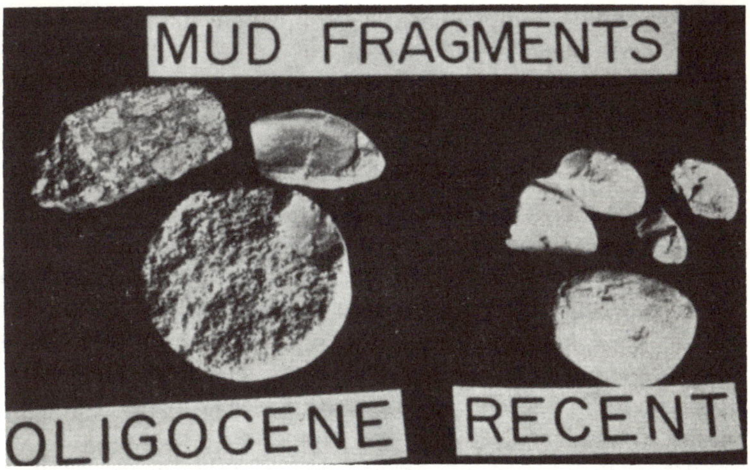


FIG. 20.—Comparison of mud balls from Oligocene sand with mud balls from Rio Grande near Brownsville, Texas.

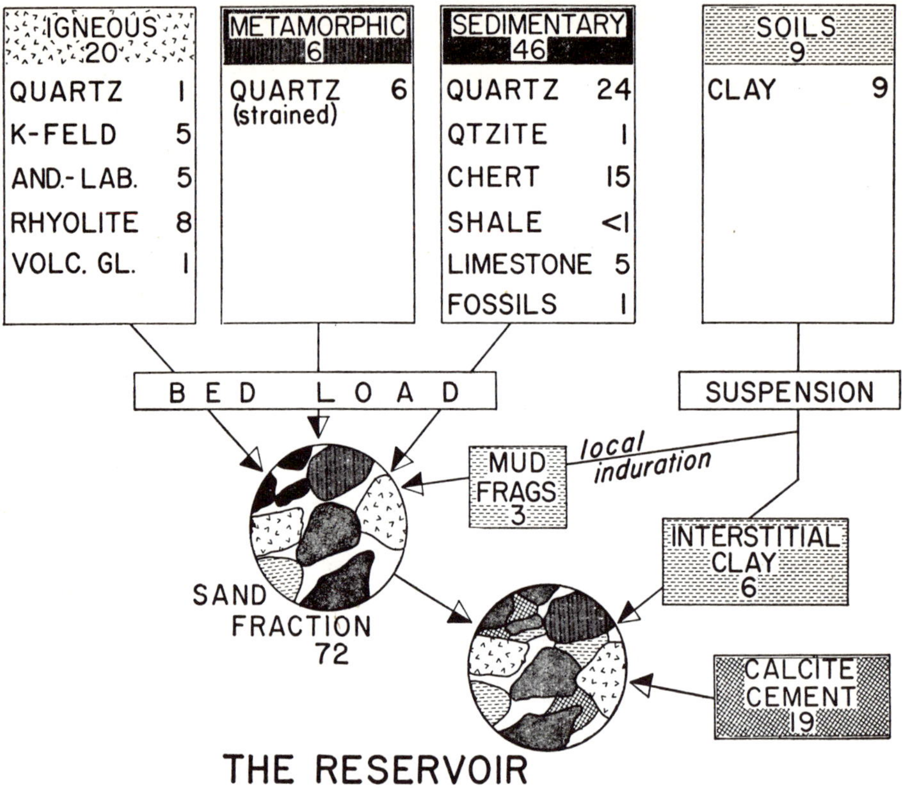


FIG. 21.—Source of materials and hypothetical mode of origin of Oligocene sand.

originating from soil erosion, and secondary calcite of unknown origin. Oil and water occupy the remaining pore spaces. These data, coupled with a detailed knowledge of the character of the older strata of the area, are useful in evaluating regional conditions at the time of deposition, and in detecting unconformities.

## SUMMARY AND CONCLUSIONS

The primary objectives of this study were to determine the detailed character and probable genesis of sand 19*B*, a typical lens-type petroleum reservoir. The study was designed to provide knowledge on the distribution of permeable beds, one of the several major factors bearing on oil accumulation. To the extent that sample distribution allowed, and present knowledge of Recent sedimentation permitted, the objectives were essentially obtained. The study demonstrates that ancient deposits can be closely compared with Recent sediments through the use of suitably exaggerated cross sections and detailed analytical work.

The major results of the study are as follows: (1) the trend of sand 19*B* is essentially at right angles to the regional depositional strike; (2) on the basis of maximum grain size, the sand body has three textural zones, with the coarsest zone forming the lower part of the sand body and the finest zone lying at the top; (3) the shales stratigraphically equivalent to the sand body are generally brown or brown and green, whereas the shales above the sand are generally green; (4) the mineralogy of the sand is complex and reflects abrasional immaturity; (5) mud fragments are common constituents of the deposit. In all these characteristics, and in the absence of fossils, sand 19*B* is similar to Recent deltaic-plain deposits of its own geologic province, and is dissimilar to any Recent beach, dune, or marine sands of the area examined to date. Since it is reasonable to assume that sedimentation in the Rio Grande embayment in Oligocene time was similar to Recent deposition in that area, the data presented in this paper point overwhelmingly to the conclusion that sand 19*B* represents the normal deltaic-plain depositional product of an Oligocene river, probably an ancestral Rio Grande. No special assumptions about a source for the oil are required. The facility with which oil can migrate from marine shales into deltaic-plain deposits is illustrated in Figure 9.

The often stated opinion that oil-field trends of the Gulf Coast, such as the Frio-Vicksburg, are "shore-line" trends has led many geologists to infer that the reservoir sands of these oil fields are strand-line deposits. The conclusions of the present study emphasize the risk of such deductions. The trends may be indirectly related to shore lines by reason of marine source beds, but many, or possibly most, of the reservoir sands are probably of continental origin. The trapping in the sand studied arises from normal faulting and gentle doming of a lenticular alluvial sand body which is oriented in the direction of regional dip.

The greatest value of these findings lies in the extrapolation and extension of this knowledge of local genesis to other parts of the stratigraphic section. In such a fashion a regional depositional and tectonic history can be outlined, centers of deposition for various geologic ages can be delimited, and areas favorable for the presence of permeable beds and oil accumulation can be approached.

## REFERENCES

Chayes, F., 1948, "A Simple Point Counter for Thin-Section Analysis," *Amer. Min.*, Vol. 34, Nos. 1 and 2, pp. 1–12.

Fisk, H. N., 1944, "Geological Investigation of the Alluvial Valley of the Lower Mississippi River," *Mississippi River Com., Corps of Engineers, U. S. Army*, Vicksburg, Miss.

———, 1947, "Fine-Grained Alluvial Deposits and Their Effects on Mississippi River Activity," *ibid.*

Krynine, P. D., 1940, "Petrology and Genesis of the Third Bradford Sand," *Pennsylvania State College Bull. 29.* 126 pp.

———, 1941, "Differentiation of Sediments During the Life History of a Landmass" (abstract), *Bull. Geol. Soc. America*, Vol. 52, p. 1915; and various other publications.

Lalicker, C. G., 1949, *Principles of Petroleum Geology*, p. 102. Appleton-Century-Crofts, New York.

Pettijohn, F. J., 1949, *Sedimentary Rocks.* Harper and Brothers, New York.

**Reprinted from:**
BULLETIN OF THE AMERICAN ASSOCIATION OF PETROLEUM GEOLOGISTS
VOL. 53, NO. 10 (OCTOBER, 1969), PP. 2150-2160, 14 FIGS., 3 TABLES

# Lower West Baden (Mississippian) Sandstone Body of Owen and Greene Counties, Indiana[1]

S. V. HRABAR[2] and PAUL EDWIN POTTER[3]
Cincinnati, Ohio 45221, and Bloomington, Indiana 47401

**Abstract** The elongate, lenticular, multistory sandstone body of the West Baden Group (Mississippian) that crops out in Owen and Greene Counties, Indiana, is an abandoned distributary finger of the late Paleozoic Michigan River system, a large river system that carried mud and sand from northeastern United States and Canada southwestward across the Illinois basin toward the Ouachita trough.

The sandstone body has two major parts: a lower, thin body probably of tidal origin, and an upper, much thicker body. Cross-bedding and ripple-bedding are the two principal bedding facies and define the reservoir units. The cross-bedded facies is very permeable, whereas the ripple-bedded facies has low permeability and inhibits vertical flow. Groundwater, flowing downward into the underlying limestone, produced permeability values near the surface that greatly exceed those of the subsurface.

A hypothetical example illustrates how permeability in a sandstone body in the subsurface, buried below an unconformity, might develop.

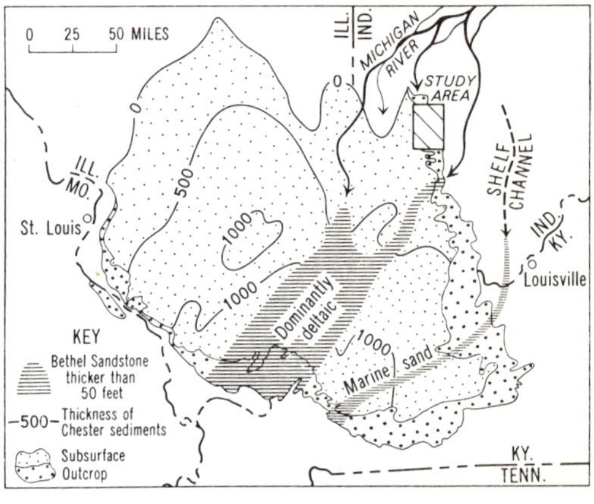


FIG. 1.—Distribution and thickness of Chesteran strata and Bethel Sandstone in Illinois basin (modified from Potter *et al.*, 1958, Fig. 1; Reynolds and Vincent, 1967, Fig. 1).

## Introduction

This study reports on a sandstone body of the West Baden Group (Mississippian) that crops out in parts of Owen and Greene Counties, west-central Indiana (Fig. 1). Primarily, this is a field study with emphasis on outcrop maps and sedimentary structures, although some subsurface data (from 2 core holes and 4 rotary wells) were used.

## Regional Setting

The Chesteran strata of the Illinois basin cover about 35,000 sq mi and consist of an alternating series of shale, limestone, and sandstone beds that thicken southwestward to a maximum of 1,200 ft. This sequence follows a major interval of carbonate deposits in the basin, which in the Indiana outcrop is called the Blue River Group. Chesteran sandstone and shale mostly represent sand and mud transported from the northeast to and across the basin by a large river system. This system, first mentioned by Weller (1927, p. 26), later was named the Michigan River system by Swann (1963, p. 12).

In the Indiana outcrop the West Baden Group is a convenient mapping unit within the Chesteran and extends from the top of the limestone of the Blue River Group to the base of the Beech Creek (Barlow) limestone. In most of west-central Indiana the group is 125–150 ft thick and normally consists of three sandstone units, two thin limestone units, and shale, shale being dominant (Fig. 2). In west-central Indiana, the group underlies the eastern part of the maturely dissected Crawford upland, where local relief commonly ranges from 100 to 150 ft.

In parts of Owen and Greene Counties most of the West Baden Group is represented by a thick lenticular sandstone body. The body is multistory; the sandstone of the Sample Formation directly overlies, and is indistinguishable from the sandstone of the Bethel Formation. Sullivan (1969) used the term "clastic belt" (Fig. 2) for the part of the West Baden Group where the sandstones of the Bethel, Sample, and Elwren Formations are indistinguishable and the normally intervening Reelsville and

[1] Manuscript received, November 11, 1968; accepted, January 11, 1969.

[2] Geology Department, University of Cincinnati.

[3] Geology Department, Indiana University.

Beaver Bend Limestones are absent. An abundance of mud and sand is believed to have inhibited carbonate deposition in the clastic belt. In early Chesteran time this fluvial-deltaic system prograded southwestward across the center of the basin and marine conditions prevailed laterally. The belt of Bethel sandstone thicker than 50 ft roughly delineates this deltaic system (Fig. 1), which contains several thin coals. Bethel sandstone is not abundant elsewhere in the basin except for a long, very narrow belt of marine sandstone that fills a channel deeply cut in the limestone of the Blue River Group (Reynolds and Vincent, 1967, p. 4). The study area is north of the main area of deltaic deposition and is separated from it.

## Characteristics of Sandstone Body

### Geometry and Bounding Lithology

The outcrop of the sandstone body in Owen and Greene Counties is approximately 16 mi long and slightly more than 2 mi wide except at Pottersville, where a small finger separates from the main body (Fig. 3). In plan, the edges of the body are gently curving, almost straight. Thickness ranges from 90 to 145 ft and averages 120 ft except at the distal, southern end, where the body thins abruptly to 20–30 ft.

The cross section illustrates the relation of the sandstone body to the bounding units (Fig.

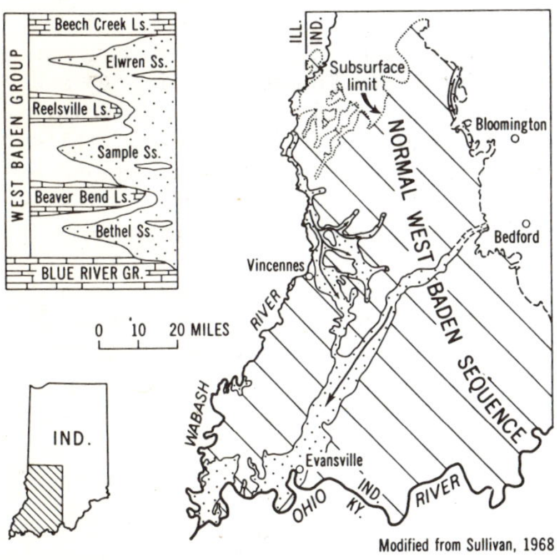


FIG. 2.—Normal West Baden sequence and clastic belt (stippled). Clastic belt defined by absence of both Beaver Bend and Reelsville Limestones (after Sullivan, 1969, Fig. 1).

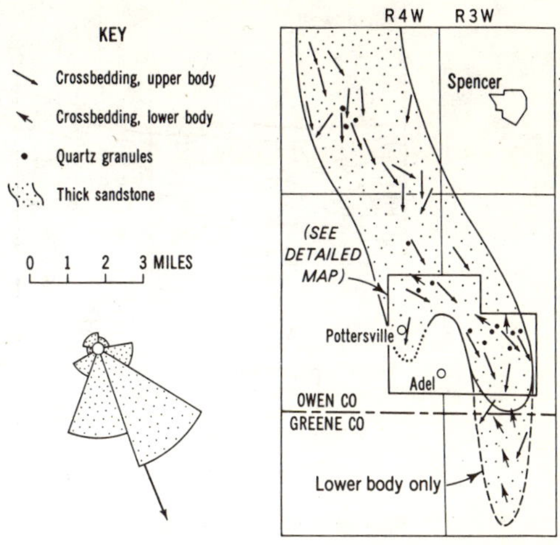


FIG. 3.—Bethel-Sample sandstone body in Owen and Greene Counties, Indiana. Broken boundary indicates where only lower body is well developed. Current rose based on 560 readings. See detailed map, Figure 9.

4). The base is unconformable on limestone of the Blue River Group, approximately 40 ft (Malott, 1952, p. 13) of which was eroded along the axis of the body's maximum thickness. Away from this axis, however, erosion of the Blue River limestone is negligible. The top of the Blue River Group rises steeply and the contact between the uppermost limestone and the clastic rocks of the Bethel Formation becomes conformable. The edges of the body are well defined by an abrupt facies change from thick, moderately well-exposed, dominantly cross-bedded sandstone to thin, poorly exposed, ripple-bedded, very fine-grained sandstone, which passes into silty shale and shale. Indiana Geological Survey drill holes (nos. 159, 162) near the margins of the body helped to document this interpretation. Both holes penetrate only minor sandstone plus limestone stringers at the position of the Beaver Bend and Reelsville Limestones; most of the section consists of gray shale containing abundant pyrite and carbonaceous debris, plus some variegated shale zones. As in the subsurface, the Beaver Bend and Reelsville Limestones are absent where the sandstone body is well developed, and it seems probable that they were not deposited, deposition having been inhibited locally by the constant influx of sand and mud. Zones of calcareous cement in the sandstone body, however, may represent the position of the Beaver Bend Limestone. In the study area the sandstone of the Elwren Formation is a readily mappable unit

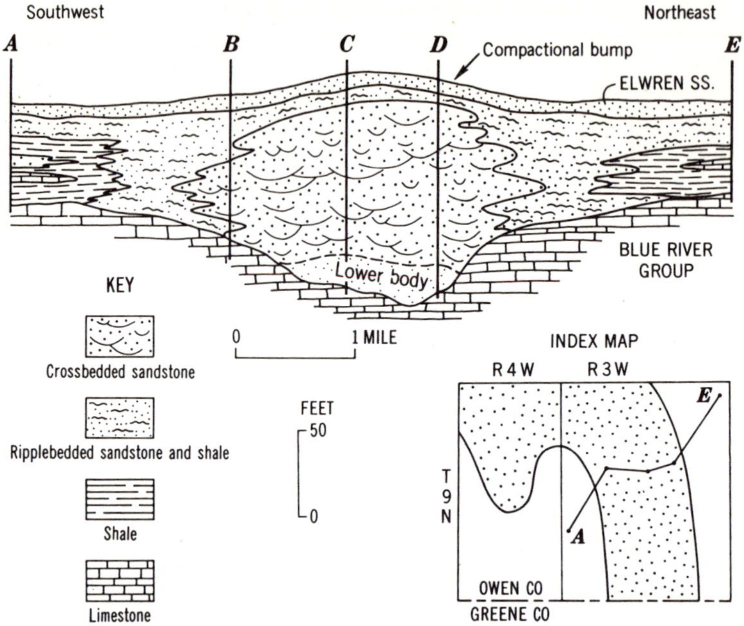


FIG. 4.—Transverse SW-NE cross section of sandstone body. Note compactional bump of Elwren Sandstone and erosion of limestone of Blue River Group at base of thickest part of sandstone body.

which extends across the Bethel-Sample body and clearly establishes its Mississippian age. The Elwren shows a compactional bump over the sandstone body.

### Internal Characteristics

The sandstone body has two major parts. An upper sandstone body, 100–120 ft thick, consists mostly of fine and medium sand, although average grain size decreases irregularly upward. Locally, however, the sandstone body is coarse grained and contains a few quartz granules and pebbles. In southeastern Illinois in Hardin County, Baxter *et al.* (1963, p. 13) also noted quartz granules and pebbles in the Bethel. Of all the Chesteran sandstone units in the Illinois basin, only the Bethel contains quartz granules and pebbles. Cross-bedding in the upper body is abundant and dips almost exclusively south and southeast. A lower sandstone body has a maximum thickness of approximately 35 ft, contains neither coarse sand nor granules, and has relatively sparse cross-bedding that dips mostly northwest. This sandstone body is not everywhere present, but is best developed at the distal, southern end of the upper body in Greene County and in southern Owen County.

Two dominant bedding facies are recognized in the Bethel-Sample body, cross-bedding and ripple mark (Table 1).

The cross-bedded facies consists mainly of trough cross-beds and some planar cosets which exceed 150 cm in thickness. The latter are found only in the thickest part of the body. Collectively, the cross-beds have an almost perfect log-normal distribution (Fig. 5). The thickest cross-bedded unit, 187 cm, can be traced for several hundred feet (Fig. 6) and is found in the central part of the body. Coarse sand and granules generally are associated with the thicker cross-beds, although commonly this facies contains fine to medium, well-sorted (0.3–0.5 $s_\phi$) grains. Some massive beds are also present.

The ripple-bedded facies is composed largely of ripple bedding with shale laminations (Fig. 7). Typically, this facies is poorly to moderately sorted (0.7–1.5 $s_\phi$) siltstone and fine sandstone. Bedding units thicker than 3 cm are uncommon. The interbedded shale is gray, silty, micaceous, and somewhat carbonaceous. Other sedimentary structures of the facies include shale-pebble conglomerate, load casts, and a few drag marks. Although as many as five zones of ripple bedding from 5 to 8 ft thick are present within the sandstone body,

**Table 1. Principal Facies and Their Characteristics**

| Facies | Lithology and Texture | Petrology (both facies) | Petrology | Porosity and Permeability |
|---|---|---|---|---|
| *Cross-Bedded* | Dominantly cross-bedded, fine to medium, well-sorted (0.3–0.5$\phi$) sand. Cross-beds from 4 to 187 cm thick. Some massive beds. | Mostly quartz (95 +percent) traces of chert and feldspar | Cement (12–17 percent) is mostly authigenic quartz, chalcedony, and iron oxide plus minor clay. | Porosity ranges from 22 to 27 percent; permeability ranges from 137 to 3,300 md and averages 600 md. |
| *Ripple Bedded* | Ripple bedding with shale laminations. Poorly to moderately sorted (0.7–1.5$\phi$) silt and fine sand. | | Sparry calcite and less-abundant clay minerals form 30–55 percent of sand. | Porosity ranges from 7 to 22 percent; permeability averages only 10 md. |

this facies forms but a small part of it. However, ripple bedding is everywhere found at the top of the body below its transition to shale; away from the thick elongate sandstone body, the ripple-bedded facies is dominant (Fig. 8). In the study area much of the sandstone of the Elwren Formation also belongs to the ripple-bedded facies.

Cross-bedding is by far the leading directional structure of the sandstone body. About 560 observations were made. Correlation with the shape of the body is excellent (Fig. 3). Moreover, in a small area (Fig. 9) where all available cross-beds were measured, the close relation between current system and sand-body shape can be seen even better—even at the bifurcation at Pottersville. This bifurcation was first suggested by seemingly aberrant due-south readings, and mapping showed that there is a definite bifurcation. Lack of a compactional bump in the Elwren sandstone just south of the southernmost exposures of the bifurcation suggests that it does not extend far into the subsurface, if at all.

Table 2 gives locality descriptions that illustrate the characteristics of the sandstone body.

Petrographically, the sandstone bodies average 87 percent quartz, most of which has unit extinction and secondary overgrowths. Feldspar constitutes 2–5 percent of the framework and is relatively fresh. Few heavy minerals are present.

No invertebrate skeletal remains have been found in the sandstone. If ever present, they must have been removed by leaching. There are, however, tracks, trails, and burrows in the ripple-bedded sandstone and siltstone, and abundant macerated carbonaceous debris in the shale.

## Depositional Environment

Origin of the sandstone body as a distributary delta finger is suggested by its gently curving but almost straight boundaries, the small downdip bifurcation at Pottersville, the southeast dip of the cross-bedding in the upper body, and the abrupt termination of the main body (Table 3).

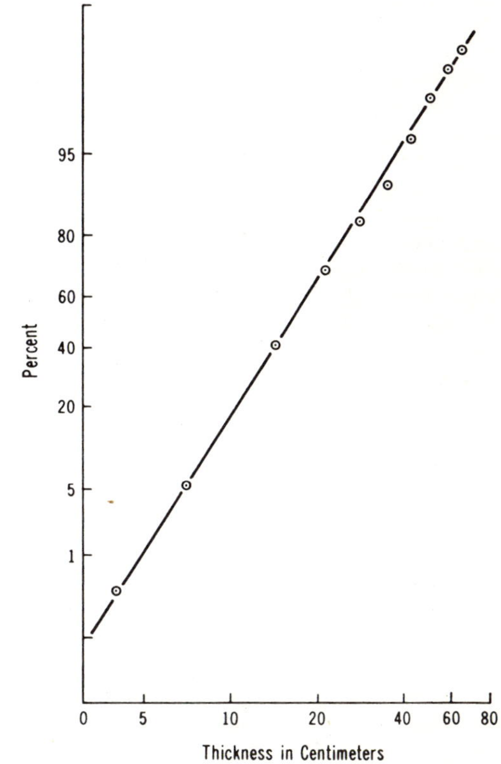


Fig. 5.—Cumulative curve of thickness of cross-beds closely approaches log normality.

FIG. 6.—Thickest cross-bed of upper sandstone body can be traced for several hundred feet along hillside. SE¼, SW¼, SW¼, Sec. 12, T9N, R4W, Owen County.

*Upper body*.—The gently curving, almost straight boundaries are characteristic of delta-finger sand bodies that develop in standing water. Figure 10 shows the course of a distributary of the Mississippi River, with similar gently curving, almost straight channel boundaries. The sand body associated with this channel has a comparable map pattern; crossbedding and ripple mark are its principal structures; the cross-bedding dips seaward; in terms of grain-size variation, the sand fines upward.

A sand body formed by a recently abandoned segment of the Mississippi River in its alluvial valley near Natchez, Mississippi, offers a striking contrast (Fig. 11). This belt of sand is 100–130 ft thick and approximately 6 mi wide. Bordering the belt with abrupt margins are backswamp mud and silt. The marked sinuosity of the sand body reflects the meandering of the thalweg as it impinged on the back swamp.

The lateral gradation of the Bethel-Sample sandstone body into a marine section is also strong evidence for its deltaic origin. In the Holocene this would be evidenced by a transi-

FIG. 7.—Typical appearance of ripple-bedded facies. East side of road in middle of S-curve in NE¼, NW¼, NW¼, Sec. 24, T9N, R3W, Owen County.

tion from natural levee and splay deposits to salt-marsh and prodelta muds. Although all these details cannot be seen, sections away from the sandstone body imply them. The Bethel shale and associated fine-grained, ripple-bedded sandstone of such sections are conformable with the marine carbonate rocks of the underlying Blue River Group and pass upward into the Beaver Bend Limestone, another marine unit.

*Lower body.*—The lower, thin sandstone body with northwestward-dipping cross-bedding, characterized by finer grain size and lack of quartz pebbles, had a different origin. The most plausible interpretation is that it is a tidal deposit filling a shallow channel cut into the limestone of the Blue River Group. This channel probably was cut by a current system related to the upper distributary finger. Sometime after the cutting, quartz sand began to fill the channel—perhaps washed into it by longshore drift—and was carried landward by tidal currents. Transport in such a tidal channel may be landward, bimodal, or seaward depending on the relative strengths of the ebb and flood ties. Such characteristics have been described in sand bodies found in the estuaries of the Rhine delta (Noorthoorn van der Kruijff and Lagaaij, 1960, p. 716, Fig. 3; Oomkens and Terwindt, 1960) and elsewhere. In Owen County this

FIG. 8.—Road cut on Indiana 46, 3½ mi east of Spencer, Owen County, showing Bethel shale overlain by ripple-bedded Bethel sandstone. Note flat basal contact of sandstone, typical of deposition away from thick lenticular sandstone body. SE¼, SW¼, SW¼, Sec. 24, T10N, R3W.

tidal body is preserved where later erosion at the base of the overlying deltaic body is minimal, and is seen best at the lower, distal end of the abandoned delta finger.

*Sequence of events.*—A postulated sequence of events emphasizes the internal consistency of the foregoing interpretations.

1. The limestone of the Blue River Group was deposited on a shallow-marine shelf.
2. An erosion channel, probably submarine, was cut into the limestone.
3. Somewhat later, fine-grained quartz sand accumulated in the channel, perhaps washed into it by long-

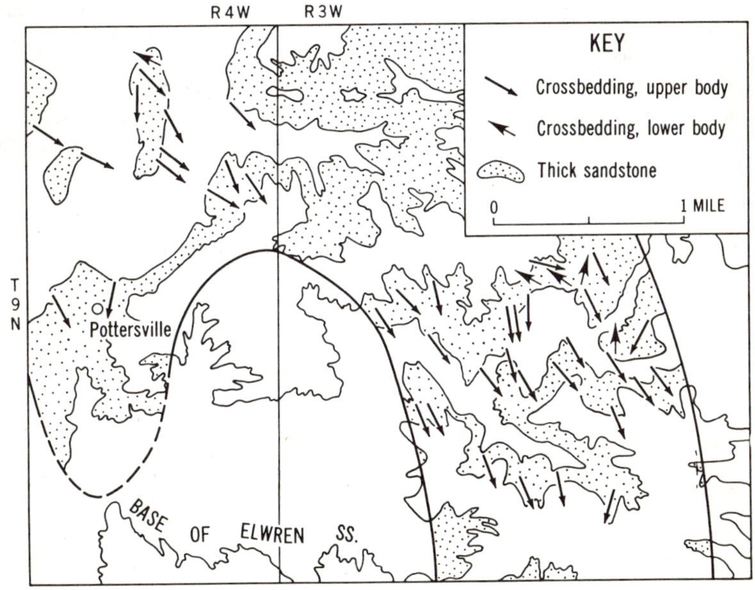


FIG. 9.—Detailed study of cross-bedding. Thick sandstone from base of Elwren Formation to top of the Blue River Group stippled.

**Table 2. Selected Localities of Sandstone Body**

| *No.* | *Location* | *Remarks* |
|---|---|---|
| 1 | SE¼, NE¼, NW¼, Sec. 29, T9N, R3W | West edge of sandstone body. Lower West Baden section mostly interbedded sandstone and shale. Elwren Sandstone exposed with Beech Creek Limestone above. |
| 2 | N½, Sec. 5, T8N, R3W | Approximately 25 ft of sandstone with northwestward dipping cross-beds exposed in abandoned quarries in south-trending valley. Considered to be part of tidal body. |
| 3 | SE¼, SW¼, SW¼, Sec. 24, T10N, R3W | Roadcut at crest of hill along S. R. 46 shows ripple-bedded sandstone (4 ft) and shale (11 ft) of "off-channel" Bethel. |
| 4 | SE¼, SW¼, SW¼, Sec. 12, T9N, R4W | West side of hillside shows largest cross-beds (see Fig. 6); quartz granules present. |
| 5 | NE¼, NW¼, NW¼, Sec. 24, T9N, R3W | Pottersville bifurcation. Ripple-bedded facies in roadcut overlies 25 ft of massive sandstone below and west of road. |
| 6 | NW¼, SE¼, SW¼, Sec. 30, T9N, R3W | Section from Blue River Group to Beech Creek Limestone exposed in long north-trending ravine along west section line. Bethel and Sample intervals mostly shale. |
| 7 | SE¼, NW¼, NE¼, Sec. 19, T9N, R3W | SW-trending ravine contains 90-ft section of sandstone that fines upward; quartz granules present. |
| 8 | SE¼, SW¼, NE¼, Sec. 20, T9N, R3W | Bluffs along Raccoon Creek. Best exposures for basal contact of sandstone body with underlying Blue River Group; tidal body (35 ft thick) overlain by delta distributary (35 ft); thick phase of distributary finger ends abruptly eastward. |
| 9 | SW¼, SE¼, SE¼, Sec. 9 T10N, R4W | Cross-bedded sandstone (65 ft thick) extends up ravine from creek level and is overlain by 6–8 ft of calcareous sandstone and more poorly exposed sandstone. Calcareous zone may represent Reelsville Limestone. |

shore currents, and then was carried landward by tidal currents.

4. The upper distributary-finger sand body prograded downdip and, as it did, erosion at its base truncated most of the lower sandstone body.

5. In the area of the main sandstone body, deltaic deposition persisted during the marine transgressions that deposited the Beaver Bend and Reelsville Limestones; the sand and mud of the delta distributary locally inhibited carbonate deposition.

6. The distributary finger was abandoned by stream diversion at some point updip before Elwren deposition; the Elwren is a shallow-marine shelf sandstone—unrelated to the older deltaic system.

7. Deposition of Beech Creek Limestone followed.

The foregoing interpretation is not without problems. For example, evidence for a shoreline in Owen and Greene Counties is absent, which suggests that wave power and tidal strength in the marine basin were minimal. In the modern Mississippi delta (Fisk, 1961), bar-finger sands form at the downdip end of the distributary, are as much as 250 ft thick, and have unidirectional, downstream cross-bedding. Neither older channels nor tidally influenced sand bodies preceded development of bar-finger sands. Another contrasting feature is the coarseness of the channel fill in the distributary of Owen County; in the modern Mississippi delta abandoned channels of distributaries fill with mud and silt, sand being restricted mostly to where the channel is abandoned. Greater competence of the Michigan River system and the fact that deposition occurred on a stable carbonate platform far from the continental margin are major differences and may be responsible for the contrasts. The prosposed interpretation emphasizes the role of units bounding a sand body above, below, and laterally. Shape and sedimentary structures are also diagnostic, especially if evaluated in terms of present-day processes and responses.

*Estimating stream size.*—Stream competence was sufficient to allow transport of some granules and pebbles across the floodplain to the edge of the strandline. These materials may have been reworked from older Mississippian or Devonian deposits in the middle and northern Appalachians or even may have been derived from the eastern parts of the present Canadian shield (Potter, 1963, p. 86). As Swann (1963, p. 14) has indicated, the Michigan River was of considerable size; the shoreline of the ocean into which it flowed shifted perhaps as much as 600 mi back and forth across the prevailing southwest-dipping paleoslope of the Illinois basin and adjacent areas. At times of

**Table 3. Essential Features of Distributary Finger of Owen County, Indiana**

*Geometry and relation to bounding units*
Elongate, narrow, and lenticular. Unconformable base; lateral facies change to siltstone and shale.

*Lithology*
Mostly permeable well-sorted cross-beds plus some impermeable, poorly sorted ripple-bedded zones.

*Paleocurrents*
Unimodal, downdip cross-bedding in distributary body; updip in tidal body.

*Physiographic setting*
Abandoned delta distributary of Michigan River system.

FIG. 10.—Mississippi River, Pass à Loutre. Note gently curving, almost straight channel. Drawn by Robert Judah from photograph by W. A. Pryor.

maximum extension the Michigan River delta system may have been near Memphis, Tennessee; at times of maximum contraction it may have been between Toledo, Ohio, and Toronto, Ontario. The thin coal beds of the Bethel Sandstone suggest a warm, humid climate with ample rainfall, perhaps comparable to that of the southern United States. Because of the areal dimensions of the drainage system and a humid climate, discharge could have been considerable. As its headwaters were far northeast, the Michigan River might have had a discharge of approximately a fourth to a third that near the mouth of the present Mississippi. If so, the Bethel-Sample distributary of Owen and Greene Counties could have had a normal discharge of perhaps 50,000-100,000 $ft^3$ per second when it was the principal outlet of the delta system.

The size of the stream also may be gauged from the thickness of the sand body. It seems reasonable to believe that the depth of scour in a thalweg during peak discharge controls the thickness of the sand body. It would be unlikely that a single-cycle sand body would extend deeper than the maximum depth of scour of its parent stream. By this reasoning, the Bethel-Sample distributary may have had a channel depth of 40–50 ft, if generous allowance is made for its multistory origin. Because the sand body was deposited initially on a carbonate shelf rather than on thick prodelta muds, local compaction virtually did not occur beneath it and hence is not a complicating factor in assessing stream-channel depth.

### Reservoir Characteristics

Bedding types defined by grain size, cementing agents, and abundance of shale laminations control porosity and permeability within the reservoir (Table 1). In the well-sorted cross-bedded facies porosity ranges from 22 to 27 percent; permeability ranges from 137 to 3,300 md and averages approximately 600 md. The highest permeability is near the base of the sandstone, just above the underlying limestone. Sorting is poor in the finer grained, ripple-bedded facies, where sparry calcite and clay matrix constitute between 30 and 55 percent of the rock. The ripple-bedded facies has an average permeability of only 10 md, and the abundant clay laminations are effective barriers to vertical flow that helped inhibit decementation. More diamond drill holes than the two available would be needed to establish the continuity of most of these zones in the reservoir; except for the uppermost zone, we suspect that the ripple-bedded zones do not have appreciable lateral continuity, primarily because of scour and channeling within the delta distributary. In any case, they and the permeable cross-bedded facies form short, fining-upward

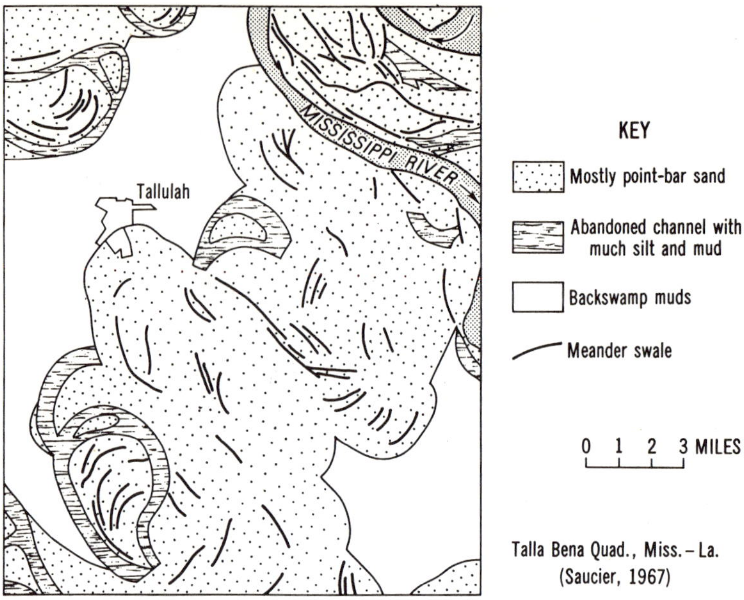


FIG. 11.—Belt of sand left behind by former course of Mississippi River, abandoned 1,100 years ago. Note scalloped edges of sand body, which is about 100–130 ft thick. Talla Bena quadrangle, Mississippi-Louisiana (after Saucier, 1967).

cyclothems and are superimposed on the larger, overall upward fining of the distributary finger shown in Figure 12. This vertical profile is comparable to that shown by Hewitt and Morgan (1965, Fig. 7) for the Robinson sandstone, a fluvial deposit of Pennsylvanian age in the Illinois basin. Although permeability correlates well with grain size and bedding type, porosity shows little dependence.

Permeability in excess of 400 md is a striking feature of the Bethel-Sample body in outcrop in Owen County. That such values *far* exceed a majority of permeability values of the Bethel in the subsurface is shown clearly by the two stepwise-cumulative curves of Figure 13.

We believe that decementation by groundwater flowing downward through the sandstone body into the limestone of the Blue River Group is the principal cause. In essence, this reservoir is without a plug at the bottom. Hence, wherever it is above creek level, it is a poor groundwater reservoir. Decementation by groundwater was most effective in the cross-bedded facies; the thin ripple-bedded zones resisted it because of more clay matrix and numerous shale laminations. Figure 13 also indicates that core analyses of outcrop materials may tell little, if anything, about subsurface equivalents.

Local relief in the study area ranges between 100 and 125 ft. Because the limestone in the Blue River Group is at or near creek level, vadose solution is maximal, and this in turn enhances groundwater flow through the overlying sandstone body. It is this combination of "plugless" reservoir and maturely dissected topography that produces the very great permeability of the exposed sand body.

The very high permeability results from the position of the sandstone body below the present erosion surface. For the petroleum industry, the Bethel-Sample sandstone body of Owen and Greene Counties is perhaps most valuable as an analog for others underlying ancient erosion surfaces now represented by unconformities. Figure 14 shows a hypothetical example of the distribution of permeability in an abandoned delta finger below an unconformity: the closer the body to the unconformity, the greater its permeability. The length of time represented by the unconformity, the intensity of the decementation process, and the extent to which the reservoir was recemented after later burial are factors controlling final porosity and permeability.

## Conclusions

1. The elongate, lenticular sandstone body of the West Baden Group in Owen and Greene Counties is an abandoned delta distributary of

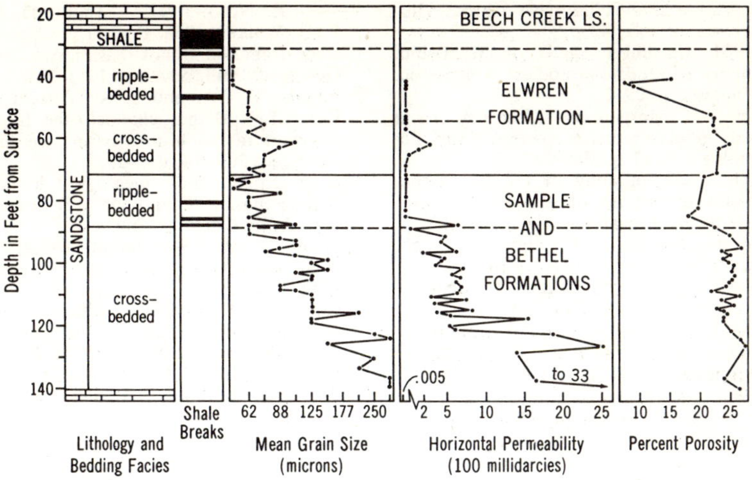


FIG. 12.—Vertical profile of bedding, grain size, permeability, and porosity, obtained from diamond drill core. Note fining upward and vertical decline of permeability. Interval extends from Beech Creek Limestone to top of Blue River Group.

the Michigan River system.

2. The delta distributary is multistory, consisting of (a) a lower, fine-grained body, perhaps of tidal origin, with cross-bedding dipping northwestward updip, and (b) an upper, coarser grained, and much thicker distributary body with abundant southeastward downdip cross-bedding.

3. Two principal bedding facies, cross-bedding and ripple-bedding, control most of the sandstone bodies' reservoir characteristics. The cross-bedded facies is well sorted, fine to medium grained, and very permeable; the ripple-bedded facies is finer grained, poorly sorted, and relatively impermeable.

4. The sandstone body is much more permeable than its subsurface equivalents, because it has been decemented by groundwater flowing downward through it into underlying limestone.

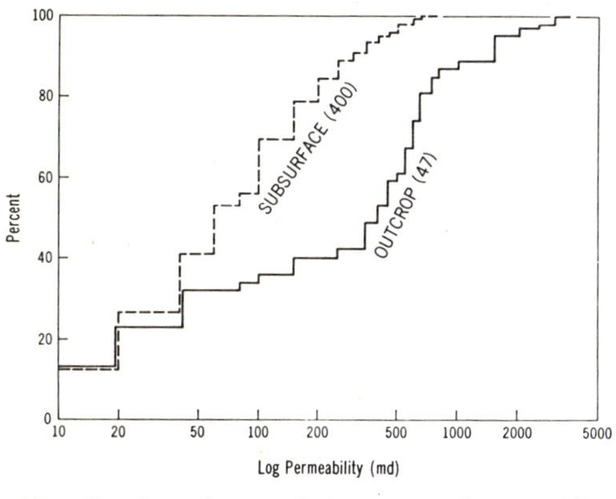


FIG. 13.—Stepwise cumulative curves of permeability values. Note marked enhancement of permeability in outcrop. The 400 subsurface samples are from Bethel Sandstone in Midland field, Muhlenberg County, Kentucky, and outcrop samples from shallow drillhole in Owen County.

## REFERENCES CITED

Baxter, J. W., P. E. Potter, and F. L. Doyle, 1963,

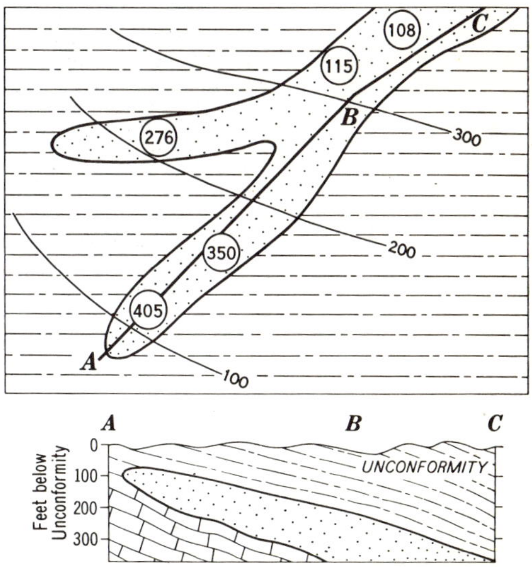


FIG. 14.—Map and cross section of hypothetical distributary finger below unconformity. Contour lines indicate depth below unconformity surface. Decementation maximal closest to unconformity.

Areal geology of the Illinois fluorspar district: Illinois Geol. Survey Circ. 342, 43 p.
Fisk, H. N., 1961, Bar-finger sands of Mississippi delta, *in* Geometry of sandstone bodies: Am. Assoc. Petroleum Geologists, p. 29–52.
Hewitt, C. H., and J. T. Morgan, 1965, The Fry *in situ* combustion test—reservoir characteristics: Jour. Petroleum Technology, v. 17, p. 337–352.
Malott, C. A., 1952, Stratigraphy of the Ste. Genevieve and Chester formations of southern Indiana: Ann Arbor, Michigan, Edwards Letter Shop. 105 p.
Noorthoorn van der Kruijff, J. F., and R. Lagaaij, 1960, Displaced faunas from inshore estuarine sediments in the Haringvleit (Netherlands): Geologie en Mijnbouw, v. 39, no. 11, p. 711–723.
Oomkens, E., and J. H. J. Terwindt, 1960, Inshore estuarine sediments in Haringvleit (Netherlands): Geologie en Mijnbouw, v. 39, no. 11, p. 701–710.
Potter, P. E., 1963, Late Paleozoic sandstones of the Illinois basin: Illinois Geol. Survey Rept. Inv. 217, 92 p.
——— E. Nosow, N. M. Smith, D. H. Swann, and F. H. Walker, 1958, Chester cross-bedding and sandstone trends in Illinois basin: Am. Assoc. Petroleum Geologists Bull., v. 42, p. 1013–1046.
Reynolds, D. W., and J. K. Vincent, 1967, Western Kentucky's Bethel channel—the largest continuous reservoir in the Illinois basin: Kentucky Geol. Survey Spec. Pub. 14, p. 2–19.
Saucier, Roger, 1967, Geological investigation Boenf-Tensas basin, Talla-Bena 15 minute quadrangle: U.S. Army Corps of Engineers.
Sullivan, D. M., 1969, Subsurface stratigraphy of the West Baden Group in Indiana: Indiana Geol. Survey Bull., in press.
Swann, D. H., 1963, Classification of Genevievian and Chesterian (Late Mississippian) rocks of Illinois: Illinois Geol. Survey Rept. Inv. 216, 91 p.
Weller, Stuart, 1927, The story of the geologic making of southern Illinois: Illinois Geol. Survey Educ. Ser., no. 1, 41 p.

---

## ERRATUM

*Bulletin,* v. 53, no. 10, pt. 1 (October 1969), Hrabar and Potter paper, p. 2150.

S. V. Hrabar and Paul Edwin Potter wish to express regret for the omission of acknowledgments in the paper "Lower West Baden (Mississippian) Sandstone Body of Owen and Green Counties, Indiana." Those who contributed importantly include Henry Gray, Dan Sullivan, Robert Blakely, and Leroy Becker of the Indiana Survey; T. G. Perry, L. J. Suttner, and H. Rudman of the Department of Geology, Indiana University; David Swann, Jack Simon, and James Baxter of the Illinois Geological Survey; and W. A. Pryor, Department of Geology, University of Cincinnati.

We also are grateful for the technical assistance of William Moran of the Drafting Department, Indiana Survey.

The Indiana Geological Survey contributed samples and data from four drill holes.

R. Alexander of Oil Field Research, Inc., Evansville, Indiana, provided the porosity and permeability analysis.

**Reprinted from:**
BULLETIN OF THE AMERICAN ASSOCIATION OF PETROLEUM GEOLOGISTS
VOL. 55, NO. 8 (AUGUST, 1971), PP. 1194-1205, 11 FIGS.

# Deltaic Stratigraphic Traps in West Tuscola Field, Taylor County, Texas[1]

J. P. SHANNON, JR.,[2] and A. R. DAHL[3]

Houston, Texas 77001, and Denver, Colorado 80201

**Abstract** Hydrocarbon occurrence in Strawn (Pennsylvanian) sandstone bodies in West Tuscola field, near Abilene in Taylor County, Texas, is the result of stratigraphic entrapment in deltaic sandstones. Cores and numerous electric logs show that the vertical succession of deltaic facies consists, from base to top, of a progradational sequence (prodelta and delta front), an aggradational unit (delta plain–marsh and interdistributary bay), and an overlying transgressive shallow-marine interval. Reservoir sandstones are present within the delta-front facies as stream-mouth-bar deposits, known locally as the "Gray" sandstone.

The stream-mouth-bar sandstones within the West Tuscola reservoir are lenticular, very irregular in outline, and have varying trends of porosity; these are characteristics to be anticipated in deltaic deposits. Such features present problems in developing effective secondary recovery methods and in predicting occurrences of other deltaic sandstones.

## Introduction

### Nature of Study

This study originated as an environmental analysis of cores from the "Gray" interval within the Strawn series of Pennsylvanian age, in and around the West Tuscola field, Taylor County, Texas (Fig. 1). Initial study of three cores from the field revealed a deltaic origin for sandstones in the "Gray" interval; subsequent drilling indicated stratigraphic entrapment of oil and provided more cores for porosity and permeability determinations and for extending the environmental studies. The amount of detailed information, including electric logs from all wells in the area, permitted determination of the size, shape, and stratigraphic relations of an ancient deltaic sandstone body and explanation of the stratigraphic complexities within the "Gray" sandstone through knowledge of modern deltaic sedimentation. From this study came the development of isopach maps depicting the trend and distribution of the deltaic sandstone, an understanding of the relation of oil and gas occurrence to sandstone distribution in West Tuscola field, and the establishment of well-to-well correlations which should prove helpful in evaluating fluid transmissibility within the field.

[1] Manuscript received, July 14, 1970; accepted, October 1, 1970.

[2] Exxon Production Research Company.

[3] Exxon Company USA.

We are indebted to Humble Oil & Refining Company for providing the materials and data which made this study possible. Many of our fellow company geologists contributed to the investigation in different ways.



### Geologic Setting

The Strawn series in north-central Texas comprises a basal limestone unit (Caddo) and a wedge of predominantly clastic rocks which thickens eastward into the area of the Ft. Worth basin (Fig. 1). The wedge terminates along a generally north-south trend which passes west of Taylor County at the margin of the eastern shelf, where Strawn rocks are mainly carbonates. The "Gray," of Strawn age, is a subsurface unit recognized in Taylor and contiguous counties, and includes a succession of sandstones and shales between the Gardner interval above and the Caddo Limestone below (Fig. 2).

Sandstones at different levels within the "Gray" interval produce oil in several fields in Taylor County (Fig. 2). West Tuscola field is one of the largest in area and amount of reserves in the county. Producing sandstones in this field lie at a depth of about 100 ft below a limestone marker near the base of the Gardner (Fig. 2). This limestone marker has been used as a datum for cross sections and as an indicator of present structure (Fig. 3); the relatively simple structural pattern shows a westward regional dip of about 35 ft/mi. The stratigraphic entrapment of hydrocarbons in West Tuscola field is due to updip (eastward) pinchout of porous sandstone (Fig. 3).

## Deltaic Sandstones of West Tuscola Field

### Field Characteristics

Estimated reservoir volume of West Tuscola field in the "Gray" sandstone is 6.7 million bbl of oil in place within an area of 1,435 acres. The recovery mechanism is solution-gas drive. Varying oil-water contacts, the occurrence of oil and gas at the same structural level in different wells, and scattered nonproductive wells within the field indicate complex local stratigraphic control within the reservoir.

Sandstones in the field occur within a strati-

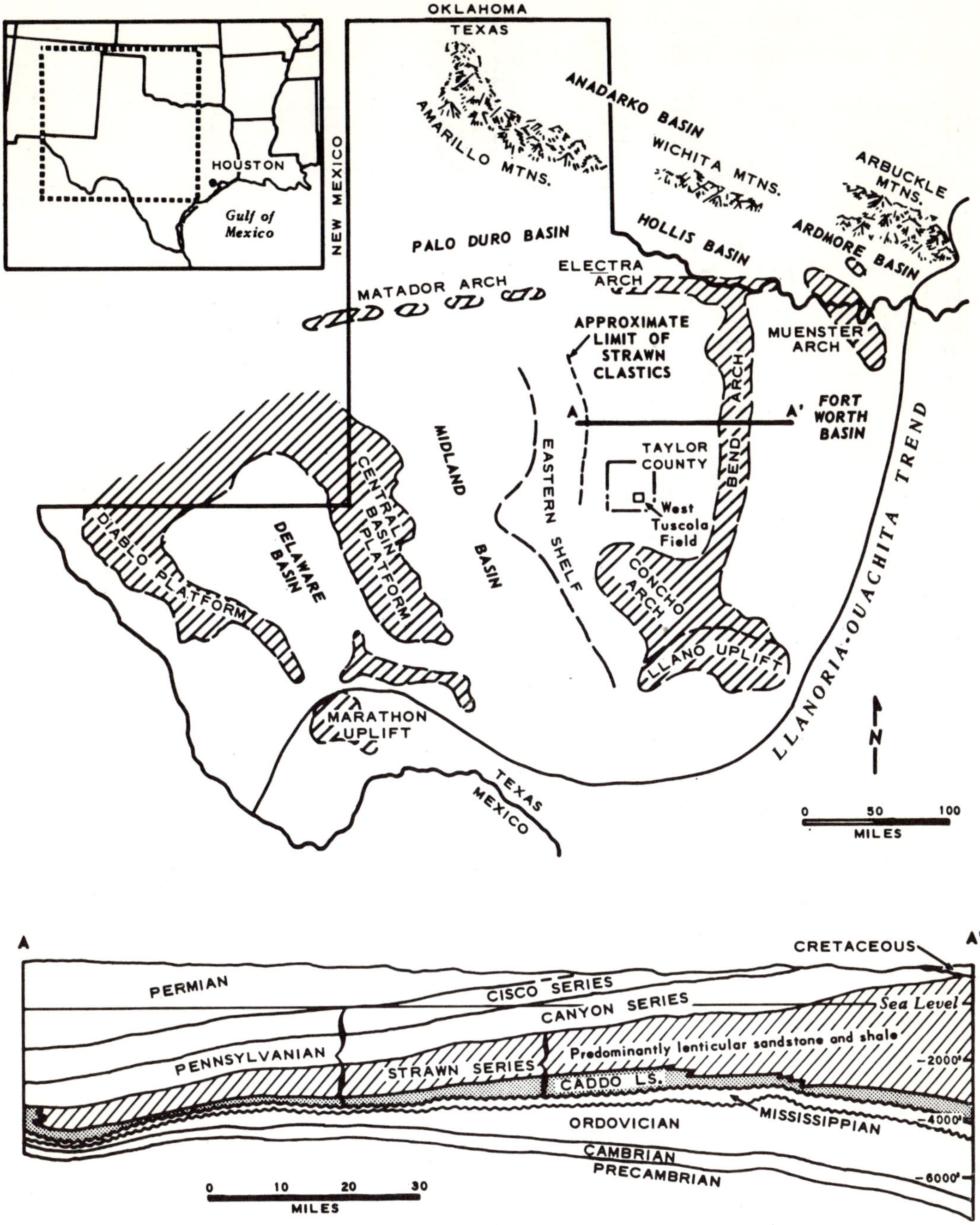


FIG. 1.—Regional setting of West Tuscola field, Taylor County, Texas.

graphic interval of 10 to slightly more than 100 ft. Individual productive sandstones (less than 25 ft thick) within this interval have an average porosity of 12 percent and permeability of 44 md. Also in the interval are units of interbedded sandstone and claystone having porosities of 5 percent and permeabilities of less than 0.1 md; these units form vertical and probably lateral permeability barriers between the producing sandstones.

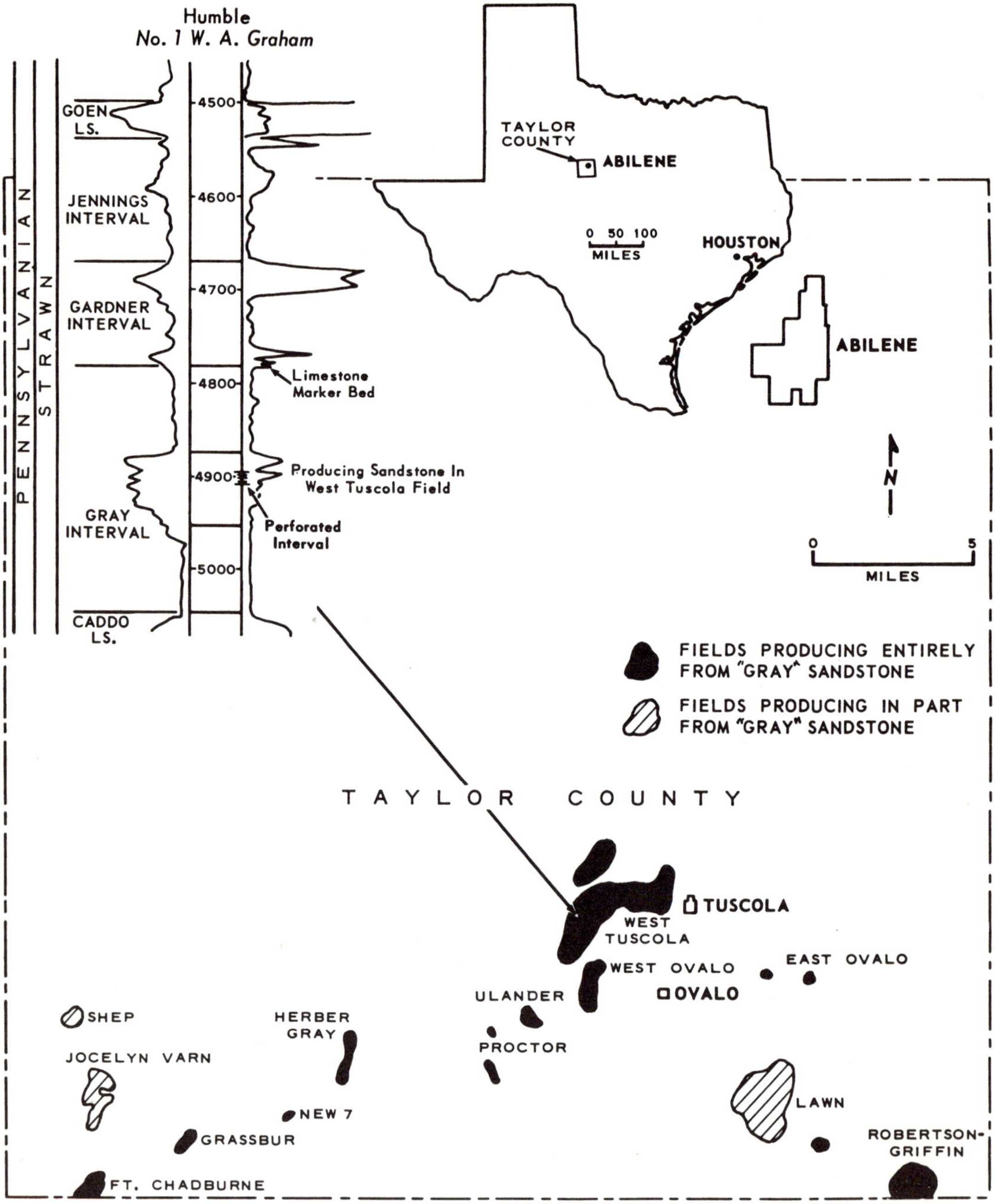


FIG. 2.—"Gray" sandstone oil production, Taylor County, Texas, with representative E-log of Strawn section in West Tuscola field.

### Facies and Environments of Deposition

The succession of facies recognized in the cores is interpreted as a prograding deltaic complex covered by shallow-marine deposits laid down during a subsequent transgression. Environmental facies within the deltaic unit comprise, from base to top, prodelta black shale; delta-front laminated carbonaceous sandstone (principal producing facies) and burrowed, intercalated carbonaceous sandstone

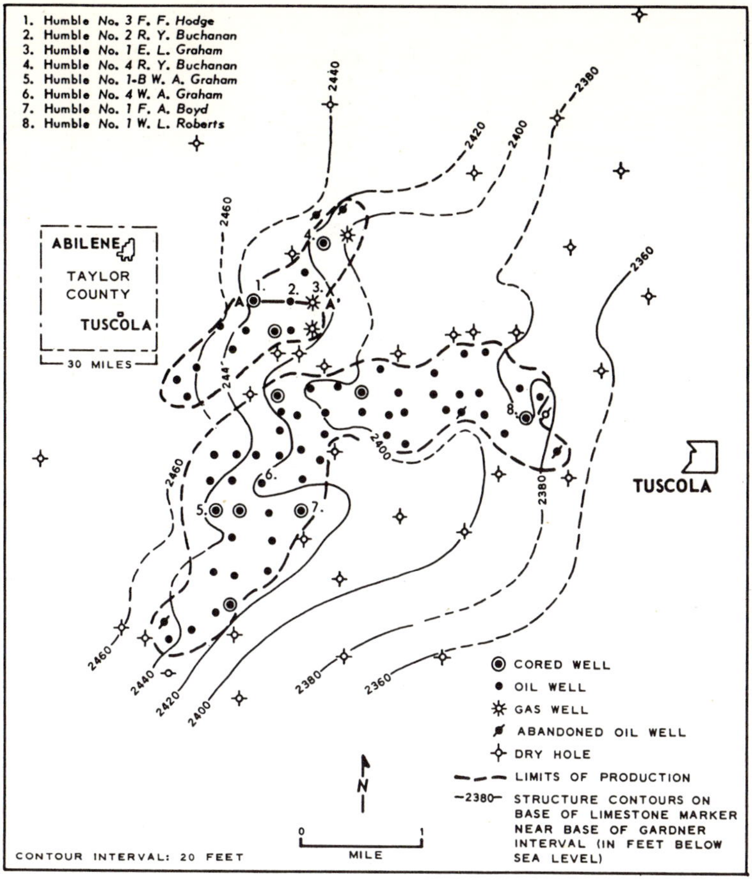


FIG. 3.—Structure map of West Tuscola field, showing limits of productive sandstone. Cross section A-A′ shown on Figure 4a.

and silty claystone; and delta-plain carbonaceous shale (marsh) and silty fossiliferous shale (bay). In cross section these facies (Fig. 4a) resemble those of a typical modern delta, shown diagrammatically with environments indicated on Figure 4b. Delta-plain and fossiliferous shallow-marine deposits overlie the delta-front facies, which grades downward into prodelta shale.

A variety of subfacies reflecting the numerous subenvironments within the delta front has been identified: stream-mouth-bar-crest laminated sandstone; bar-slope burrowed intercalated sandstone and silty claystone; distributary-channel crossbedded sandstone; and calcareous (fossiliferous) spit sandstone.

## Sandstone-Body Geometry

In order to establish the geometry of productive deltaic sandstones at West Tuscola, deltaic and related facies were identified in cores and on electric logs from throughout the field area, and this information was assembled in a series of detailed cross sections and maps. Sandstones in the field have a common top about 100 ft below a thin limestone near the base of the Gardner interval, which is used as a time-stratigraphic marker. The distribution and thickness of delta-front deposits whose tops occur at this level are shown on Figure 5. Figure 6 depicts the thickness of the clean sandstone zone and more sharply defines the field boundaries and the trends of delta growth.

The inferred direction of growth of the delta into the West Tuscola area is from the east along a trend of thick delta-front deposits which extend westward across the field (Figs. 5, 6). Diverging directions of delta growth in the western part of the field created thick delta-front sections on the north and south. Broad productive sandstone trends on the south thin markedly and terminate along a well-controlled

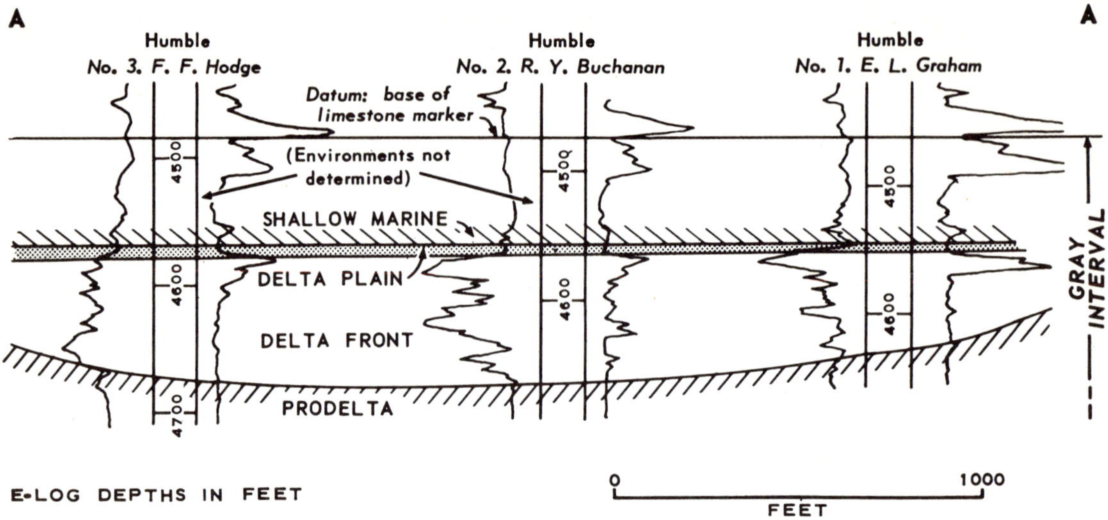


**a**

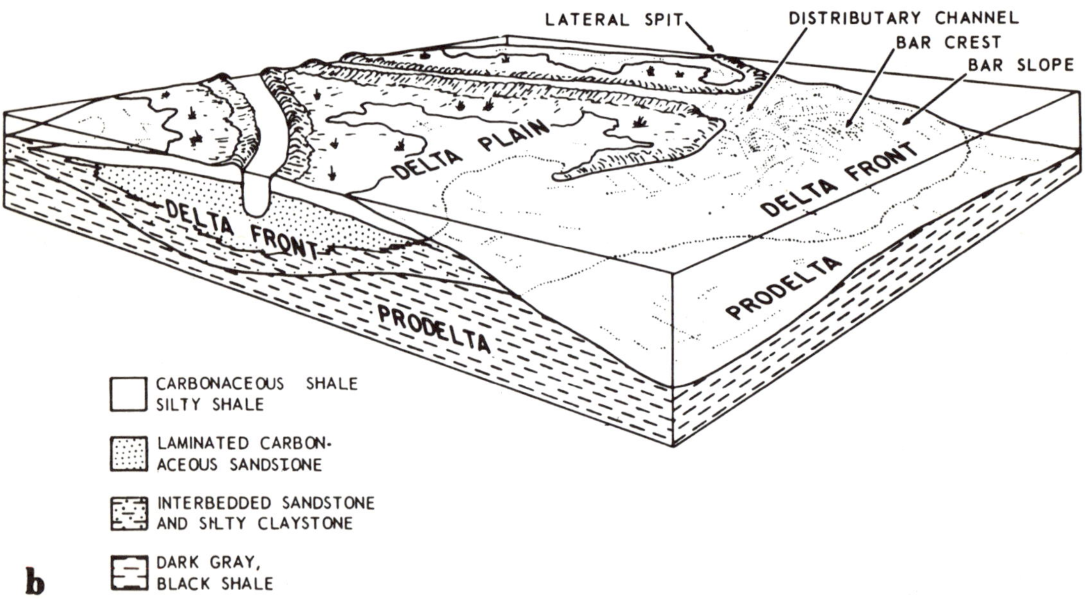


**b**

FIG. 4.—**a.** Facies recognized in "Gray" interval, West Tuscola field. For location of section, see Figure 3. **b.** Facies and environments of typical modern delta.

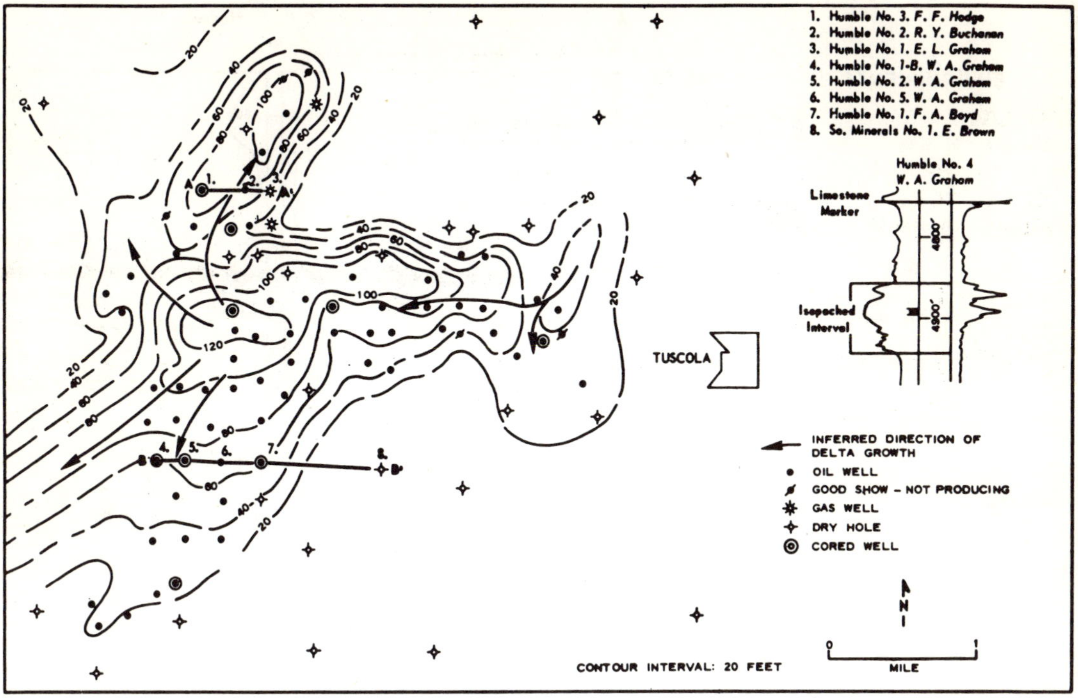


FIG. 5.—Isopach map of delta-front facies, West Tuscola field.

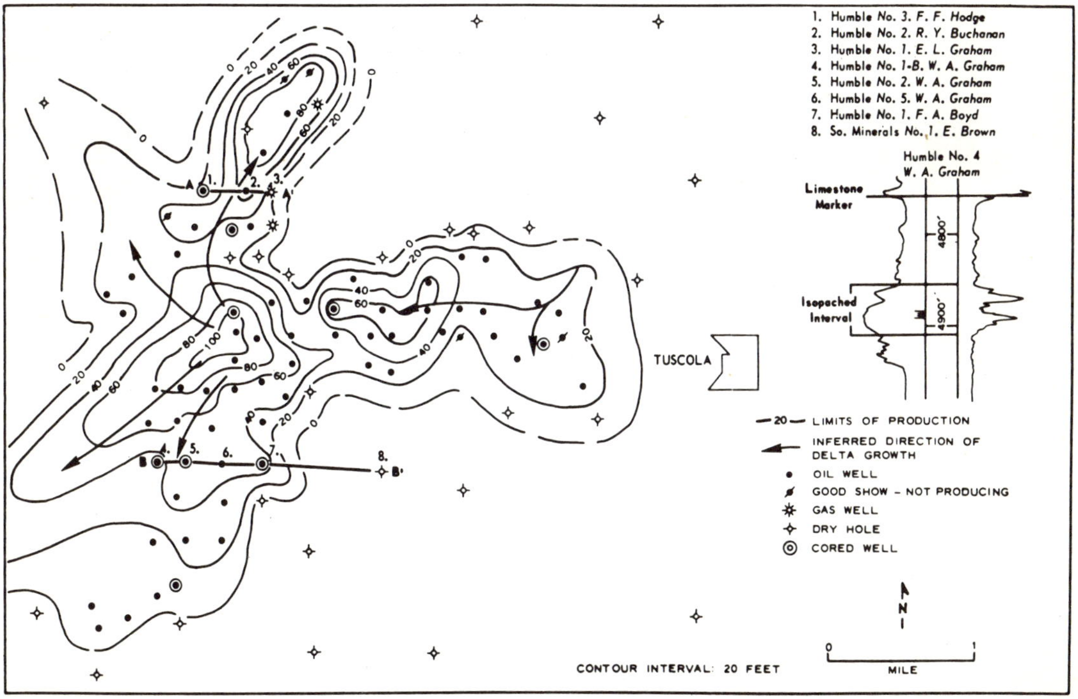


FIG. 6.—Isopach map of clean sandstone zone, West Tuscola field.

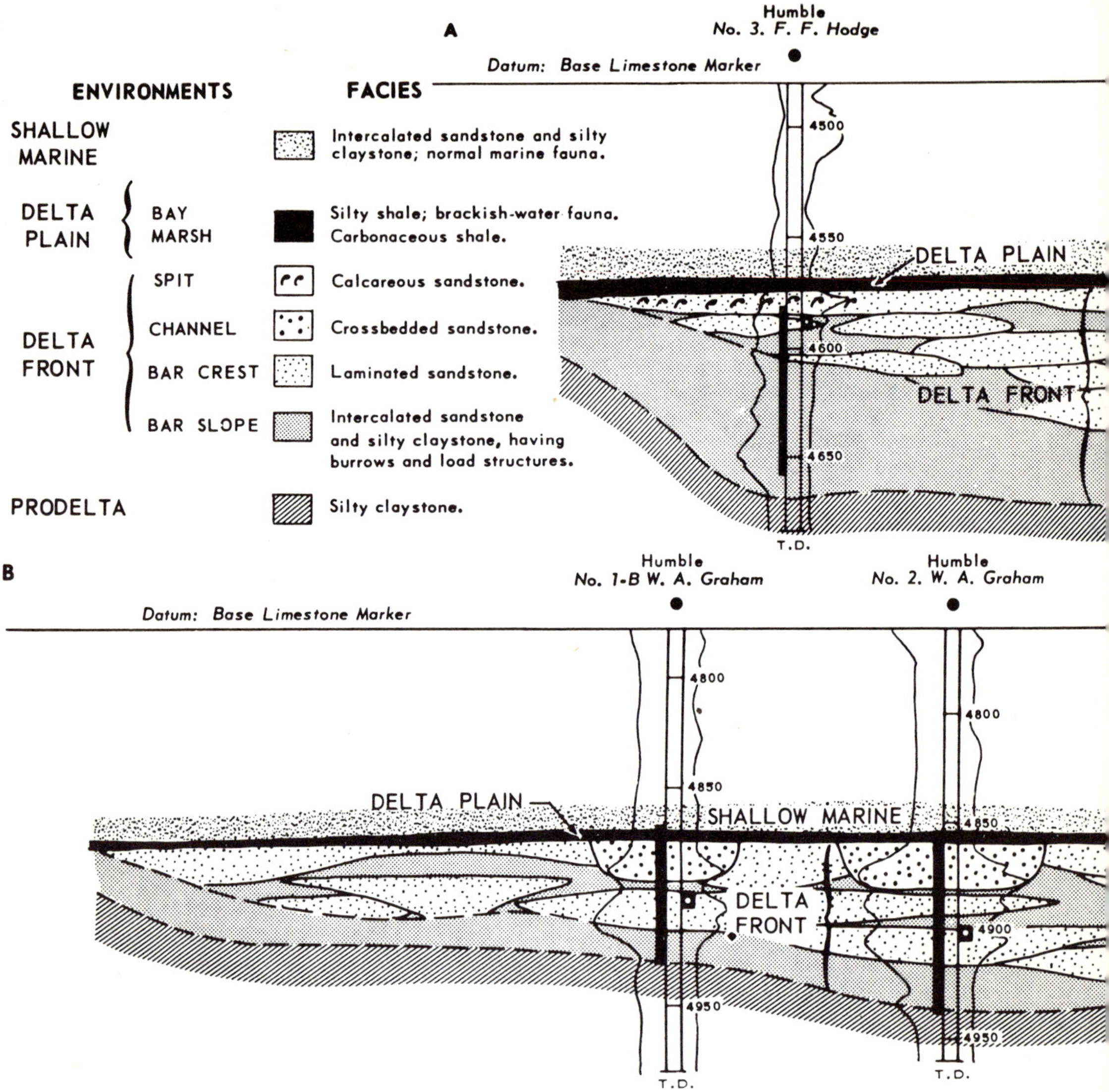


FIG. 7.—West-east cross sections A-A′ and B-B′, showing detailed facies relations, West Tuscola field. For location of sections set Figure 6.

zero line of porosity; possible continuation of sandstone toward the southwest is indicated by the presence of stratigraphically equivalent water-wet deltaic deposits at the western edge of the map (Fig. 6).

The eastern limits of sandstone occurrence are well defined by dry holes near the town of Tuscola and along the edges of the northern and southern deltaic extensions. These dry holes reflect decreased porosities in thin bar-crest deposits on the flanks of the delta. The eastward pinchout of porous sandstone, a postulated "shaling up" of any channel extensions eastward, and the present regional dip of 30–40 ft/mi toward the west (as indicated by the overlying limestone marker, Figs. 5, 6) create the stratigraphic trap. Local erratic distribution of porosity within the field is indicated by scattered dry holes which penetrated stacked successions of very thin bar-crest sandstones.

Distribution and nature of fluids within the reservoir indicate local stratigraphic irregulari-

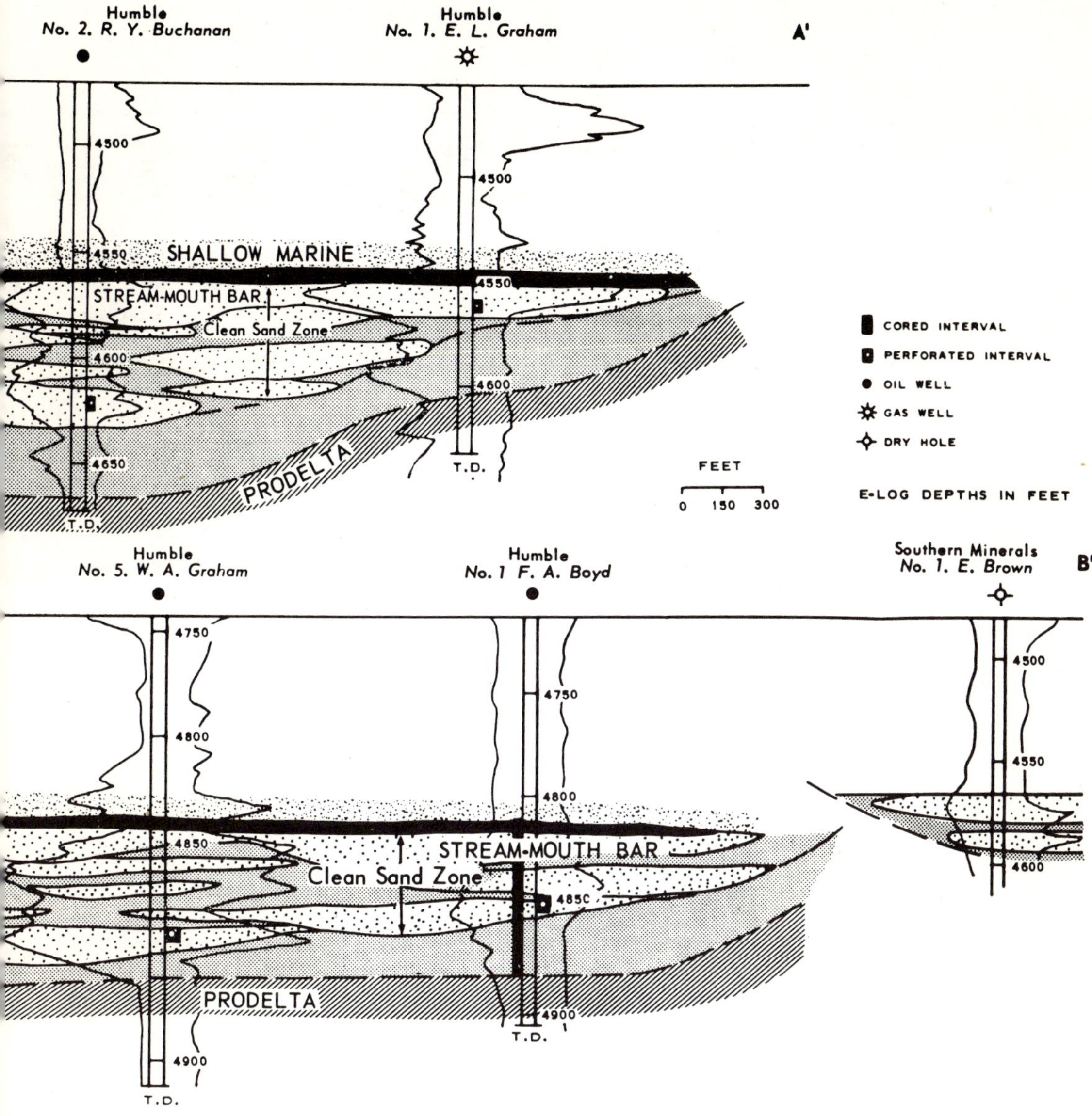


ties. Gas production is limited to a few wells on the eastern edge of the northern deltaic extension (Fig. 6), whereas only oil has been recovered far updip near the town of Tuscola. Water has been recovered from several wells on the western edge of the field, and oil-water contacts have been observed on logs and in drill-stem tests. Average levels of oil-water contacts were established for two areas in the field (Fig. 6); these averages, however, are themselves different and, within each area, local variations occur. Although the downdip limit of porous sandstone is uncertain owing to lack of control, the oil-productive part of the reservoir is fairly well defined.

### Internal Characteristics and Genesis of Reservoir Sandstone

Detailed facies relations of the producing interval in the West Tuscola field are shown in Figure 7. Both sections, constructed from the base of a limestone datum approximately 100 ft above the producing deltaic sandstone, are transverse to the general trends of a series

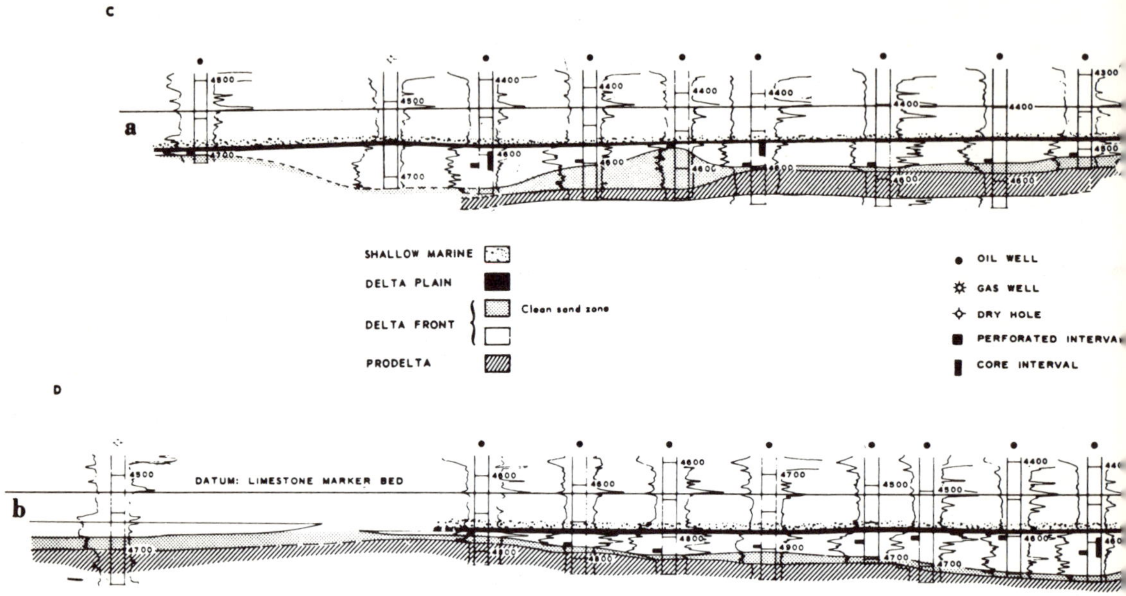


FIG. 8.—**a.** West-east cross section C-C′. **b.** South-north cross section D-D′. Deltaic deposits, West Tuscola field and adjacent areas.

of individual, partly interconnected, lenses enveloped by bar-slope deposits. This irregular pattern of bar-crest deposits, based on the marked variation in log patterns from well to well, is typical of the field. The varied levels of oil and gas production, as shown on the sections (Fig. 7), are due in part to perforating as low as possible; the erratic gas/oil/water relations, however, attest to local stratigraphic controls within the field.

Two of the cored wells on cross section B-B′ (Fig. 7), Humble No. 1-B W. A. Graham and Humble No. 2 W. A. Graham, penetrated distributary channel deposits which represent the filling of minor channels that coursed to, or partly through, the stream-mouth bar at the time of its development in this area. Tight sandstone at the top of the delta-front unit in Humble No. 3 F. F. Hodge (Fig. 7, cross section A-A′) is interpreted as a calcareous lateral spit or beach. Marsh deposits, marking the final stages in outbuilding of a delta lobe, are best developed in cores from the southern part of the area (Fig. 7, cross section B-B′). Bay and shallow-marine deposits, which form an impermeable seal over the porous deltaic sandstones, reflect marine incursion over the subsiding, abandoned deltaic complex.

Cross sections of deltaic deposits in West Tuscola field and adjacent areas (Fig. 8) show gross facies relations and several significant features of the deltaic sandstone body and the overlying datum. The top of the delta-front facies is essentially planar and parallel with the limestone datum, which iself was formed on a relatively flat surface parallel with sea level. (These two surfaces have since been tilted in accord with the regional structural pattern.) The top surface of the sandstone interval is the result of settling of deltaic deposits to a position of equilibrium parallel with and near sea level at the time the delta was prograding. The base of the clean sandstone zone is very irregular, however, indicating variations in supply of bed-load sands to different parts of the deltaic complex. The intricacy of the porous sandstone pattern within the clean sandstone zone is evident from the lack of continuity of sandstone beds throughout the field and the numerous levels of production. This lack of continuity can affect strongly fluid transmissibility within the reservoir.

The development of a thick interval of delta-front deposits and the stacking and irregular distribution of bar-crest sandstones within that interval are interpreted as being the result of interaction of various sedimentary processes and variations in stream load at a river mouth

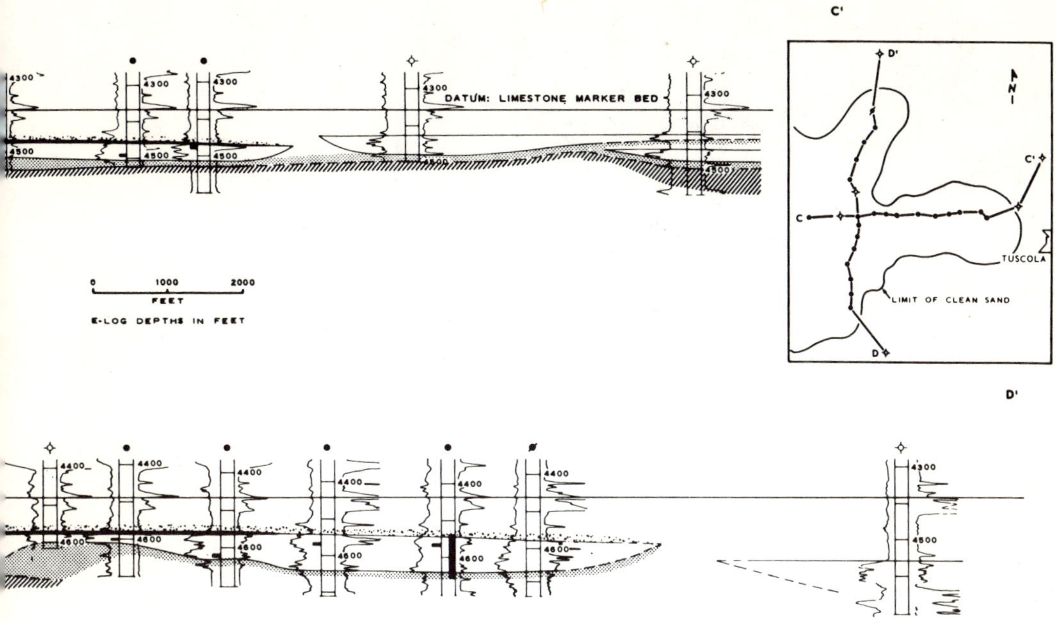


(Fig. 9a). During times of normal river discharge, stream bed load is carried through bifurcating subaqueous channels to the shoaling crest of a stream-mouth bar, where sand is deposited, then reworked and spread laterally by wave action into a crescentic bar (Fisk, 1961). Finer sediments settle from suspension on the slope of the bar. As river flow increases (annually, during spring flood, or catastrophically), the channels enlarge and cut through the bar crest, allowing bed-load sand to spread as a bulge onto the frontal bar slope. The sand settles into the finer material on the bar slope and, as river stage drops, is covered by the normal bar-slope silt and clay (Fig. 9b). In this manner, pods of sand virtually enclosed in bar-slope deposits are created at the extremities of each of the subaqueous channels.

Development of a thick, stacked succession of bar-crest sandstones (recognized in the "Gray" interval in West Tuscola field) is the result of settling of bar-crest sands into a thick clay platform by displacement and compaction. Areas of maximum settling lie just seaward of the bar crest on the front slope of the bar (Morgan, 1961). Repeated switching of subaqueous channels (in response to river-stage variations) and feeding of bed-load sands into this unstable zone result in a thick alternating succession of bar-crest and bar-slope deposits (Fig. 9c). Areas lateral to the subaqueous channels receive thick accumulations of bar-slope deposits which become the stratigraphic equivalent of the bar-crest sands (Fig. 7).

The building of a near-sea-level stream-mouth bar is necessary for deltaic progradation to take place. When accumulation of bar-crest sands has compacted the substrate until the weight of the sand no longer exceeds the strength of underlying silt and clay, temporary equilibrium is reached and the bar crest stands high (commonly to within a few feet of sea level). The delta plain then builds seaward over the platform of stream-mouth–bar-crest sands. Progradation is continuous and must be preceded by the formation of a stable stream-mouth bar; hence it is likely that the uppermost bar-crest sands (marking the positions of successive bars) have continuity throughout much of the delta lobe, in contrast with the distribution of sands low in the complex, which are discrete pods commonly connected, however, by lateral overlap. These relations are demonstrated in West Tuscola field (Fig. 7), where widespread clean sandstones lie at a consistent level at the top of the delta-front facies, but occur only sporadically below that level. Lateral extensions of the uppermost sandstones (Hum-

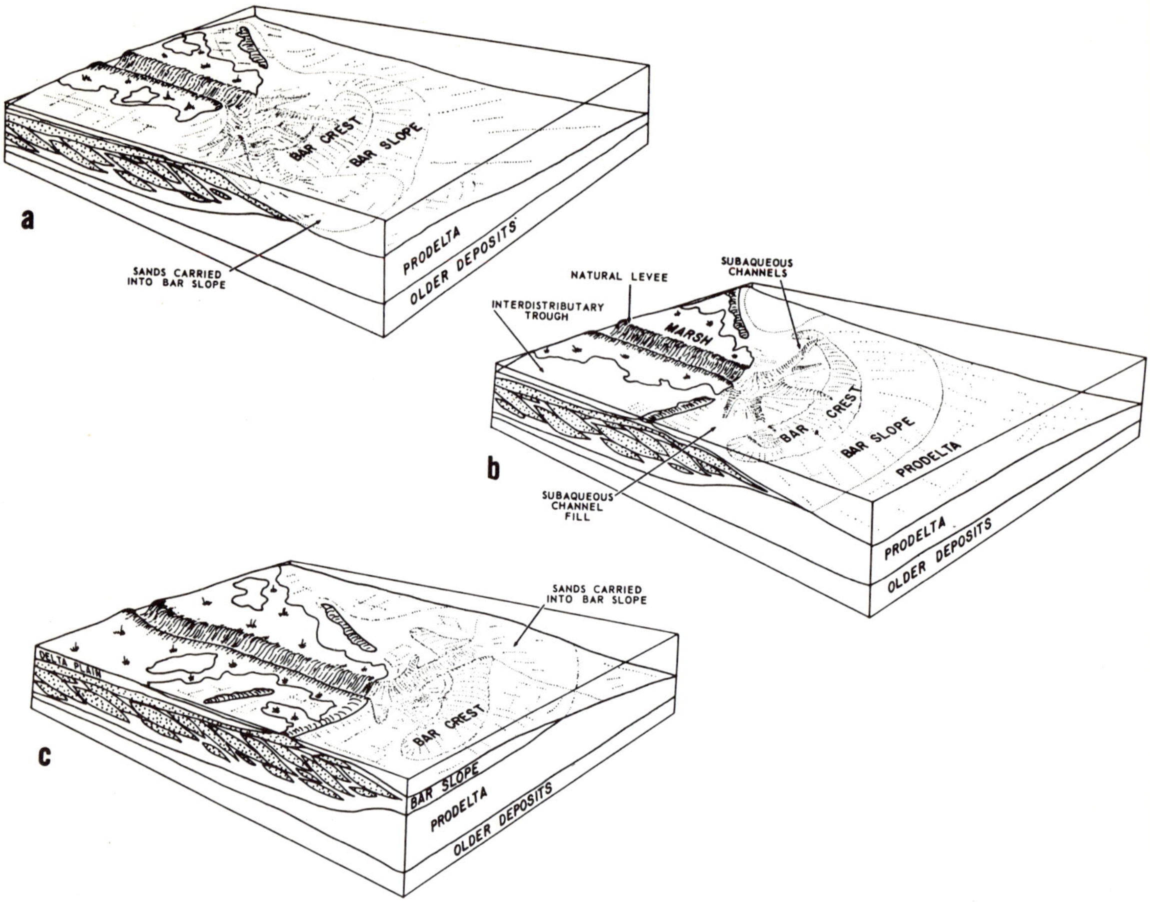


FIG. 9.—Stages in development of delta-front unit.

ble No. 3 Hodge, Fig. 7a) are spits and bars developed during the progradational phase off the flanks of the delta (Fig. 9c).

### Development of Strawn Deltaic Complex

Successive progradation and abandonment of individual delta lobes have led to the development of discrete sandstone bodies within the Strawn succession of Taylor County (Fig. 10). Upstream diversion of Strawn channels led to formation of deltaic complexes separated in space and time (Fig. 11); such relations are postulated for the "West Tuscola" system and systems adjacent to it. Thus, the potentially productive sandstone bodies can occur at various stratigraphic levels and can be separated by impervious units—bar-slope and/or delta-plain material (marsh and bay deposits). In detail, the individual deltas build as fingerlike extensions seaward, flanked by brackish interdistributary areas receiving fine-grained sediments. As progradation continues and the deltaic mass settles to an equilibrium position near sea level, marsh deposits of the delta plain develop locally over delta-front sands. Upon abandonment of a channel, diversion of stream flow to an adjacent area, and continued subsidence of the abandoned deltaic lobe, that area becomes the site of brackish bays lateral to the newly formed delta.

Continued shifting of the stream channels and seaward extension of deltas as sedimentation exceeds regional subsidence result in a widespread deltaic complex. It is probable that such a complex makes up much of the Strawn succession in Taylor and surrounding counties. Individual deltas within the complex may have a variety of porosity trends; hydrocarbon entrapment in these trends would depend on their orientation to present regional strike, the posi-

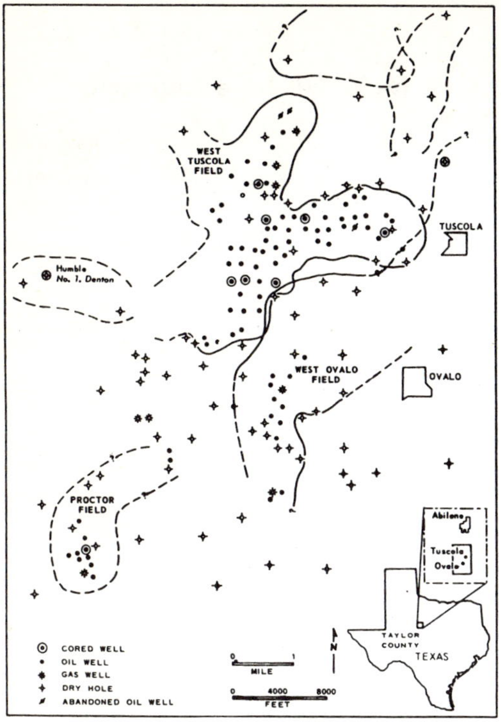


FIG. 10.—Distribution of deltaic sandstones, Tuscola and Ovalo areas.

tion of updip pinchout, and/or the degree of sinuosity which could localize oil pools within the deltaic sandstone bodies. The documented occurrence of deltaic sandstones as stratigraphic traps within the "Gray" interval emphasizes the potential of the Strawn sandstones over broad areas of north Texas (Fig. 1).

## REFERENCES CITED

Fisk, H. N., 1961, Bar-finger sands of Mississippi delta, *in* J. A. Peterson and J. C. Osmond, eds., Geometry of sandstone bodies: Am. Assoc. Petroleum Geologists, p. 29–52.

Morgan, J. P., 1961, Genesis and paleontology of the Mississippi River mudlumps—Pt. 1, Mudlumps at the mouths of the Mississippi River: Louisiana Geol. Survey Geol. Bull. 35, 116 p.

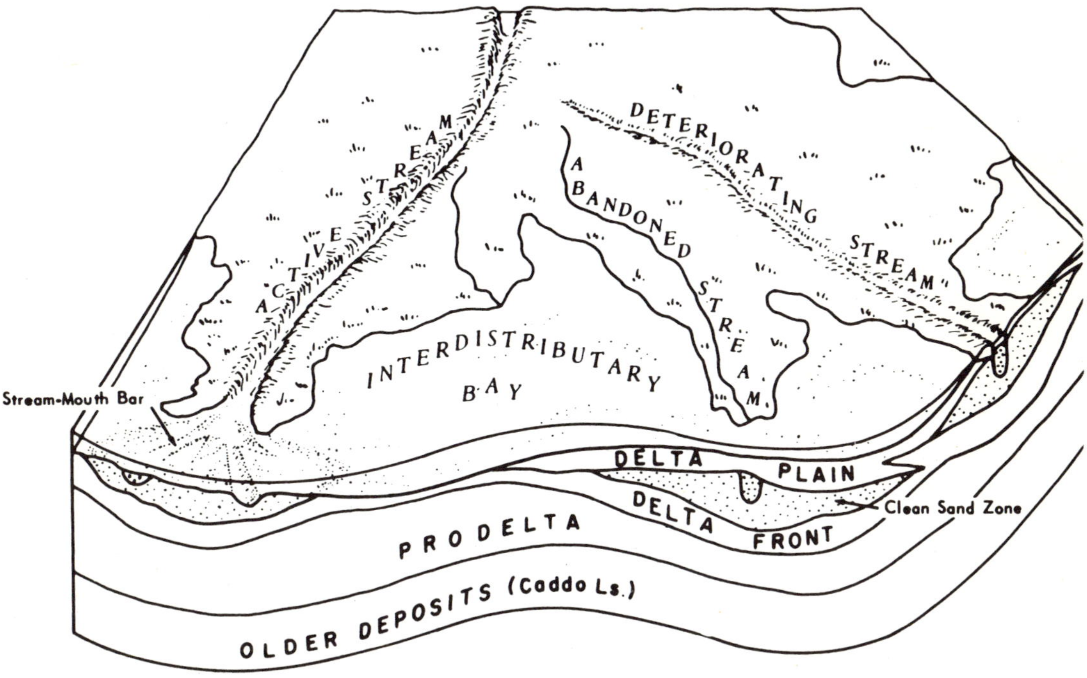


FIG. 11.—Diagrammatic representation showing development of several delta lobes in Tuscola area.

**Reprinted from:**
BULLETIN OF THE AMERICAN ASSOCIATION OF PETROLEUM GEOLOGISTS
VOL. 56, NO. 3 (MARCH, 1972), PP. 554-571, 16 FIGS., 2 TABLES

# "$G_2$" Channel Sandstone, Main Pass Block 35 Field, Offshore Louisiana[1]

**J. A. HARTMAN**[2]

New Orleans, Louisiana 70160

**Abstract** The Miocene "$G_2$" sandstone, Main Pass Block 35 field, is a classic example of oil production from a stream channel. The channel has a maximum width of about 0.5 mi, and its edges are closely defined by dense well control. Nineteen wells or sidetracks have been completed in this sandstone, and ultimate recovery will exceed 12 million bbl of oil.

## Introduction

The Main Pass Block 35 field is in the Gulf of Mexico about 50 mi southeast of New Orleans in 10 ft of water (Fig. 1). The Miocene "$G_2$" sandstone, the largest single reservoir in the field, is a classic example of an oil production from a stream channel. Data for this study were obtained from examination of all electrical logs in the field, from examination of numerous sidewall samples, and from one wireline core.

FIG. 1—Main Pass Block 35 field, offshore Louisiana.

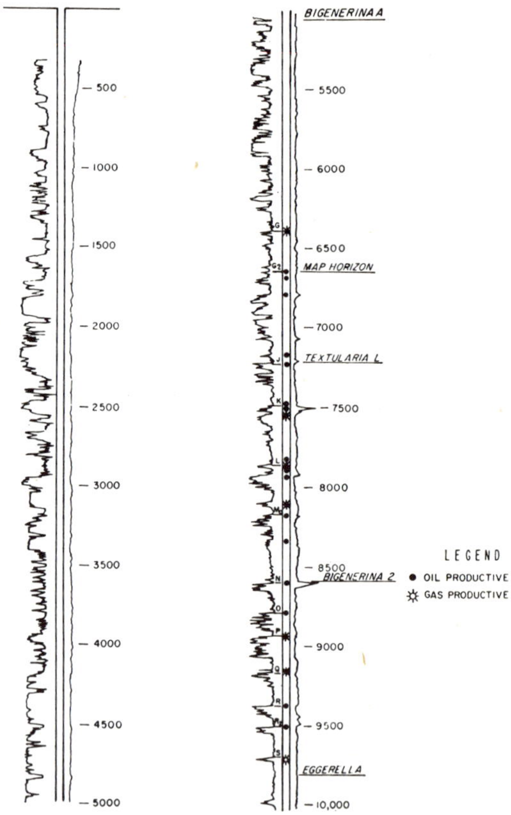


FIG. 2—Composite type log, Main Pass Block 35 field.

## History of Field

Main Pass Block 35 field was discovered in 1951 by Shell Oil Company with the successful completion of State Lease 1961 No. 1. As of January 1, 1971, 154 wells had been drilled in the field resulting in 134 completed wells and 20 dry holes. Eighteen of the completed wells were duals, initially. Commercial hydrocarbon accumulations have been proved in 23 Miocene sandstones, as indicated on Figure 2, a type log for the field. Cumulative oil production is about 70 million bbl, and ultimate recovery is estimated at about 100 million bbl. The geology of the field has been discussed briefly by Hartman (1967), and a paleostructural study of the "N" sandstone, one of the main producing zones, has been published by Paynter (1970).

[1] Manuscript received, March 4, 1971; accepted, June 10, 1971.

[2] Shell Oil Company.

I thank J. Braunstein and D. D. Paynter for their critical review of this paper and Shell Oil Company for permission to publish it.

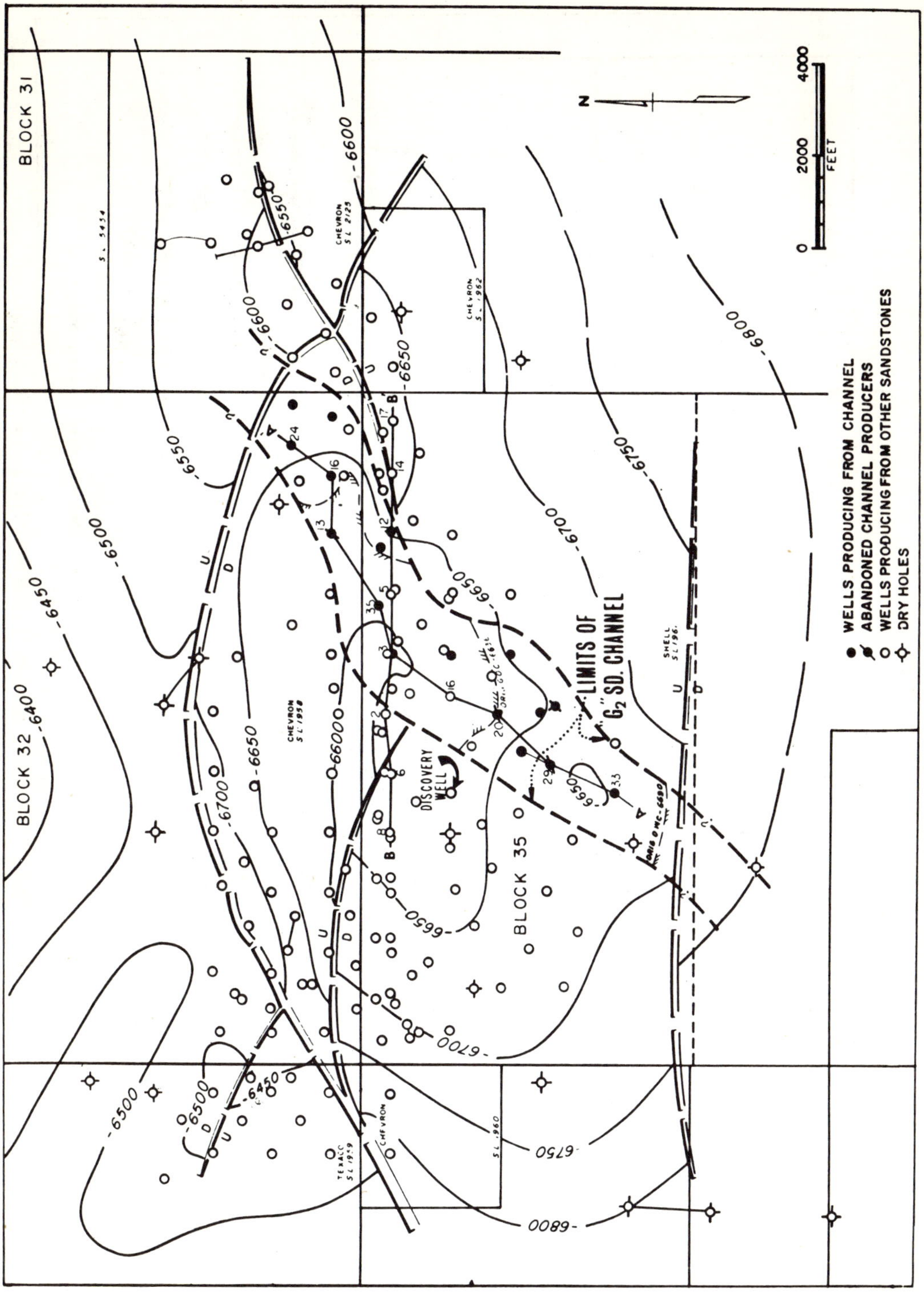


FIG. 3—Structure map of Main Pass Block 35 field. Datum is top of "$G_2$" sandstone.

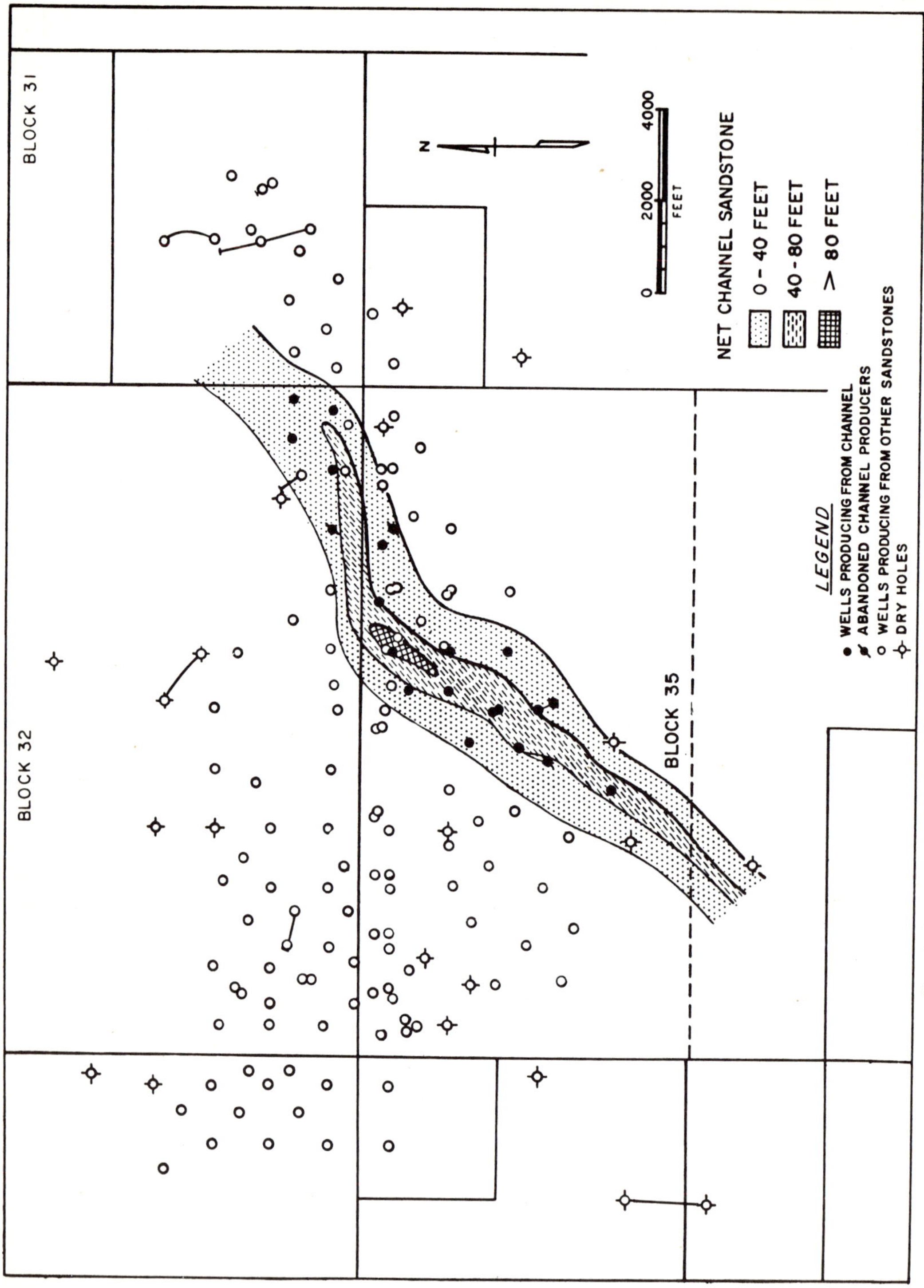


FIG. 4—Isopach net sand map, "$G_2$" channel sandstone, Main Pass Block 35 field.

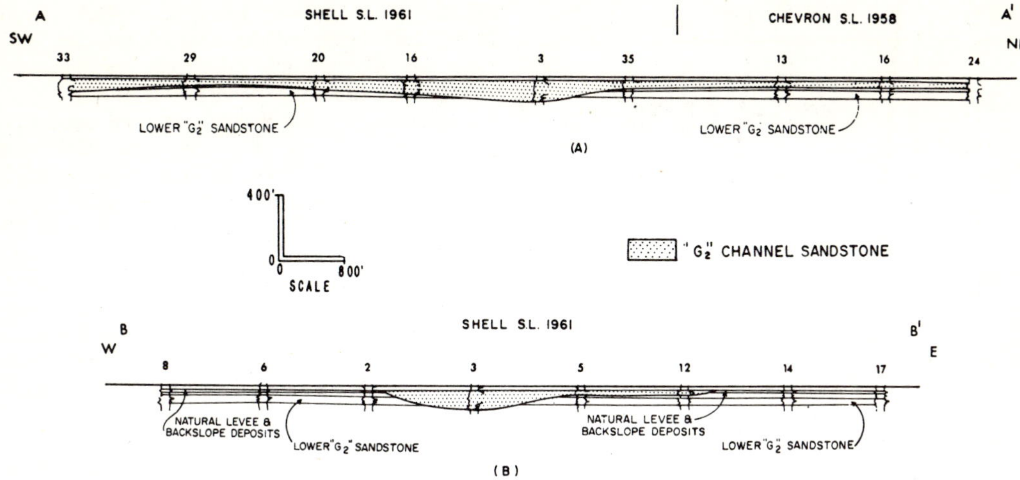


FIG. 5—Longitudinal (*A*) and transverse (*B*) cross sections through "$G_2$" channel sandstone, Main Pass Block 35 field.

## "$G_2$" CHANNEL SANDSTONE

In the discovery well, the 6,500-ft "$G_2$" sandstone was penetrated on the west side of the channel in the natural levee and backslope deposits. The channel facies was penetrated first in the Shell State Lease 1961 No. 3 in what was shown by subsequent development to be the deepest part of the channel. To January 1, 1971, 31 wells or sidetracks of old wells have penetrated the channel facies. Thus, control within the productive area of the field is good, and the limits of the channel sand are closely defined. The gas-oil contact at 6,636 ft subsea and the oil-water contact at 6,690 ft subsea also are clearly defined by well control.

Figure 3 is a structural interpretation of the "$G_2$" sandstone. The general structure is that of a rollover anticline in the downthrown block of a south-dipping contemporaneous fault which at the "$G_2$" level has about 200 ft of displacement. In addition to the major fault, several others intersect the "$G_2$" sandstone but have no effect on the hydrocarbon accumulation in the channel sandstone.

Wells which are or have been completed in the "$G_2$" channel sandstone are shown as black dots on Figure 3. These wells are in a northeast-southwest linear trend crossing the structure. The limits of the channel also are indicated on Figure 3. Figure 4 is a net sandstone isopach map of the channel sandstone. Figure 5A is a longitudinal cross section, and Figure 5B is a transverse cross section through the channel. Cumulative production to June 1, 1970, is 8.4 million bbl of oil, and ultimate primary recovery is estimated at about 12 million bbl of oil. Prior to a recent increase in voidage, this reservoir was producing under strong waterdrive conditions. Recent water influx, however, has not been sufficient to match increased withdrawals, and the reservoir pressure is declining. To arrest this decline a pressure-main tenance project was initiated in early 1971.

## DESCRIPTION OF "$G_2$" SANDSTONE, NATURAL LEVEE, AND BACKSLOPE DEPOSITS

The "$G_2$" channel sandstone consists of a clean, well-sorted, fine-grained quartz sandstone with up to 20 percent feldspar and minor amounts of clay and carbonate. Average porosity is about 34 percent and average permeability is more than 3 darcys. A typical capillary pressure curve for this deposit is shown on Figure 6. Bottomhole-pressure data indicate that continuity within the channel is excellent.

In contrast to the channel sandstone, the natural levee and backslope deposits are very fine-grained sandstone or siltstone. Average porosity is about 26 percent, and average permeability is about 75 md. A typical capillary pressure curve also is shown in Figure 6. The difference in the reservoir quality of these two deposits is readily apparent from a comparison of the two capillary pressure curves.

## GEOLOGIC HISTORY

The "$G_2$" sandstone series is part of an overall regressive sequence. Paleontologic data

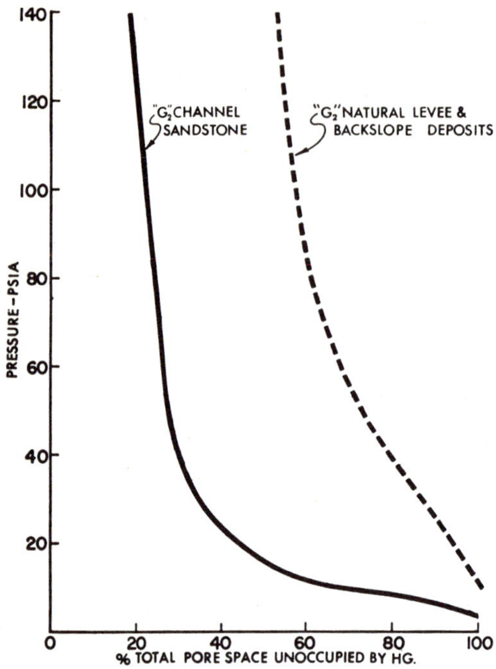


FIG. 6—Average Hg-air capillary pressure curves for "$G_2$" channel sandstone and "$G_2$" natural levee and backslope deposits, Main Pass Block 35 field.

from an unpublished Shell Oil Company report by Shelton and Parrott indicate that the lower "$G_2$" series consists of very fine-grained sandstone, siltstone, and shale deposited in a shallow marine fringe environment. These sediments were deposited in front of a delta system which, based on interpretation of the directional data obtained from a wireline core, was advancing from the northeast toward the southwest. The course of one of the distributaries from this delta system traveled through the Main Pass Block 35 area. This channel incised its course completely through the lower "$G_2$" sediments in the vicinity of State Lease Well No. 3 and deposited the sediments which now form the "$G_2$" channel sandstone. As the water supply exceeded the capacity of the channel, the stream overflowed its banks and deposited the finer grained natural levee and backslope deposits (Fig. 4). These sediments are assumed to be separated from the channel sandstone by some sort of permeability barrier because the commercially producible hydrocarbons are confined to the channel area. This barrier may be a clay drape, or it may be only a function of the finer grain sizes in the levee and backslope deposits. The poorer reservoir quality of the levee and backslope deposits is shown by the capillary pressure curves (Fig. 6). Following deposition of the channel sand, the delta abandoned this channel, and the entire area subsided. The overlying shales then were deposited in a marine environment.

Paynter (1970) published a paleostructural analysis to explain the anomalous distribution of hydrocarbons in the "N" sandstone. Figure 5 (p. 786) of that paper is an isopach map of the $K_m$ structure at the time of "$G_2$ deposition". This map shows a northeast-southwest-trending low area on the northern part of State Lease 1961, and this low closely coincides with the position of the "$G_2$" channel. Paynter (personal commun.) believes that the "$G_2$" channel flowed in this low area to avoid the adjacent highs.

The extension of the channel northeast and southwest cannot be determined with existing control, for none of the flank wells penetrated the channel facies.

## REFERENCES CITED

Hartman, J. A., 1967, Main Pass Block 35 field, Plaquemines Parish, Louisiana, *in* Oil and gas fields of southeast Louisiana, v. 2: New Orleans Geol. Soc., p. 119–124.

Paynter, D. D., 1970, Main Pass Block 35 field, Louisiana: Paleostructural analysis: Am. Assoc. Petroleum Geologists Bull., v. 54, p. 783–788.

**Reprinted from:**
BULLETIN OF THE AMERICAN ASSOCIATION OF PETROLEUM GEOLOGISTS
VOL. 44, NO. 9 (SEPTEMBER, 1960), PP. 1473-1504, 19 FIGS., 8 TABLES

# CRETACEOUS SEDIMENTATION IN UPPER MISSISSIPPI EMBAYMENT[1]

WAYNE ARTHUR PRYOR[2]
Pittsburgh, Pennsylvania

## ABSTRACT

The Cretaceous (Gulfian) sediments in the Upper Mississippi Embayment are the result of deltaic deposition. The deltaic system (the "McNairy Delta") centered in the northeastern end of the embayment and dispersed the sediments toward the southwest. Three major environments of deposition are represented by the embayment Gulfian sediments: (1) the fluviatile and upper deltaic environment, represented by the non-marine, arenaceous Coffee and McNairy formations; (2) the pro-delta shelf environment, represented by the marine, calcareous, and argillaceous Selma and Owl Creek formations; and (3) the transitional delta platform environment, represented by the marine, argillaceous Coon Creek formation and the marine parts of the arenaceous McNairy and Coffee formations.

Several regional trends are apparent from subsurface studies—the sedimentary body thickens from less than 100 feet in the north to more than 1,100 feet in the southwest; the clastic ratio decreases from greater than 2:1 in the north to less than 1:2 in the south; the sediments range from totally non-marine in the north to totally marine in the southwest.

McNairy formation sands are first-cycle protoquartzites. The heavy-mineral assemblages of the McNairy formation and the adjacent Coffee, Selma, Coon Creek, and Owl Creek formations are identical and are composed chiefly of kyanite, staurolite, tourmaline, zircon, and sillimanite, with minor amounts of other minerals.

Major sand bodies in the McNairy formation trend southwest, parallel with the paleoslope and the regional transport direction. Local trends in the McNairy formation can be predicted from studies of cross-stratification and fabric.

The Gulfian sediments in the Upper Mississippi Embayment were derived chiefly from the high-rank metamorphic rocks of the Blue Ridge and Piedmont Plateau of the southern Appalachians and were transported to the embayment by a single, large stream system. Humid, temperate to subtropical climatic conditions were prevalent in both the source and depositional areas. The depositional history involves two transgressive-regressive cycles of late Cretaceous seas, separated by an erosional interval. The initial transgressive-regressive cycle resulted in the deposition of the Coffee-Selma-Coon Creek-McNairy sequence, and the later cycle, apparently more rapid and widespread, resulted in the deposition of the Owl Creek-Prairie Bluff sequence.

A depositional model, postulated for the Cretaceous (Gulfian) Upper Mississippi Embayment, includes as primary elements (1) *basin geometry:* a shallow, open-end, intracratonic basin, plunging southwest; (2) *depositional pattern:* deltaic with longitudinal filling; (3) *paleoslope:* parallel with basin axis, toward the southwest; (4) *depositional strike:* normal to paleoslope, northwest-southeast; (5) *transport direction:* parallel with paleoslope and normal to depositional strike, toward the southwest; (6) *sand trends:* parallel with transport direction and paleoslope and normal to depositional strike, toward the southwest; (7) *sedimentary association:* protoquartzite grading down paleoslope into impure carbonates; and (8) *tectonic activity:* mild subsidence and uplift.

[1] Presented at the S.E.P.M. meeting in Dallas, Texas, March, 1959. Manuscript received, January 2, 1960. Publication authorized by the Illinois State Geological Survey, Urbana, Illinois. Part of this material was used as a dissertation submitted to the Graduate School, Rutgers University, New Brunswick, New Jersey, in partial fulfillment of the requirements for the degree of Doctor of Philosophy, May, 1959.

[2] Formerly assistant geologist, Illinois State Geological Survey. Present address, Gulf Research and Development Company, Pittsburgh, Pennsylvania. The investigation was conducted as part of a program of basic research in ground-water geology at the Illinois State Geological Survey. Paul Edwin Potter of the Illinois Survey and J. H. C. Martens of Rutgers University made many helpful suggestions. Appreciation is expressed to the many people of the Illinois State Geological Survey at Urbana, Illinois, and the U. S. Geological Survey at Paducah, Kentucky, and at Memphis, Tennessee, who offered assistance and suggestions. Special acknowledgments are extended to Miss Margaret Castle, who drafted the illustrations and to Paul Edwin Potter, who critically read and edited the manuscript.

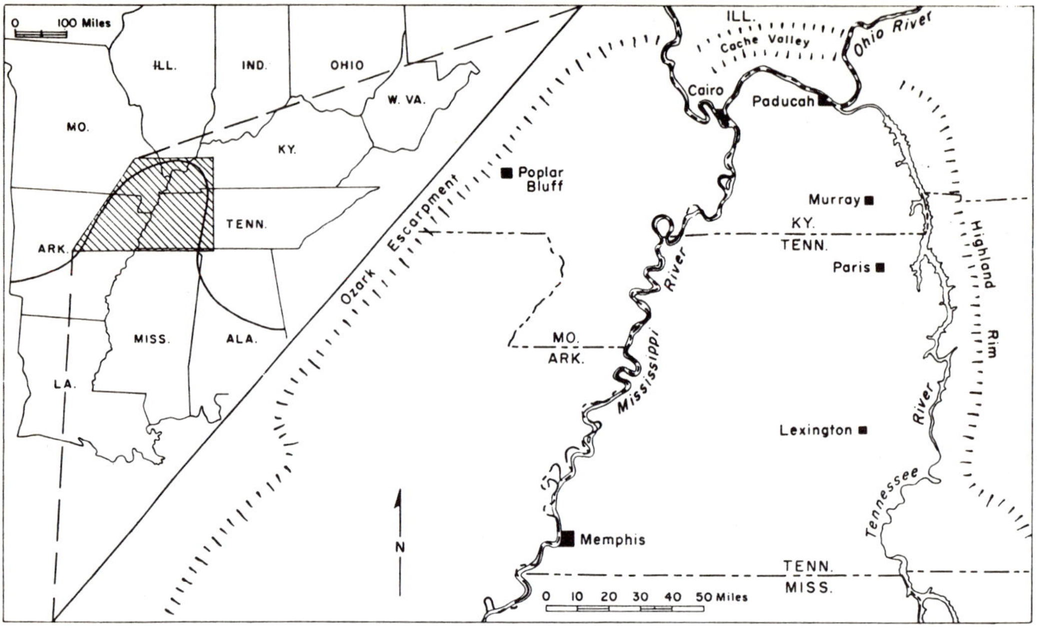


FIG. 1.—Index map showing location of Upper Mississippi Embayment. Shaded section is area studied.

## Introduction and Objectives

That there are a relatively limited number of sedimentary associations and relatively few types of sedimentary basins has been increasingly stressed in recent years (Dapples *et al.*, 1948, p. 1943; Krumbein *et al.*, 1949, pp. 1876–77; Pettijohn, 1957, pp. 610–47; and others). This viewpoint holds that the vast majority of sedimentary rocks belong to relatively small numbers of recurring patterns of sedimentation and that both the patterns of deposition and accompanying sedimentary associations are closely related to the tectonic control and type of depositional basin. The results, therefore, of systematic study of a relatively few properly chosen depositional basins should be applicable to similar sediments in other basins.

The Mississippi Embayment is an intracratonic type of depositional basin that has been common to the North American craton for long periods of geologic time. The purpose of this paper is to depict the gross geometry, determine the patterns of sedimentation and depositional environments, show relations of sediment transport to depositional strike, determine the provenance, and interpret the geologic history of the late Cretaceous Upper Mississippi Embayment. The ultimate objective is to contribute to the formulation of a descriptive model for this type of deposition.

The late Cretaceous (Gulfian) sediments were chosen for this investigation because their present limits apparently differ little from the original depositional extent; they exhibit a wide range of lithic types over relatively short distances; they are well exposed; and subsurface control is available.

The strata reported in this paper occur wholly within the Upper Mississippi Embayment, here defined as the area of the embayment north of the latitude of the Tennessee-Mississippi state line (Fig. 1). The late Cretaceous sediments are exposed in an arcuate band 10–50 miles wide extending continuously from Crowleys Ridge in southeastern Missouri, through the area south of the Cache River Valley in Illinois, and into western Kentucky and Tennessee, generally west of the Highland Rim and the Tennessee River. A limited outcrop band of these sediments, less than one mile wide and 10–15 miles long, occurs in northeastern Arkansas, adjacent to the Ozark Escarpment. The Cretaceous sediments occur in subsurface throughout northeastern Arkansas, southeastern Missouri, extreme southern Illinois, western Kentucky, and western Tennessee.

### Previous Investigations

Most of the earlier work on the Cretaecous Mississippi Embayment sediments dealt with general stratigraphy and paleontology. Harper in 1857 subdivided the Cretaceous sediments of Mississippi into two general units, a lower "Glauconitic group" and an upper "Calcareous or Lime group." Hilgard (1860) recognized four units, the Eutaw group, the Tombigbee group, the Rotten Limestone group, and the Ripley group. In 1864 Safford extended these units into Tennessee and noted the northward change in lithologic character of the calcareous sediments. Glenn (1906) was the first to present a unified picture of a large part of the Mississippi Embayment, with his ground-water study of the Tennessee, Kentucky, and Illinois parts of the embayment. He also was the first to recognize the widespread character of Hilgard's Ripley group.

L. W. Stephenson (1914, 1955), one of the most prolific workers on Coastal Plain geology, established important interregional correlations of the Coastal Plain sediments. He and Monroe (1938, p. 1642) showed the interfingering relations of the Cretaceous clastic and carbonate sediments in Mississippi and Alabama. Stephenson *et al.* (1942, Chart 9) recognized seven stratigraphic units in the Upper Mississippi Embayment, which in ascending order are the Tuscaloosa formation, the Eutaw formation, the Coffee sand, the Selma chalk, the Coon Creek tongue and the McNairy sand member of the Ripley formation, and the Owl Creek formation. Early in his investigations of the Gulf Coastal Plain, Stephenson (1914, p. 17) suggested a near-shore, estuarine, and in part fresh-water origin for the McNairy sand. In 1917, Berry suggested a deltaic environment of deposition for the Tuscaloosa formation in Mississippi and Alabama.

Recent investigations by Stearns and Armstrong (1955) and Stearns (1958), based largely on sample studies and geophysical well logs, show the Gulf series thickening toward the south and sand-shale ratios decreasing in that direction. Stearns (1958, p. 1095) suggested deltaic deposition in the northeastern part of the embayment. He presented an interpretation of the tectonic and depositional history of the embayment that involved several major transgressions and regressions of the Cretaceous seas with no major breaks in sedimentation.

The only comprehensive petrographic study of Cretaceous sediments in the Mississippi Embayment is an investigation of the heavy minerals in the Tombigbee sand in Mississippi and Alabama by Needham (1934), who found epidote to be the major heavy mineral in the suite. Dutton (1928) studied the heavy minerals of a limited number of sand samples from the Cretaceous of southern Illinois.

### Stratigraphy

Detailed discussions of the stratigraphy are presented by Stephenson and Monroe (1938), Mellen (1958), and Pryor (1959 and in preparation).

Figure 2 demonstrates the general stratigraphic relations of the Upper Cretaceous (Gulf series) rock units in the Mississippi Embayment. Shown are four groups of sediments, separated by three major breaks in sedimentation, resting on a floor of Paleozoic rocks.

The Tuscaloosa formation, a major unit in Mississippi and Alabama, is present on the eastern periphery of the embayment as a series of limited outliers. It is composed mainly of clayey chert gravels, derived from the siliceous Paleozoic rocks of adjacent areas. The Eutaw formation is present only in the extreme southern part of the area as a thin series of glauconitic and fossiliferous sands and clays.

Disconformably overlying the Tuscaloosa formation on the north and the Eutaw formation on the south, is a group of sediments, composed of two major facies—a coarse clastic facies in the north and a fine clastic and carbonate facies in the south. The basal unit of this closely related group of sediments, the Coffee formation, is composed of lignitic sands and clays in the north, grading into sandy, calcareous clays toward the south, where it is transitional with the lower Selma formation chalks. In southern Tennessee and Mississippi the Selma formation is made up of as much as 400 feet of impure, clayey chalks, which thin toward the north and grade into calcareous, glauconitic, and fossiliferous clays. In the vicinity of Lexington, Tennessee, these materials merge with the overlying Coon Creek for-

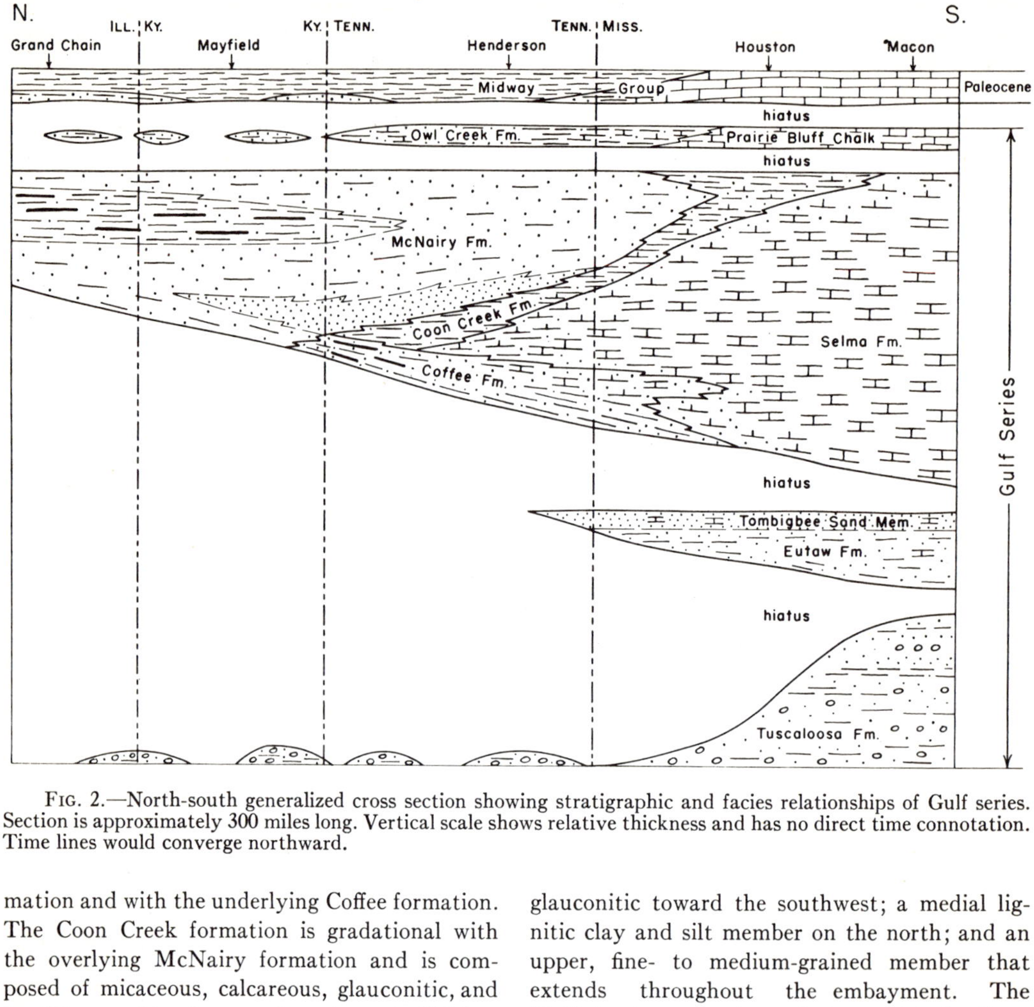


FIG. 2.—North-south generalized cross section showing stratigraphic and facies relationships of Gulf series. Section is approximately 300 miles long. Vertical scale shows relative thickness and has no direct time connotation. Time lines would converge northward.

mation and with the underlying Coffee formation. The Coon Creek formation is gradational with the overlying McNairy formation and is composed of micaceous, calcareous, glauconitic, and fossiliferous silts, clays, and sands up to 200 feet thick. The Coon Creek thins gradually toward the north, and in the vicinity of the Kentucky-Tennessee state line merges with the McNairy and Coffee formations.

The McNairy formation is the most widespread and lithologically consistent unit in the Upper Mississippi Embayment. It is as much as 600 feet thick, and in the northern part of the area is composed of lignitic, fine to medium, very micaceous, cross-stratified sands, with numerous clay laminations and clay beds. Southwestward it grades into calcareous, glauconitic sands. There are three distinct members present in the McNairy formation—a lower, very fine-grained member on the south that grades into the Coon Creek formation and becomes calcareous and glauconitic toward the southwest; a medial lignitic clay and silt member on the north; and an upper, fine- to medium-grained member that extends throughout the embayment. The McNairy formation is gradational with the Coon Creek formation and with the Coffee formation. The cross section (Fig. 2) shows the coarse, clastic facies (McNairy formation) and the fine clastic-carbonate facies (Selma formation) grading into each other, these gradational sediments being represented by the Coon Creek formation.

Separated from the McNairy, Coon Creek, and Selma formations by a sharp sedimentary break and in some places by an angular unconformity, is the Owl Creek formation. It is composed of up to 50 feet of glauconitic, fossiliferous, calcareous, and micaceous sandy clays that grade southward into chalks and limestones (Prairie Bluff chalk). The top of the Owl Creek formation is marked by a pronounced unconformity, on which a soil zone has locally developed. Overlying this unit in most of the area are the sediments of the Paleocene Midway stage (Clayton and Porters Creek formations).

## Methods

An isopachous map of the Gulf (late Cretaceous) sediments was constructed, based on published geophysical logs and well-sample studies. Sand-clay ratios and the percentage of marine beds were calculated from the well-sample studies and a map showing lithofacies and percentage of marine beds was drawn. An isopach and lithofacies map of the McNairy formation was constructed from the well-sample studies.

The fossiliferous outcrops were examined and population counts of megafossils were made. Samples were examined for microfossil content and population counts were made to aid in interpretation of environment.

Textural analyses of the sands were made and a map was constructed to show median diameter distribution. The size data were plotted on arithmetic probability paper. The curves were compared with the textural investigations of Doeglas (1946, pp. 19–40; 1950, pp. 238–47), and van Andel and Postma (1954, pp. 80–89) whose environmental interpretations were used in mapping the resulting sand types.

Outcrops were examined for sedimentary structures and bedding thickness. Cross-stratification directions were measured at each locality. A circular dip direction indicator (Pryor, 1958, p. 230) was inserted along the plane of stratification and the azimuth of maximum inclination was obtained with a Brunton pocket transit. No tilt correction was necessary for these essentially flat sediments.

As the large area involved in this study precluded visiting all available outcrops and measuring all of their directional properties, a systematic sampling procedure was designed to combine economy in field sampling with measures of reliability for regional and local mean directions. The procedure used in general followed the approach to data sampling and treatment used by Potter and Olson (1954) and Potter *et al.* (1958, pp. 1019–20). The outcrop area was divided into 43 six-mile intervals, perpendicular to the strike of the outcrop belt. This system of intervals permitted an even distribution of sampling points and facilitated the processing of the data. Potter and Siever (1956, p. 233), using the multi-level sampling procedure, found that 1–5 outcrops per interval and 2 directional observations per outcrop yielded satisfactory results in their regional sampling of the Caseyville sandstones. Potter *et al.* (1958, p. 1019) in sampling the Chester sandstones modified this method slightly and found that 5 observations per outcrop, if that many were possible, was the optimum number. A sampling density of 2–9 outcrops per interval and 2–10 directional observations per outcrop was adopted and followed for this study.

The Tukey orientation test (Tukey, 1954; Harrison, 1957, pp. 98–105; Rusnak, 1957, pp. 53–55; Potter *et al.*, 1958, pp. 1022–24), which combines the vector method of deriving resultant mean orientations with the chi-square test of statistical significance and reliability, was used. The simplicity of the test permits rapid handling of large amounts of directional data. This test, based on 9 class intervals of 40° each, was used for each of the individual outcrops, each of the sampling intervals, and for the entire area, using all of the observations.

Point counts were made on thin sections of the McNairy formation sands to estimate the percentage of each constituent present and to classify the sediment. In addition, unlithified sands were mounted, etched, and stained, and percentage estimates of their constituents were made. In both thin section and etched mount, 200 point counts per slide were made. Additional 200 grains of quartz per thin section were observed to differentiate the unstrained, strained, and polycrystalline varieties. Estimates of granular angularity of 200 quartz grains per thin section and stained mount were made.

Heavy minerals from the sands and the argillaceous and calcareous sediments were mounted and a line count of 200 grains per slide made. Approximately 100 grains of tourmaline and of zircon per slide were counted and degree of roundness was estimated with the Powers (1953) chart. "Normal analysis" (van Andel, 1950, pp. 41–43) was used and heavy-mineral assemblages from sands exhibiting the same magnitude of median diameter and degree of sorting were analyzed to cope with the effects of granular variation.

X-ray spectrometer analyses of the minus 5-micron fraction of clays were made, and the relative amount of each clay mineral per sample was estimated.

## Subsurface Investigations

### GENERAL

Regional variations in thickness, lithofacies, and environmental relations yield a picture of the configuration of the depositional basin and of the distribution of the Cretaceous sediments in the Upper Mississippi Embayment.

Accurate and useful subsurface information in the northern part of the Mississippi Embayment is sparse. Most of the available information consists of very generalized drillers logs. The informative records are few, scattered, and consist of geophysical well logs and well-sample studies. The subsurface records used in this study are chiefly sample studies obtained from the limited literature concerned with the area and the writer's studies of the samples he collected.

### THICKNESS

The Gulf series is thickest in the southeast and thins markedly west and gradually north. The isopach map (Fig. 3) shows a range in thickness from zero at the truncated borders and less than 100 feet at the western limit of outcrops to more than 1,100 feet in southwestern Tennessee.

Stearns and Armstrong (1955, pp. 19–20) found that the Upper Cretaceous sediments thickened markedly southward, reaching more than 3,000 feet in northern Mississippi. Mellen (1958, p. 37) noted more than 3,500 feet of Gulfian sediments in central Mississippi. In the northern part of the area, localized thinning to as little as 25 feet is present, in response to high relief developed on the erosional surface of the underlying Paleozoic rocks (Pryor, in preparation). Erosion and truncation are important as control factors in present thickness in only a very small area along the periphery of the embayment. The deepest downwarp is shown by the area of thickest accumulation of sediments, in the southeastern part of the embayment. The axis of this depositional and structural trough is in the east-central part of the embayment and trends northeast-southwest.

### LITHOFACIES AND ENVIRONMENTAL RELATIONS

A lithofacies map (Fig. 4) of the Gulf series was made, based on geophysical well logs and well-sample studies.

Four sand-clay ratios, >2:1, <2:1 to >1:1, <1:1 to >2:1, and <1:2, were used in Figure 4 to divide the Gulf series into major lithic zones. Coarse clastics are dominant in the north and fine clastics in the central and southern part of the area. The fine clastics become gradually more calcareous southward and the decrease

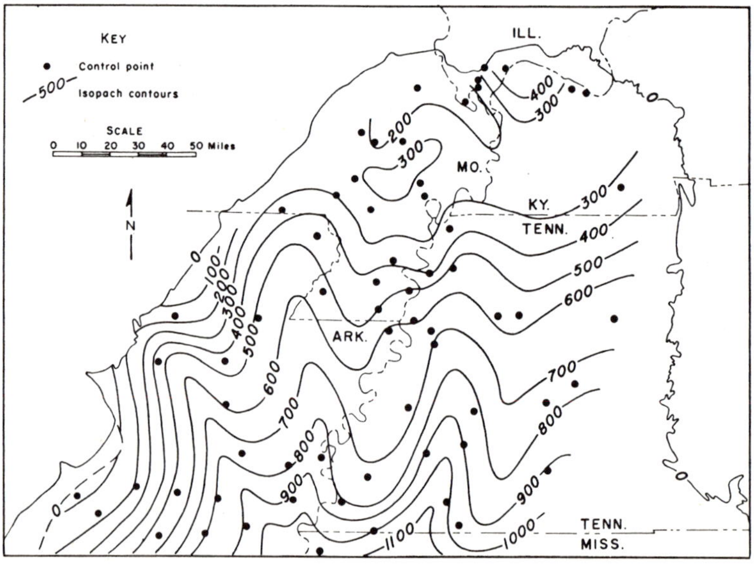


Fig. 3.—Isopach map of Gulf series. 100-foot contour interval.

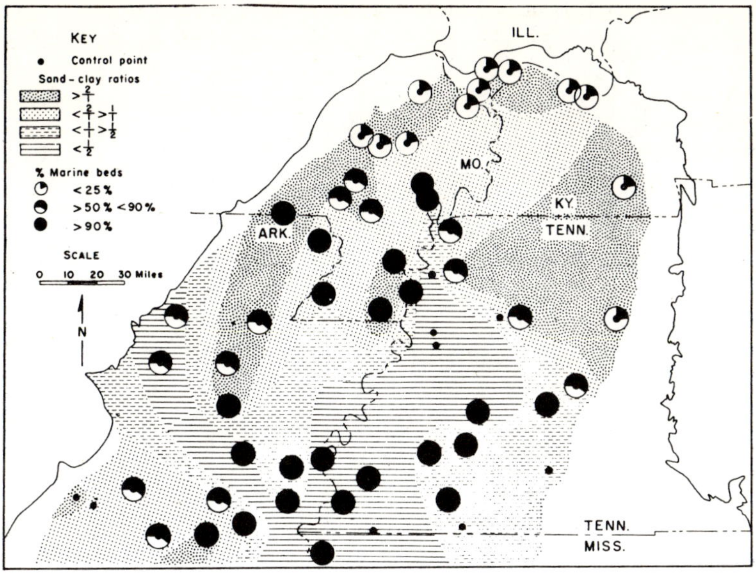


FIG. 4.—Map of Gulf series showing lithofacies and percentage of marine beds in Upper Mississippi Embayment.

in sand-clay ratios in the southerly direction can be attributed to the thickening of the finer sediments below the McNairy formation (Fig. 2).

Examination of well samples and published sample studies suggested several general environmental relationships, and criteria based on well-sample studies were used to interpret these relations. For the marine environment, criteria consisted of the presence of (1) glauconite, (2) megafossils and microfossils, (3) chalk and limestone, and (4) calcareous sand and clays. The criteria for the non-marine environment consisted partly of negative evidence (*i.e.*, the absence of criteria for the marine environment) and the presence of (1) lignite beds, (2) leaf imprints, and (3) highly carbonaceous clays and sands.

The percentage of marine sediments, based on thickness, was estimated for each control well where sample studies were available. These percentages are divided into three major groups, less than 25 per cent, 50–90 per cent, and more than 90 per cent, and are plotted in pie diagrams in Figure 4.

The dominantly non-marine (<25 per cent) occurs in the northern and northeastern part of the area, whereas the dominantly marine zone (>50–<90 per cent) occurs on the south and southwest, and the most dominantly marine zone (>90 per cent) occurs in the extreme south (Fig. 3).

Subsurface investigations yield a picture of the gross geometry of this depositional basin as being an elongate, southward-plunging trough, whose structural axis trends southwest toward the area of maximum downwarp and sediment accumulation.

The map showing lithofacies and percentage of marine beds indicates that coarse non-marine clastics occupy the northern end of the embayment and grade southward into fine marine clastics, suggesting that the sediments entered the basin of deposition from the north or northeast and were distributed southward.

## PALEONTOLOGY

Paleontological observations were made to establish lines of correlation and to aid in environmental interpretations. The paleontological literature of the area was the chief source of information, but it was supplemented by samples collected from a limited number of outcrops. Complete descriptions of the paleontology of the Selma, Coon Creek, and Owl Creek formations are given by Stephenson (1914), Wade (1926), Pierce (1942), Cushman (1946), Hayes (1951), Stephenson (1955), and Loeblich and Tappan (1957). The organic constituents of the Upper

TABLE I. INCIDENCE OF MEGAFAUNA IN GULFIAN SEDIMENTS

| *Formation* | *Sample No.* | *Exogyra ponderosa* | *Ostrea falcata* | *Ostrea larva* | *Ostrea pluma* | *Gryphaea vesicularis* | *Cucullaea vulgaris* | *Sphenodiscus pleurosepta* | *Exogyra costata* | *Exogyra cancellata* | *Inoceramus proximus* | *Baculites ovatus* | *Cardium lowei* | *Turritella trilari* | *Turrilites alternatus* |
|---|---|---|---|---|---|---|---|---|---|---|---|---|---|---|---|
| Owl Creek | 3 | | | | | | | x | X | | | x | x | x | |
| | 120 | | | | | | | | X | | | | | | x |
| Coon Creek | 215 | | | | | | x | | x | | x | | | | |
| | 216 | | | x | | | | | | x | | | | | |
| | 217 | | | | | | | | | | | | | | |
| | 218 | | | | | | | | | | | | | | |
| | 219 | | x | | x | x | x | | X | x | | | | | |
| | 221 | | | | | | | | | | | | | | |
| | 222 | | | x | x | | x | | x | | | | | | |
| | 223 | | x | | | X | | | x | x | x | | | | |
| | 224 | | x | x | x | X | x | | | X | | | | | |
| | 225 | | x | x | x | x | X | | X | x | x | x | | | |
| Selma | 226 | | | x | x | X | x | x | x | X | | | | | |
| | 227 | | | x | | x | | | x | X | | | | | |
| | 228 | x | x | | x | x | x | | X | | | | | | |
| | 229* | | | | | | | | | | | | | | |
| | 230* | | | | | | | | | | | | | | |
| | 231* | | | | | | | | | | | | | | |
| | 232 | | | x | x | x | | | X | x | | | | | |
| | 233 | | x | | | x | | x | X | X | | | | | |
| | 234 | x | x | x | x | x | | | x | x | | | | | |

* Core sample.
X Abundant.
x Present.

Cretaceous sediments in the embayment consist of plant remains, megafossils, and microfossils.

Plant remains in all stages of preservation are common in the Coffee and McNairy formations and were observed at most outcrops. No correlations were attempted on the basis of plant remains. However, the abundance of terrestrial fossils suggests non-marine environments for the sediments in which they are found. A comprehensive study of the Cretaceous plant remains of the area was presented by Berry (1919).

Eighteen outcrops were visited and sampled. The more abundant, readily apparent, and well preserved megafossils were identified and their relative abundances estimated. Table I shows the results of these observations. The samples collected at these outcrops and three well cores were disaggregated and treated to recover the microfossils. The most abundant microfossils were identified and their relative abundances were estimated. Table II shows the results of the microfossil analysis.[3]

Three major faunal zones are represented by the samples listed in Table I: the *Exogyra ponderosa* zone, the *Exogyra costata* zone, and the *Exogyra cancellata* zone. Stephenson (1914, Pl. X) defined these faunal zones and used them in regional correlations of the Cretaceous sediments of the Atlantic Coastal Plain and the Gulf Coastal Plain. Stephenson (1914, Pl. X) and Stephenson *et al.* (1942, Chart 9) showed the *E. ponderosa* zone extending from the base of the Eutaw formation to the upper part of the Selma formation; the *E. costata* zone extending from the upper part of the Selma formation to the top of the Owl Creek formation; and the *E. cancellata* zone extending from the top of the *E. ponderosa* zone to the top of the Coon Creek formation.

[3] Exact locations of sampled outcrops are recorded in Pryor, 1959.

Fossils of the *E. ponderosa* zone were found in only two of the outcrops. The other fossiliferous outcrops contained fauna of the *E. costata* zone and the *E. cancellata* zone. Fossils of the *E. costata* zone were found in outcrops of the Selma, Coon Creek, and Owl Creek formations, and fossils of the *E. cancellata* zone were found in the Selma and Coon Creek formations. The occurrence of *E. ponderosa* and *E. costata* in the same zone at outcrops suggests either convergence of the two zones or reworking of the lower zone into the upper zone, but detailed study of the northern extension of the Selma formation is needed to determine the cause of this concurrence.

Table II shows the results of a population count of the most common genera and species of microfossils found in the outcrop samples. Complete lists of microfauna from these units are available in the literature. The microfossils are here divided into benthonic and planktonic types and the relative abundance of each is noted for each outcrop.

The microfauna observed in these sediments is common to the *Globotruncana* zone of Loeblich and Tappan (1957), which is late Cretaceous (Gulf) in age and correlative with the Maestrichtian of Europe. The most abundant planktonic foraminifera present belong to the genus *Gümbelina*. Although present in most samples they are abundant in only a few.

Ostracods, scolecodonts, shark teeth, fish scales, and arthropod appendages constitute a small part of the microfossil assemblages of many of the samples and are more abundant in the Coon Creek formation than in either the Owl Creek or Selma formations.

In the outcrop area, marine fossils are restricted to the Selma, Coon Creek, and Owl Creek formations. Marine fossils are absent in the northern part of the Coffee and McNairy formations but are present on the south, where the formations merge with the Selma and Coon Creek formations.

The Coon Creek formation has a larger population of benthonic foraminifera and ostracods than either the Selma or Owl Creek formation, which suggests its deposition in a near-shore environment (Phleger and Parker, 1951, p. 81). The Selma and Owl Creek formations also have larger populations of planktonic foraminifera, which Phleger and Parker found, in their study of recent sediments of the Gulf of Mexico, as characteristic of the fauna in offshore areas.

In summary, paleontologic evidence suggests deposition of the Tuscaloosa formation and northern parts of the Coffee and McNairy formations in a non-marine environment and deposition of the Selma, Coon Creek, and Owl Creek formations and the southern parts of the Coffee and McNairy formations in a marine environment.

## Textural Analysis

Size analyses of 107 samples of the McNairy formation sands were made and the quantitative data and derived parameters were recorded. The data were plotted on arithmetic probability paper in order to permit comparison of the curves with the large amount of grain-size data collected and studied by Doeglas (1946, 1950), van Andel and Postma (1954, pp. 80–89), and Kruit (1955, pp. 37–48), and, if possible, apply their interpretations.

Doeglas (1946, p. 39) suggested the occurrence of a limited number of curve shapes when the data were plotted in this manner. These grain-size types are based on the shapes of the curves, not on the actual grain size of the deposit. He proposed that three main types of frequency distributions, R, S, and T, occur in sedimentary deposits as a result of the sorting action of the transporting medium. Van Andel and Postma (1954, p. 89) adopted this method of sediment classification in their study of the Recent sediments of the Gulf of Paria and added two further types to the three suggested by Doeglas. They redesignated the Doeglas types the F, M, and B sands and named the new types the C and S clays. The work by Doeglas (1950) and by van Andel and Postma (1954) has suggested that these types or groups of similar curves are characteristic for specific environments in the Recent sediments they studied. Figure 5 shows eight size analyses plotted in this manner.

The sand types of the McNairy formation (Fig. 5) most closely fit the R type of Doeglas (1950) or the F and F+S types of van Andel and Postma (1954, p. 81). Doeglas (1950, p. 310) stated, "R-distributions are characteristic of stream channels where fine particles are washed out, due to periodical increase of capacity." Van Andel and Postma (1954, p. 89) stated, "The F-sands are

TABLE II. INCIDENCE OF MICROFAUNA IN GULFIAN SEDIMENTS

| Formation | Sample No. | Benthonic | | | | | | | | | | | | | | | | | | | | | | | | | | | | | | Planktonic | | | | | |
|---|---|---|---|---|---|---|---|---|---|---|---|---|---|---|---|---|---|---|---|---|---|---|---|---|---|---|---|---|---|---|---|---|---|---|---|---|---|
| | | *Robulus münsteri* | *Lenticulina coffeensis* | *Marginulina watersi* | *Marginulina pseudomarki* | *Vaginulina substriata* | *Vaginulina selmaensis* | *Palmula rugosa* | *Guttulina adhaerens* | *Lagena substriata* | *Bulimina quadrata* | *Anomalina selmaensis* | *Robulus culatpaensis* | *Dentalina pauperata* | *Globulina lacrima* | *Anomalina involuta* | *Planulina taylorensis* | *Cibicides stephensoni* | *Robulus cultratus* | *Textularia ripleyensis* | *Dentalina subarcuata* | *Nodosaria alternata* | *Rotalia aequilateralis* | *Anomalina nelsoni* | *Anomalina tennessensis* | *Cibicides excolata* | *Robulus navarroensis* | *Dentalina confluens* | *Dentalina lorneiana* | *Nodosaria obscura* | *Frondicularia cuspidata* | *Gümbelina reussi* | *Gümbelina plummerae* | *Gümbelina globulosa* | *Gümbelina semicostata* | *Gümbelina costulata* | *Globotruncana cretacea* |
| Owl Creek | 3 | | | | | | | | | | | | | | | | | | | | | | | | | | x | | | | | x | | | | | |
| | 120 | | | | | | | | | | | | | | | | | | | | | | | | | | x | | | | | x | | X | | | x |
| Coon Creek | 215 | | | | | | | | x | | | | X | | | x | | x | x | | | x | | x | X | X | | x | x | | x | | | | | | |
| | 216 | x | | | | | | | | | | | x | x | | | | x | | x | | | x | x | | x | | | x | | x | | | | | | x |
| | 217 | | | | | | | | | | | | | | | | | | | | | | | | | | | | | | | | | | | | |
| | 218 | | | | | | | | | | | | | | | | | | | | | | | | | | | | | | | | | | | | |
| | 219 | | x | | | | x | | | | | | x | x | x | | x | x | x | | | | | | X | | x | x | | x | | | | x | | x | |
| | 221 | | | | | | | | | | | | x | | x | X | x | | x | x | x | x | | X | x | x | | | x | | X | | | | | | |
| | 222 | | | | | | | | | | | | x | x | | x | x | | X | | | | | x | x | X | | x | | X | | | | | | | |
| | 223 | | | | x | | | | | x | | | x | | x | x | | x | x | x | | x | | | x | x | | | x | | | | | x | | x | x |
| | 224 | x | | | | | | | | | | | | x | x | | x | | x | | | | x | x | x | x | x | | | X | x | | | | | | |
| | 225 | | | x | | | | | | | | | x | | x | | x | X | x | | x | | x | X | x | | | | | | | | | | | | |
| Selma | 226 | X | | x | | x | | x | x | | x | x | | x | x | x | | x | x | | | | | | | | | | | | | | x | X | x | | x |
| | 227 | | x | | x | | | | x | | x | x | | | | | x | x | | | | | | | | | | | | | | | | x | X | x | X |
| | 228 | | x | | x | | x | | | | x | x | x | | | x | | | | | | | | | | | | | | | | | x | | x | | x |
| | 229* | | x | x | | x | | | x | x | | x | | | x | x | x | | | | | | | | | | | | | | | | | X | | x | x |
| | 230* | x | | x | | x | x | | x | x | | x | | | | | | | | | | | | | | | | | | | | | | x | x | | X |
| | 231* | x | x | | x | | | | x | | | x | | | x | x | | | | | | | | | | | | | | | | | X | x | | x | x |
| | 232 | | x | | x | x | | x | | x | x | x | | x | x | | x | x | | | | | | | | | | | | | | | | | x | | x |
| | 233 | x | x | | x | | | | x | x | | x | | | x | | x | | | | | | | | | | | | | | | | | x | x | | |
| | 234 | x | | x | | x | | x | | x | | | | | | x | | x | | | | | | | | | | | | | | | | X | | x | X |

* Core sample.
**X** Abundant.
x Present.

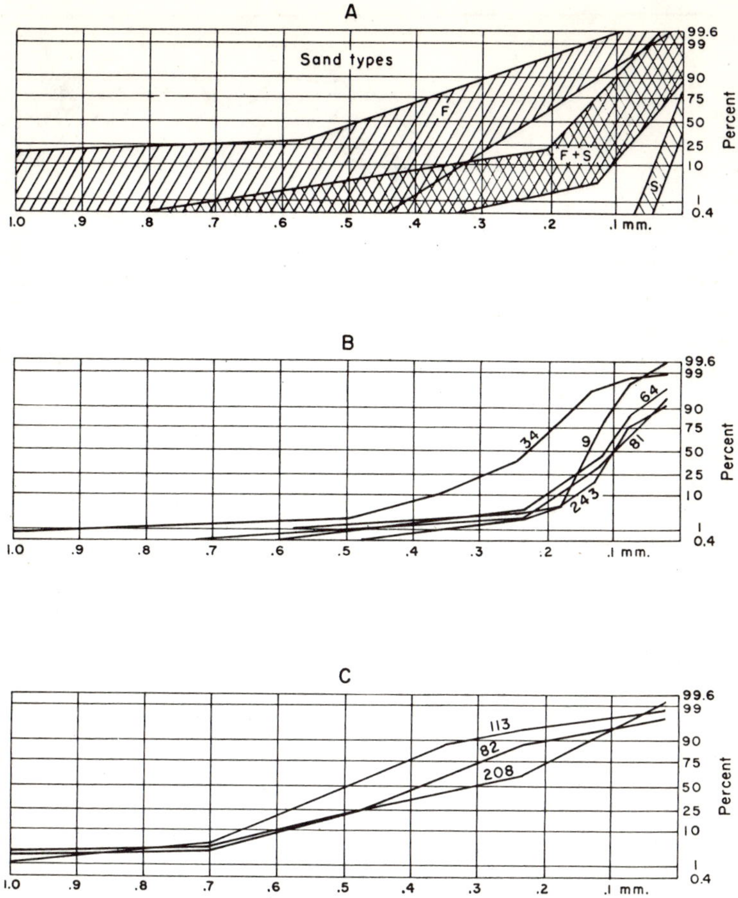


FIG. 5.—Size distribution of McNairy formation sands plotted on an arithmetic-probability scale. (a) Sand types of van Andel and Postma (1954); (b) McNairy formation F+S type sands; (c) McNairy formation F type sands. Sample numbers are shown adjacent to curves.

the principal sand type in the upper delta, where tidal influence is dying out and a normal river regime may be assumed." They summarized the relations between the F type of grain-size distribution and the environments of deposition in the Gulf of Paria by stating, "F-types: upper delta with sufficient supply of coarse and suspended material and predominance of river regime." The S type of van Andel and Postma (1954, p. 89) is a clay containing up to 50 per cent silt and is characteristic of the lower delta and estuary environments. The F+S type of distribution is a mixture of the upper and lower delta environments. The F sand is the most common type found in the McNairy formation, which also contains minor amounts of F+S types. Application of the Doeglas and the van Andel and Postma interpretations suggest that the sands of the McNairy formation are similar in size distributions to sands deposited in an upper delta environment with a predominance of the river regime.

Along the outcrop belt of the McNairy formation the F and F+S distribution types have a significant pattern of occurrence. Figure 6 shows a relatively concentrated occurrence of the F+S type of sand along the northern limb and along the southern half of the eastern limb of the outcrop belt. The F sand is dominant in the central part between the two extremities.

To derive the various parameters, cumulative size frequency curves for 107 sand samples from McNairy formation were plotted, by using phi units. Curves for 8 of these analyses are shown in Figure 7.

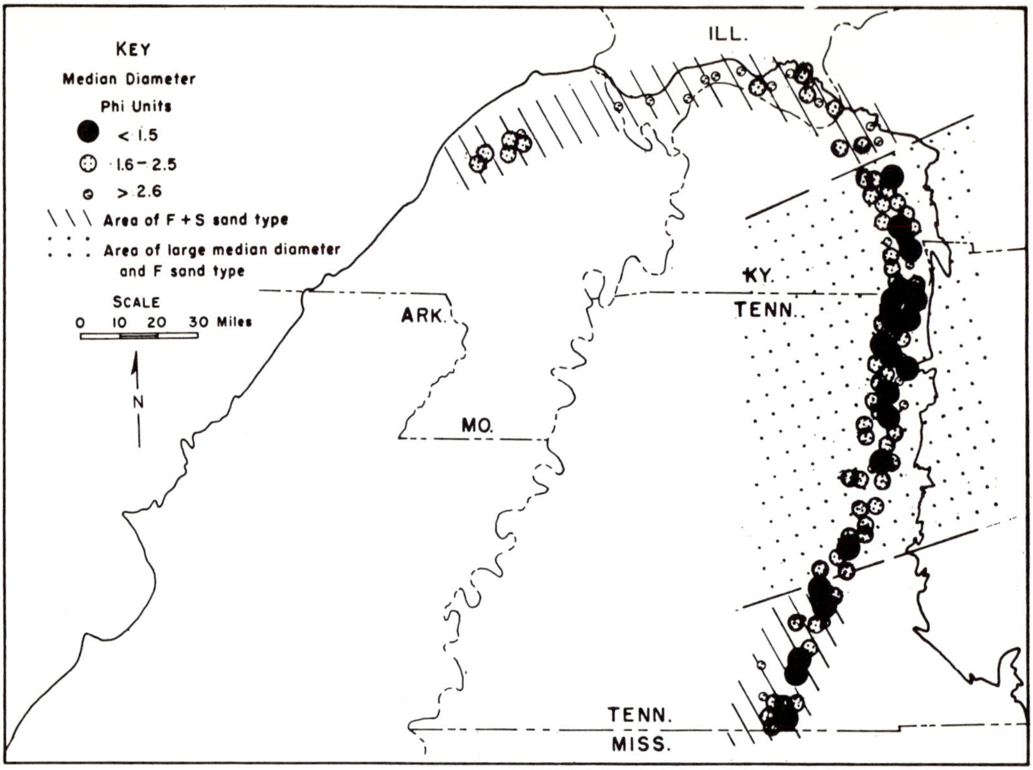


FIG. 9.—Distribution of F+S and F sand types and median diameters along outcrop belt of McNairy formation.

The parameters derived from the plotted distribution curves were median diameter, decile deviation, coefficient of sorting, coefficient of geometric quartile skewness, and quartile kurtosis. The median diameters of the McNairy formation sands range from 0.75$\phi$ (0.58 mm.) to 3.68$\phi$ (0.078 mm.), coarse to very fine, respectively. The median diameters are graphically plotted on the map in Figure 6, the dark circles representing the largest medians. The plotted data show a significant regional variation in this parameter. There is an area in Kentucky and northern Tennessee where the median diameters generally are larger than in adjacent areas in Illinois, Missouri, and southern Tennessee. Variations in median diameter reflect variations in velocities of the currents that move the materials to their sites of deposition and this well defined zone of coarse median diameters probably marks an area where higher velocities persisted during the deposition of the McNairy formation sands.

The coefficients of decile deviation vary from 0.24 to 1.22, the latter indicating poorer sorting than the former. Regional plotting of this parameter shows no striking patterns, although there is a poorly defined area of slightly better sorting, or of lower coefficients of decile deviation, in essentially the same area as the larger median diameters. The coefficients of sorting for the McNairy formation sands range from 1.08 to 1.88 for a well sorted series of sands. There are no well defined regional patterns of variation in the coefficients of sorting for the McNairy formation sands. There is a well developed symmetry present for most of the sands studied, the coefficients of skewness generally approaching 1.0. No strongly developed regional variations in this parameter were observed. Kurtosis values for McNairy formation sands range from 0.12 to 0.34, indicative of a strong degree of peakedness, with high proportions of the materials in two or three grade sizes. No strongly developed regional variations in this parameter were observed.

Textural analyses of McNairy formation sands suggest that the coarse clastics in the northern part of the Mississippi Embayment were depos-

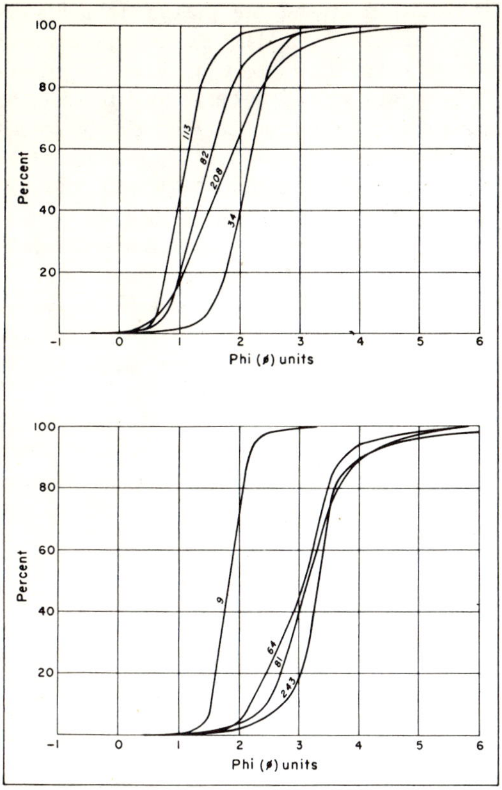


FIG. 7.—Cumulative size curves at McNairy formation sands, shown in Figure 5, plotted on an arithmetic percentage—phi scale. Sample numbers are shown adjacent to curves.

ited principally by streams whose competence diminished toward the southwest. The distributional pattern of sand types suggests the prevalence of fluviatile and upper deltaic environments in the northeastern part of the area that grade into lower deltaic and estuarine environments to the south and west. Although most of the textural parameters are of little apparent significance, the distribution of the higher median diameters is significant and coincides with the area of the F (fluviatile-upper deltaic) sand.

## SEDIMENTARY STRUCTURES

### GENERAL

The sedimentary structures of the late Cretaceous sediments are most prominently associated with the coarse clastics of the McNairy formation and to less extent those of the Coffee formation. The major sedimentary structures in these units include cross-stratification, bedding, clay galls, clay pebbles, ripple marks, and cut-and-fill.

### CROSS-STRATIFICATION

The most common sedimentary structure present in the McNairy formation is cross-stratification. It is abundant but less common in the argillaceous Coffee formation. The cross-stratified units may be separated by one or several layers of normally stratified sands or clays, but commonly they occur as sets of two or more in vertical succession. The cross-stratified units range in length from less than 5 feet to more than 30 feet and in width from less than 1 foot to more than 10 feet. The foreset strata generally are separated by planes occupied by mica concentrations and abundant clay pebbles and generally are less than $\frac{1}{2}$ inch thick. The foreset laminations commonly are tangential with the underlying strata, but in some places are not. Some of the cross-laminated units apparently have uniform thicknesses, but most are slightly wedge-shaped. Aspects of shape (wedge and uniform) and type (simple, planar, and trough) depend in large part on the size of the unit, in both long and short dimensions, and on the section or view exposed at the outcrop.

Numerous attempts have been made to associate types of cross-stratification with specific environments, but Pettijohn (1957, pp. 166–67) doubted that there are any genetic types. He concluded that only the mean direction, variability, and scale (thickness) appear to be significant for interpretations of environment from cross-bedding.

The directions of maximum inclination of the foreset beds in the McNairy formation were measured in order to ascertain the current directions involved in the transportation and deposition of the sediments and to determine the dispersal pattern of the sediments.

Directional data for each outcrop were sampled in a random manner. Beginning at the base or the top of each outcrop, whichever was the most accessible, the direction of maximum inclination and thickness of the first ten (if present) sets of cross-stratification were measured. From 175 outcrops 920 observations of direction of cross-stratification[4] inclination were recorded. The results of the directional measurements and their

[4] The detailed cross-stratification data are recorded in Pryor, 1959.

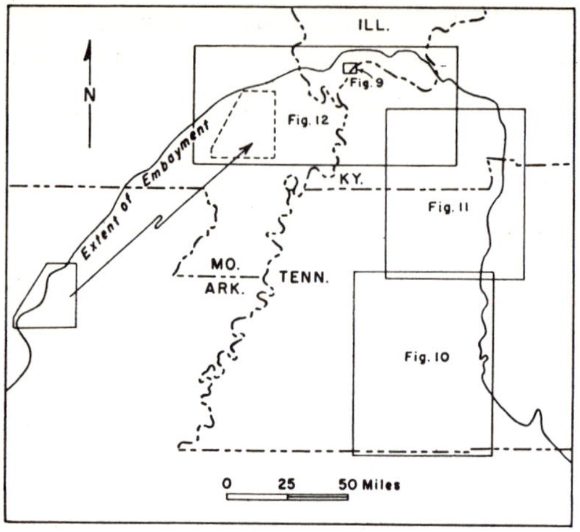


FIG. 8.—Index to detailed maps of McNairy formation cross-stratification.

statistical treatment are presented in Figures 8–12, and in Table III. An area of about 3 square miles along the Ohio River in Illinois (Fig. 9) was studied in detail to determine the variability of direction between closely spaced sampling points. Eighty-one observations at 19 outcrops showed a preferred orientation toward the southwest. The windrose diagram of nine 40° segments shows a strong southwestern orientation. The Tukey test shows a grand mean for the area of 242°, which is significant ($X^2_{obs}=11.41$; $X^2_{0.995}=10.60$). Figure 9 shows that there are no significant variations of direction in a series of closely spaced points.

The mean for the southern Tennessee area (Fig. 10 is 232° ($X^2_{obs}=184.21$; $X^2_{0.995}=10.60$); for the northern Tennessee-Kentucky area (Fig. 11) it is 242° ($X^2_{obs}=243.63$; $X^2_{0.995}=10.60$); and

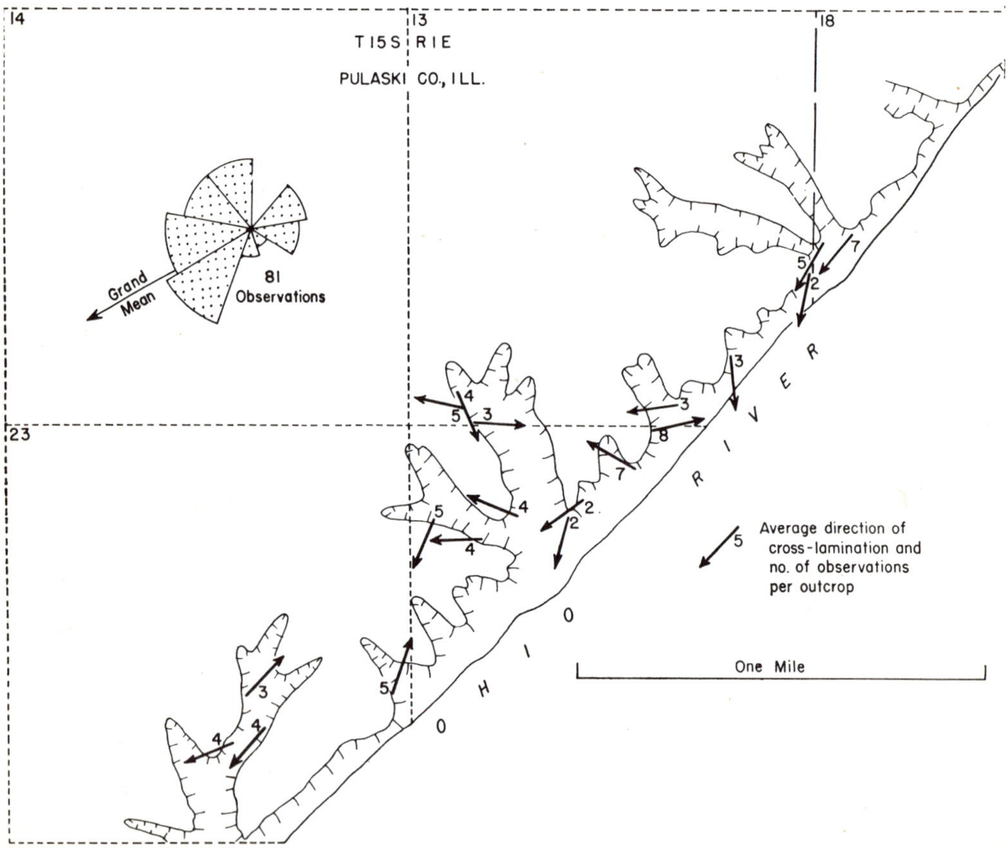


FIG. 9.—Cross-stratification in McNairy formation along Ohio River in Illinois.

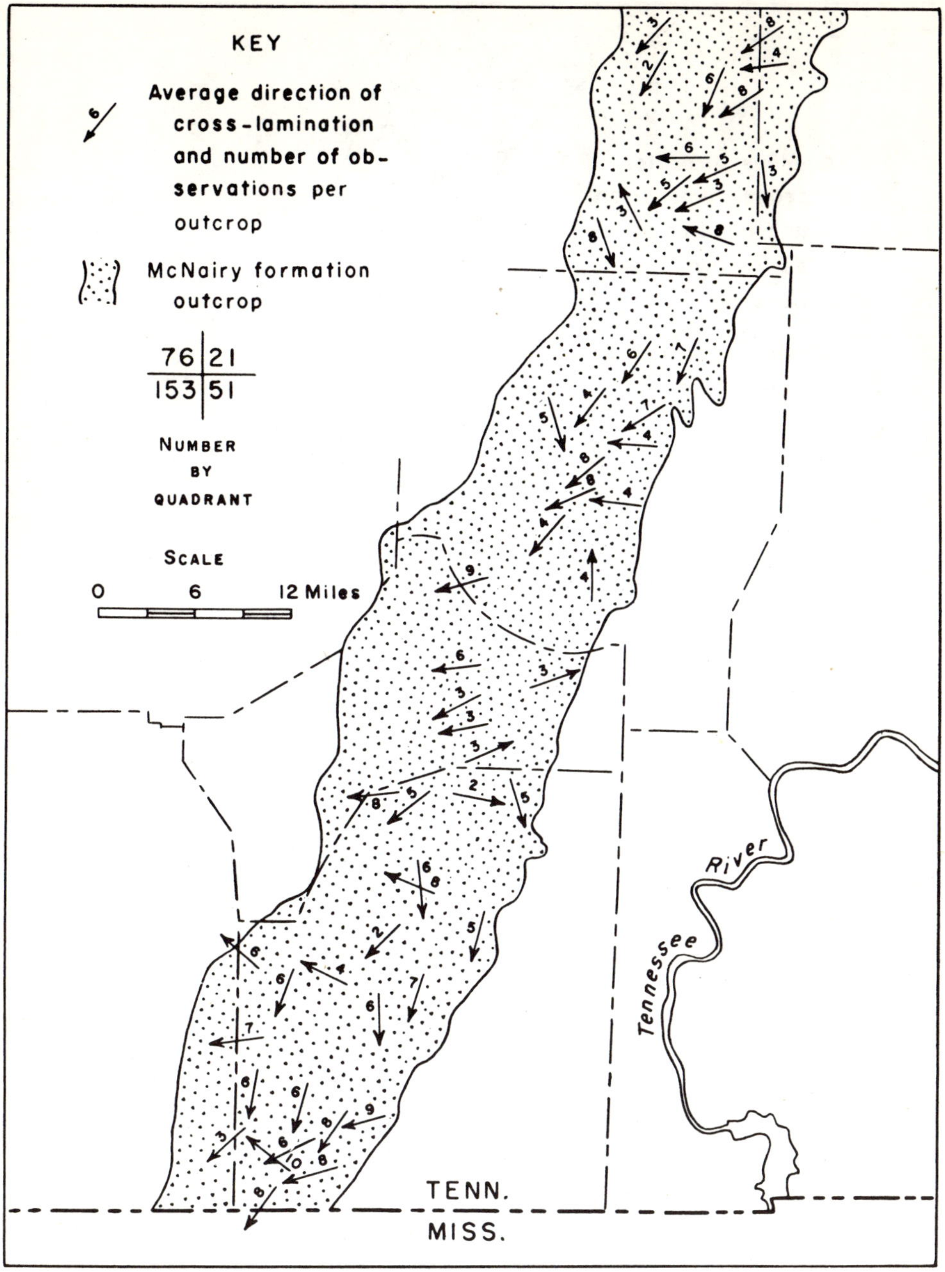


FIG. 10.—Cross-stratification in McNairy formation in Tennessee.

for the Illinois-Missouri-Arkansas (Fig. 12) area it is 200° ($X^2_{obs}=70.77$: $X^2_{0.995}=10.60$).

The directional data obtained from measuring cross-strata is summarized in Figure 13. The means for the sampling intervals are shown, as is the "grand mean," for cross-stratification directions in the McNairy formation. The grand mean azimuth for all of the 920 observations is 236° and is highly significant ($X^2_{obs}=464.83$; $X^2_{0.995}=10.60$). The mean directions of the sam-

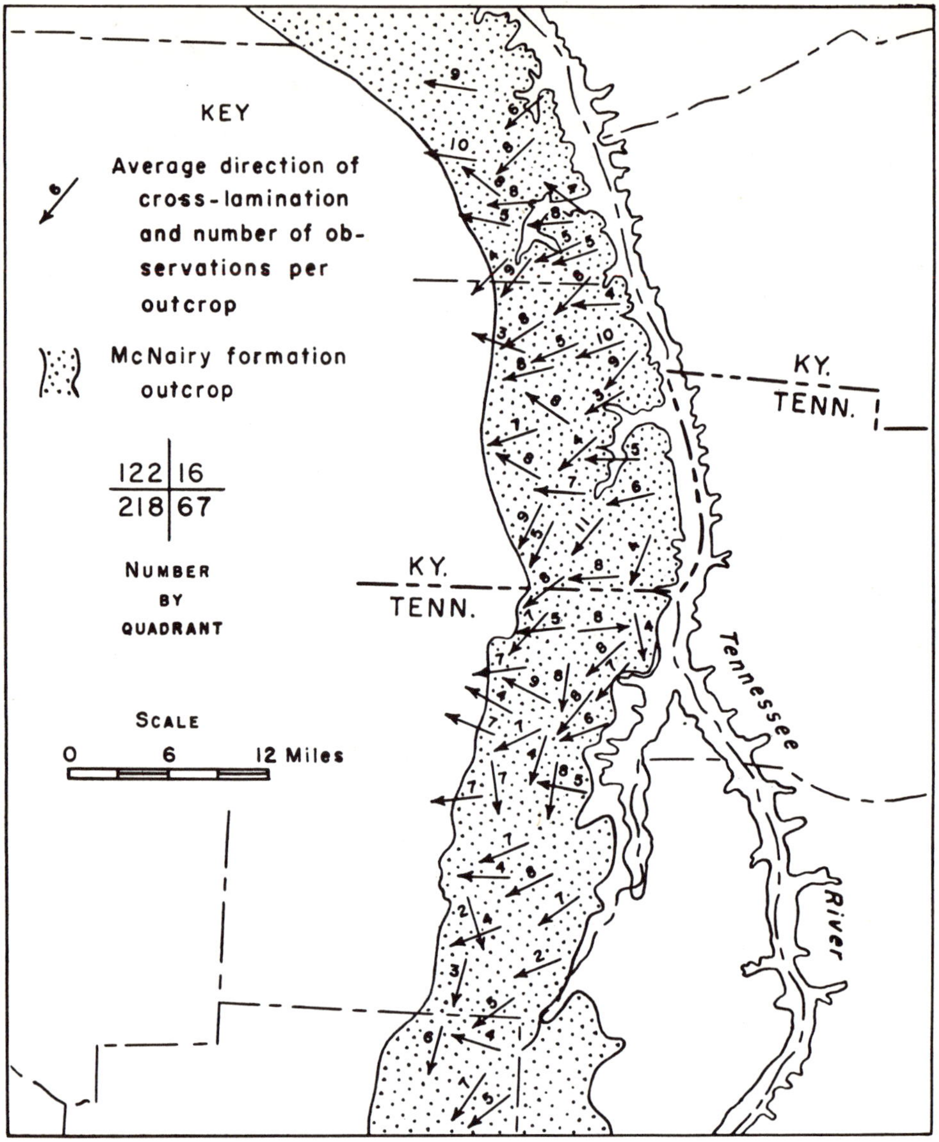


FIG. 11.—Cross-stratification in McNairy formation in Tennessee and Kentucky.

pling intervals show little variation over the entire embayment area, the directions being consistently to the south and west. The windrose diagram shows the cross-stratification to be essentially unidirectional, with the large majority of observations being in one hemisphere. The grand mean is parallel with the structural axis of the Mississippi Embayment and subparallel with the present-day course of the Mississippi River.

Pettijohn (1957, p. 167) suggests that the distribution of cross-stratification azimuths may have a genetic significance. McDowell (1957, pp. 28–29) compared the windrose diagrams of the cross-stratification in the Mississagi quartzite to similar diagrams of fluvial deposits and tidal deposits and on the basis of these comparisons concluded that the cross-stratification distribution had a fluviatile origin. Pelletier (1958, p. 1059) presents a windrose for the cross-stratification azimuths of the fluvial Pocono

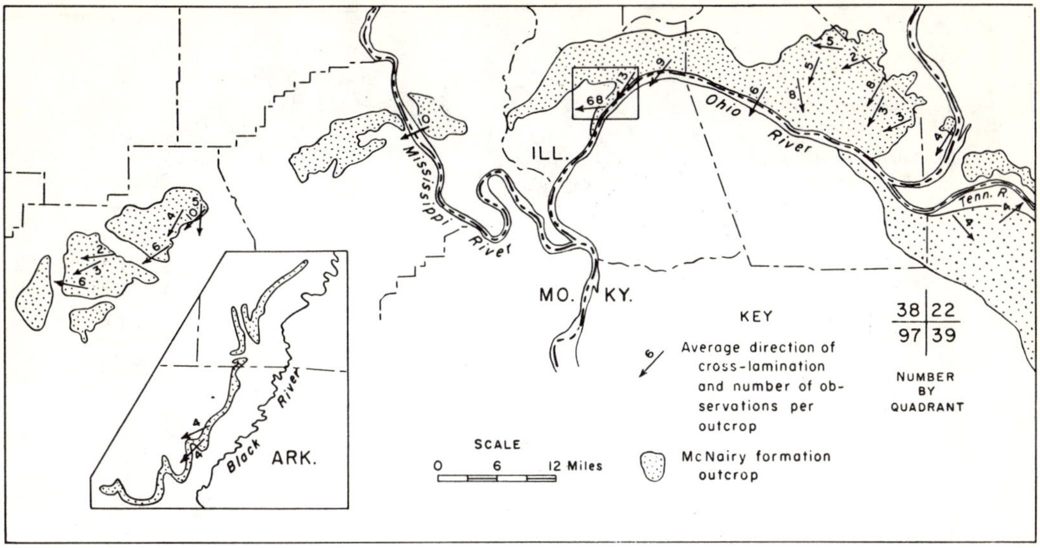


FIG. 12.—Cross-stratification in McNairy formation in Illinois-Missouri-Arkansas area.

formation, and Hamblin (1958, pp. 54, 92, and 110) presents several windroses for the fluvial basal Cambrian formations of the Upper Michigan Peninsula. These compare closely with the cross-stratification windrose for the McNairy formation. In contrast to this type of dispersion for fluvial materials, the littoral, shallow-marine, late Pennsylvanian and early Permian sandstones of Oklahoma show two opposing means on a diagram by Tanner (1955, p. 2482) and the shallow-marine shelf Franconia sediments of southwestern Wisconsin show widely dispersed patterns (Farkus, 1958, p. 34).

The variances and standard deviations (Dixon and Massey, 1951, p. 21) of cross-stratification in six ancient sedimentary rocks are shown in Table IV. The marine Hawthorne (Emrich, in preparation), Glenwood-St. Peter (Potter and Pryor, in preparation), and Franconia (Farkus, 1958, p. 34; Hamblin, 1958, p. 54) have cross-

TABLE III. SUMMARY OF CROSS-STRATIFICATION DATA ON UPPER CRETACEOUS SEDIMENTS OF MISSISSIPPI EMBAYMENT

(Numbers of observations in class intervals)

| *Fig. No.* | | *Class Interval (Azimuth)* | | | | | | | | | *Totals* |
|---|---|---|---|---|---|---|---|---|---|---|---|
| | | *0–40* | *41–80* | *81–120* | *121–160* | *161–200* | *201–240* | *241–280* | *281–320* | *321–360* | |
| 10 | | 5 | 14 | 7 | 17 | 52 | 82* | 61 | 45 | 18 | 301 |
| 11 | | 6 | 9 | 10 | 30 | 55 | 104 | 107* | 70 | 32 | 423 |
| 12 | | 3 | 14 | 12 | 15 | 33 | 50* | 38 | 19 | 12 | 196 |
| 13 | Total | 14 | 37 | 29 | 62 | 140 | 236* | 206 | 134 | 62 | 920 |
| 13 | Per Cent | 1.5 | 4.0 | 3.2 | 6.7 | 15.2 | 25.7 | 22.4 | 14.6 | 6.7 | 100.0 |

* Modal cross-lamination class interval.

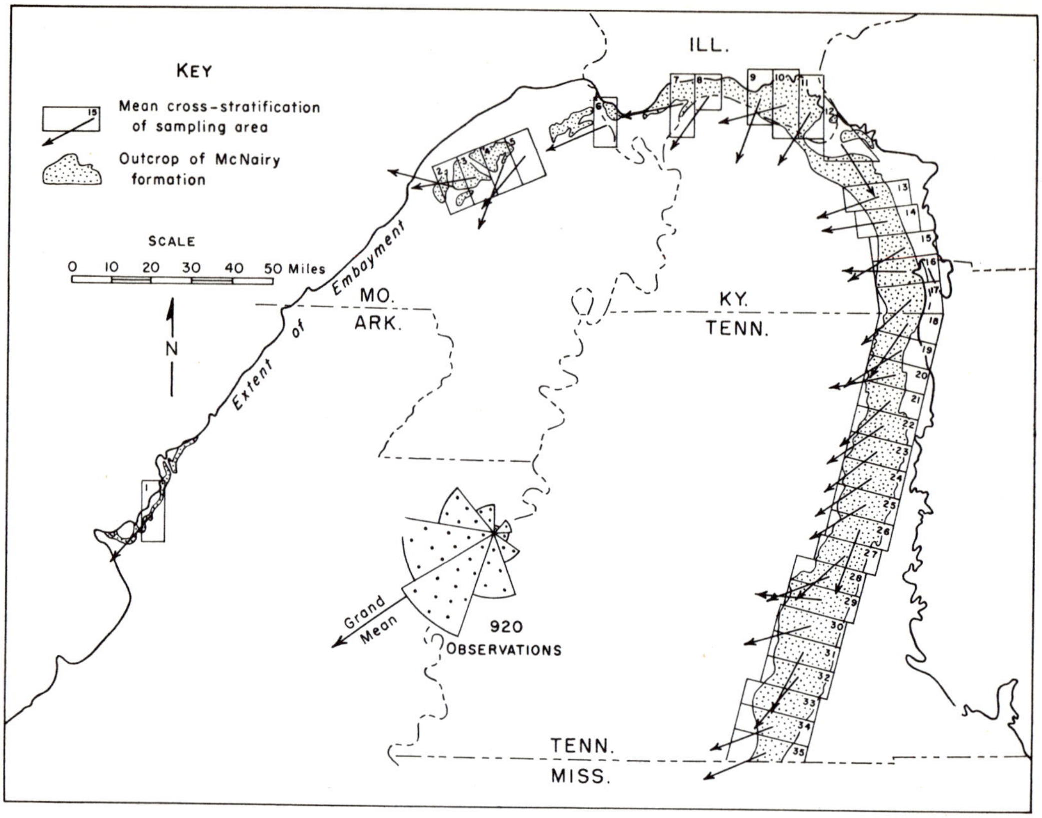


FIG. 13.—Directions of cross-stratification in McNairy formation, summarized by sampling interval, for Upper Mississippi Embayment. Windrose is based on nine 40° intervals.

stratification variances ranging from 7,000 to 8,000 and standard deviations of 84° to 90°. However, the non-marine fluvial basal Pennsylvanian (Potter and Siever, 1956), Pocono (Pelletier, 1958) and Mississagi (McDowell, 1957) have cross-stratification variances ranging from 4,900 to 5,800 and standard deviations from 70° to 76°. Judged from the comparisons in Table IV, the McNairy cross-stratification variance and standard deviation are in the range of that for the non-marine fluviatile sediments.

Pelletier (1958, pp. 1044–45) discussed the possibility of using standard deviation to distinguish environmental types of cross-stratification at the local level and concluded that, at present, it can not be done on the basis of statistics alone. Although there is an indication (Table IV) that variances and standard deviations of regional cross-stratification patterns may be environmentally diagnostic, several factors require further investigation. The differences may be only apparent and may be in partial response to differing sizes of sampling areas and varying numbers of observations per unit area, in addition to environmental controls.

Probably the most important factor controlling the variance and standard deviation of dispersal patterns is the amount of slope of the depositional surfaces—the greater the amount of slope, the smaller the variance and standard deviation. Hence, sediments deposited in a fluvial environment generally will have a dispersal pattern with a smaller variance than one deposited on a more gently sloping continental shelf. Before variance and standard deviation of dispersal systems are accepted or rejected as being environmentally diagnostic, however, many more examples of ancient and recent dispersal systems

TABLE IV. VARIANCE AND STANDARD DEVIATION OF VARIOUS MARINE SHELF AND NON-MARINE FLUVIAL CROSS-STRATIFICATION PATTERNS

| Formation | Statistical Variance | Standard Deviation | Number of Observations |
|---|---|---|---|
| *Marine Shelf* | | | |
| Hawthorne (Miocene) | 7,391 | 85.9° | 284 |
| Glenwood-St. Peter (Ordovician) | 8,094 | 89.9° | 140 |
| Franconia (Cambrian) | 7,120 | 84.4° | 1,511 |
| *Non-Marine Fluvial* | | | |
| Basal Pennsylvanian | 5,151 | 71.7° | 792 |
| Pocono (Mississippian) | 5,227 | 72.3° | 388 |
| Mississagi (Precambrian) | 5,791 | 76.1° | 1,260 |
| *Mississippi Embayment Cretaceous* | | | |
| McNairy | 5,018 | 70.8° | 920 |

than are presented in Table IV should be analyzed and the factor of numbers of observations and size of sampling area considered.

In comparing the distribution patterns of cross-stratification in ancient rocks with the distribution of channel directions in a series of Recent sediments, the map of the post-Pleistocene delta complex prepared by Fisk (1944) was used. Fisk shows seven overlapping deltas and associated river and distributary channels. Figure 14 is an adaptation of this map showing random directional observations of the channels. A grid system, based on township and range divisions, was superposed on Fisk's map and random directional observations were plotted wherever the channels crossed the grid lines. The "mean" for the 509 observations was 144° and is highly significant ($X^2_{obs}=352.91$; $X^2_{0.995}=10.60$). The windrose diagram shows a strong southeastern orientation. The part of the Mississippi River Valley and delta depicted in Figure 14 has approximately the same area as the Upper Mississippi Embayment. The variance and standard deviation are 3,418 and 50.5°, respectively. Compared with the ancient fluvial sediments (Table IV), the values

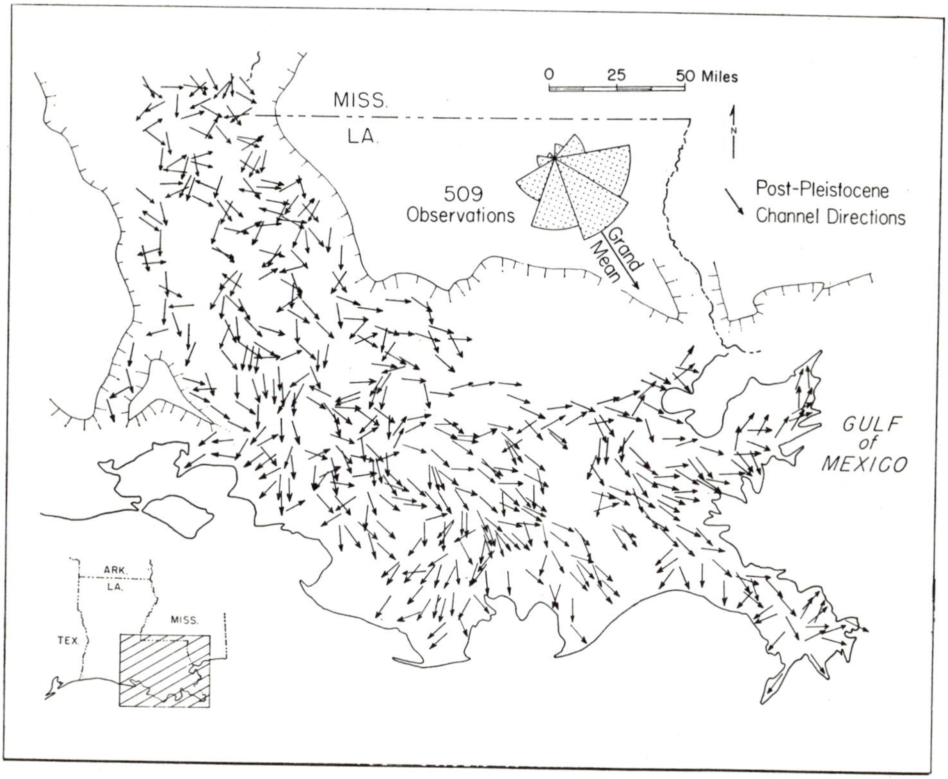


FIG. 14.—Directions of channel trends in post-Pleistocene Mississippi Delta complex.

are somewhat low. However, as the measurements are of channel directions, they do not take into consideration the variations in orientation of the megaripples and sand bars within the channels. If the cross-stratification in these were measured, the variance and standard deviation probably would be larger.

A comparison of Figure 13 with Figure 14 shows a similar orientation of channel and cross-strata trends. The windrose diagrams exhibit perhaps the most striking similarities of general outline and are essentially mirror images of each other.

It is suggested that the fan-shaped, unidirectional windrose diagrams of cross-stratification directions with regional variances of 5,000 to 6,000 and standard deviations of 70° to 76° are common to fluviatile sediments and imply a deltaic dispersion pattern. The dispersal pattern of cross-stratification is therefore consistent with a deltaic interpretation of the sedimentation for McNairy formation sands.

## BEDDING

Cross-stratification was measured and an average thickness was obtained for each unit. The distribution of the thickness of the measured sedimentary units is shown in Figure 15. The average thicknesses range from 6 to 12 inches and, whereas Schwarzacher (1953, p. 326) and Pelletier (1958, p. 1055) observed thickening of

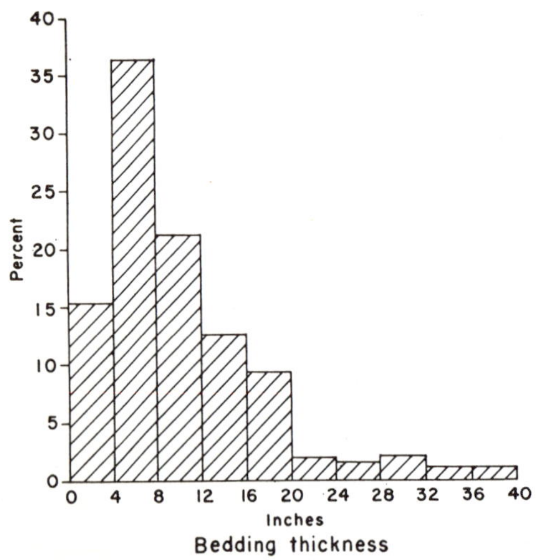


FIG. 15.—Frequency distribution of bedding thicknesses in McNairy formation.

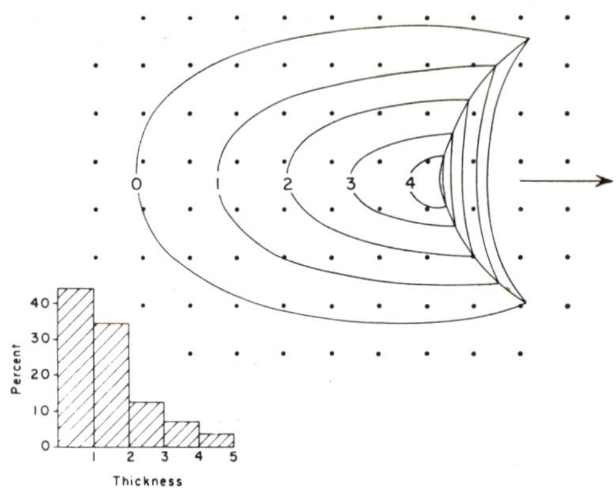


FIG. 16.—Diagram showing relation of systematic sampling of subaqueous dune to frequency distribution of bedding thickness. Dots are sampling points and arrow denotes transport direction.

cross-stratified units toward the source areas, no gradients were observed in the thickness of cross-stratified units of the McNairy formation. Most of the variations in observed thickness can be attributed to variations in the true thickness of units, but it also can be shown that random observations on a single unit will yield the same kind of frequency distribution. Figure 16 shows an isopach of a diagrammatic dune, a configuration suggested by Potter and Siever (1956, p. 231), Hamblin (1958, p. 50), Niehoff (1958, p. 281), and Wurster (1958, pp. 323, 337, 342) for cross-stratified units. Superposed on this dune is a systematic sampling design, indicated by dots. Random measurements of thickness of this single unit yields the frequency distribution in Figure 16, the same type of distribution exhibited in Figure 15, and similar to that shown by Pelletier (1958, p. 1043).

Recent investigations by Brush (1958, p. 1542) and others have demonstrated a relation between the thickness of cross-strata and mean current velocity. However, the factor of sampling should be considered prior to estimating current velocities responsible for the deposition of ancient sediments on the basis of bedding thickness.

## CLAY GALLS AND CLAY PEBBLES

Clay galls commonly are associated with cross-stratification in the McNairy and Coffee formations. They generally are found on the foreset lamination planes and were observed at every cross-stratified exposure. They are tabular in shape, a fraction of an inch to several inches in length and width and less than ½ inch thick. Many are subrounded, but most are angular to subangular, denoting very short distances of transport. They are most certainly derived by the bottom erosion of nearby clay laminations. Clay pebbles and clay pebble conglomerates are present in places but are not abundant. The pebbles range from tabular to subspherical in shape and always exhibit some degree of rounding. They range in size from about one inch to more than 4 inches in diameter, generally are associated with local erosional unconformities, and were probably derived from the erosion of nearby clay beds. Clay pebbles and clay galls commonly are associated with fluviatile sediments.

#### RIPPLE MARKS

Although ripple marks probably are abundant, their recognition in unconsolidated sediments is difficult. They were recognized in less than 10 per cent of the several hundred outcrops visited. Where recognizably present, ripple marks invariably were associated with clay laminations or sediments partly cemented by iron oxides. The types of ripple marks observed were asymmetrical (current), symmetrical (oscillation), and interference. Asymmetrical and interference ripple marks are the most common. Very small-scale cross-stratification commonly was associated with the ripple marks.

#### CUT AND FILL

Cut-and-fill structures commonly are associated with cross-stratification and are the sedimentary structures second in abundance to cross-stratification. They generally are trough-shaped and plunge in the direction of adjacent cross-stratification inclination.

#### OTHER STRUCTURES

Sedimentary structures such as current lineation, loadcasts, current fluting, drag grooves, and slumps may be present but are difficult to observe in unconsolidated sediments. With the exception of several occurrences of slump structures, none of these was identified with certainty.

### PETROGRAPHY

Petrographic studies of the Cretaceous units, concentrating on the more readily studied McNairy Formation, were conducted to determine regional variations in mineralogy and the provenance of the sediments. Mineral analyses of the Tuscaloosa, Coffee, Selma, Coon Creek, and Owl Creek formations were made to determine their provenance relations to the McNairy formation.

In order to study these sediments systematically and to present a balanced geographic density of petrographic data, an attempt was made to collect samples on 6-mile centers.

Samples of unconsolidated McNairy formation sands from 35 well spaced outcrops were mounted, etched with hydrofluoric acid, and stained, following the techniques suggested by Willman (1942, p. 10). Impregnation with a rapidly polymerizing plastic was conducted in the field at seven other outcrops and thin sections were made from these samples. The thin sections and stained mounts were studied and point-count percentages of the constituents were estimated. Tables V and VI are the results of these analyses.

Quartz makes up 76–98 per cent of the sands but averages about 90 per cent. Secondary overgrowths were noted on only a few of the grains on several of the slides and thin sections. The individual grains of all sizes exhibit an extreme

TABLE V. THIN-SECTION ANALYSIS OF MCNAIRY FORMATION SANDS
(Point-count percentage)

| *Sample No.* | *Quartz* | *Muscovite* | *Feldspar* | *Chert* | *Graphite* | *Matrix* | *Percentage of Quartz Angularity* |
|---|---|---|---|---|---|---|---|
| 47 | 88.00 | 9.50 | 1.50 | .50 | — | .50 | 95 |
| 65 | 85.00 | 8.00 | 1.00 | 2.50 | .50 | 2.00 | 98 |
| 108 | 86.00 | 7.00 | 2.00 | 2.50 | — | 2.50 | 98 |
| 138 | 87.50 | 10.50 | .50 | .50 | .50 | .50 | 94 |
| 156 | 84.50 | 6.50 | 1.50 | 2.50 | — | 5.00 | 100 |
| 157 | 76.00 | 8.00 | 1.00 | 5.00 | — | 10.00 | 99 |
| 182 | 82.00 | 9.50 | 3.50 | 2.00 | .50 | 2.50 | 93 |
| Average | 84.14 | 8.43 | 1.57 | 2.22 | .22 | 3.28 | 97 |

TABLE VI. AVERAGE MINERAL COMPOSITION OF McNAIRY FORMATION SANDS
(By grain count)

| *Per Cent by Number of Grains in Sand Fraction* | | | | | *Weight Per Cent, grm.* | | *Percentage of Quartz Angularity* |
|---|---|---|---|---|---|---|---|
| *Quartz* | *Muscovite* | *Feldspar* | *Chert* | *Graphite* | *Silt-Clay* | *Heavy Minerals* | |
| 93.90 | 3.04 | 1.70 | 1.00 | 0.35 | 2.99 | 1.34 | 95 |

angularity (89–100 per cent angular) and low sphericity; most of them are prolate and tabular. The grains are up to 2 mm. in diameter, averaging between 0.25 mm. and 0.5 mm.

The following varieties of quartz were noted in the thin sections and counted: (1) unstrained quartz, (2) strained quartz, and (3) polycrystalline quartz. The unstrained particles compose 40–55 per cent of the quartz, whereas the strained particles and polycrystalline particles compose 40–53 per cent and 5–19 per cent of quartz, respectively. Pettijohn (1957, pp. 512–14) suggested that the variety of quartz is a guide to the nature of the source area. Bailey *et al.* (1958, p. 1402) have demonstrated the unreliability of optical varieties of quartz for differentiating between metamorphic and igneous rocks. They conclude however that strained quartz is the result of deformed rocks and for this reason the ratios of unstrained-strained quartz can show changes in provenance.

The amount of interstitial silt and clay was determined chiefly by sieve analyses and in thin section was counted as matrix. This material makes up 1–10 per cent of the sands, as determined in thin section. No attempt was made to differentiate the various clay minerals in thin sections. The silt grains are composed chiefly of quartz, with minor quantities of heavy minerals.

Muscovite mica is the chief accessory mineral of the sands and is particularly noticeable in the field, where laminae and beds up to 3 inches thick composed entirely of muscovite have been observed. Muscovite occurs in amounts ranging from 1 to 10 per cent. Because of the shape factor, the point-count method in the loose grain mounts tends to exaggerate the amount of muscovite. Muscovite flakes have about one-fifth the weight of a quartz grain of the same diameter. The factor 0.20 therefore was used to arrive at the percentages of muscovite given in Table VI. The individual grains of muscovite have a fresh appearance with slightly frayed edges. Muscovite flakes occur in all size fractions, but are most abundant in the coarser fractions, averaging about 1 mm. in diameter. Frequently observed as inclusions in the muscovite flakes are euhedral graphite particles. Micaceous minerals other than muscovite are extremely rare in the samples but include biotite and chlorite. All the mica particles appear to be detrital.

Potash feldspar was the only feldspar type observed in thin section and etched mount. It occurs in quantities ranging from less than one per cent to five per cent and occurs in sizes up to 0.5 mm. in diameter. The individual grains show stages of weathering ranging from fresh to highly weathered. All observed grains are here considered as detrital. No authigenic feldspar was recognized in thin section, although its presence was suggested in several X-ray spectrometer analyses.

Detrital grains of chert, averaging 0.25 mm. in diameter, were found in all thin sections and etched mounts. It occurs in amounts ranging from 1 to 3 per cent. The grains generally have an irregular shape, a high degree of angularity, and exhibit a microcrystalline character under crossed nicols.

Graphite, a rare accessory mineral in sedimentary rocks, was most readily identified in the etched mounts and is present in amounts ranging from 0 to 1 per cent. Shrode (1954) reported its occurrence in the McNairy formation sands in southern Illinois and the present writer has observed that it occurs widely in the Upper Mississippi Embayment sediments. Graphite occurs chiefly as free grains with subhedral to irregular outlines up to 0.125 mm. in diameter. It also occurs as inclusions in muscovite flakes, where it is generally euhedral, and very rarely as inclusions in quartz grains.

From point-count percentages of the thin sections, in which chert grains and polycrystalline

TABLE VII. AVERAGE HEAVY-MINERAL COMPOSITION, BY FORMATION, OF GULFIAN SEDIMENTS
(Line-count percentage)

| *Formation* | *Kyanite* | *Staurolite* | *Tourmaline* | *Zircon* | *Sillimanite* | *Rutile* | *Monazite* | *Hypersthene* | *Andalusite* | *Epidote* | *Garnet* | *Chloritoid* | *Xenotime* | *Ilmenite* | *Leucoxene* | *Magnetite* | *Percentage Angular Tourmaline* | *Percentage Angular Zircon* |
|---|---|---|---|---|---|---|---|---|---|---|---|---|---|---|---|---|---|---|
| McNairy | 43 | 14 | 10 | 8 | 8 | 6 | 3 | 1 | 1 | 1 | tr. | 1 | tr. | 3 | 1 | tr. | 92 | 79 |
| Owl Creek | 34 | 18 | 3 | 15 | 5 | 6 | 5 | 1 | 1 | 2 | 5 | tr. | tr. | 4 | 1 | tr. | 96 | 87 |
| Coon Creek | 21 | 14 | 7 | 25 | 5 | 6 | 6 | 2 | 2 | 2 | 3 | 1 | 3 | 2 | tr. | 1 | 97 | 88 |
| Selma | 16 | 10 | 6 | 28 | 10 | 7 | 7 | tr. | 1 | 2 | 4 | 1 | 4 | 3 | tr. | 1 | 96 | 93 |
| Coffee | 25 | 15 | 9 | 15 | 5 | 8 | 6 | 1 | 1 | 1 | 1 | 1 | 1 | 6 | 3 | 2 | 99 | 96 |

quartz grains were summed as rock fragments (no argillaceous rock fragments were observed), and silt, clay, muscovite, and graphite were summed as matrix, the McNairy formation sands were classified as slightly argillaceous, very micaceous, and sparingly feldspathic sands that most closely fit the description "protoquartzite" of the lithic sandstone family (Pettijohn, 1957, p. 291).

Heavy minerals generally constitute less than 2 per cent of the sediments, but may range up to 30 per cent, by weight. They consist of kyanite, staurolite, tourmaline (brown, brownish green, colorless, and blue), zircon, sillimanite, rutile, monazite, hypersthene, andalusite, epidote, chloritoid, xenotime, ilmenite, leucoxene, and magnetite. Nearly all the grains are angular and have diameters generally less than 0.25 mm.

Kyanite, staurolite, and sillimanite are the most abundant heavy minerals and constitute up to 81 per cent of the entire suite. Tourmaline and zircon are next in abundance, ranging up to 32 per cent in the sands. Table VII[5] presents the averaged results of the heavy-mineral analyses, by unit. Thirty-seven samples from the McNairy formation, 2 samples from the Owl Creek formation, 6 samples from the Coon Creek formation, 4 samples from the Selma formation, and 2 samples from the Coffee formation were analyzed by line counts of 200 grains per sample. The differences between the units are variations in ratios of specific minerals, which are attributed to variations in granular median diameter of the samples. However the Tuscaloosa formation is nearly devoid of heavy minerals and the limited suite consists almost wholly of well rounded tourmaline and zircon grains, similar to those in the underlying and adjacent Paleozoic rocks.

[5] Detailed heavy-mineral analyses are presented in Pryor, 1959.

Tourmaline grains range from 80 to 100 per cent angular. Most angular grains are euhedral or subhedral, whereas the round grains are extremely well rounded and generally have frosted and pitted surfaces. Zircon grains range from 53 to 100 per cent angular, most of the angular grains being euhedral. Several tourmaline grains exhibited a rounded, central core of tourmaline with slightly rounded euhedral authigenic overgrowths. The brown and brownish green varieties of tourmaline are the most abundant; the colorless and blue varieties are present, but rare.

Needham (1934) studied the heavy minerals of the upper part of the Eutaw formation in northeastern Mississippi and reported epidote and garnet as the most abundant heavy minerals, and staurolite and sillimanite rare. This difference between the heavy-mineral assemblages of the Eutaw formation and the overlying Cretaceous units in the Mississippi Embayment suggests slightly different provenance characteristics.

Clay samples from 120 widely spaced localities were analyzed by X-ray diffraction techniques, and the presence of three main clay minerals, kaolinite, mica (illite), and montmorillonite, was ascertained. The clays of the Tuscaloosa, Coffee, and McNairy formations are composed chiefly of kaolinite; those of the Selma and Owl Creek formations are chiefly montmorillonite; and those of the Coon Creek formation are nearly equal mixtures of kaolinite, mica, and montmorillonite. A close relationship between clay mineralogy and depositional environments is suggested by these data. Kaolinite is associated chiefly with non-marine sediments and montmorillonite is chiefly associated with marine sediments (Pryor and Glass, in preparation).

The petrographic studies suggest that the dominant rock types in the source area consisted of high-rank metamorphic rocks such as mica schists, gneisses, and crystalline and contact metamorphic rocks. Minor quantities of acid and basic igneous rocks, pegmatites, and pre-existing sedimentary rocks also are considered to have contributed to the sediments. The Tuscaloosa formation was derived directly from the adjacent Paleozoic rocks of the area.

Although low percentages of feldspar can be the result of a long history of abrasion, its rarity in these sediments is believed, because of the immaturity of the sediments, to be the result of lack of feldspar in the source area or intense weathering conditions in the source area, or both. The immaturity of these sediments is well expressed by the extreme angularity observed in the quartz, tourmaline, and zircon grains, and suggests a first cycle sediment with a short history of transportation and abrasion.

## Sedimentation

### Geometry of Basin

The late Cretaceous Mississippi Embayment plunges southwest in the form of a slightly asymmetrical trough, with the lesser slope on the east. This basin of deposition widens toward the south, and eventually becomes part of the east-west-trending Gulf Coast geosyncline. The Cretaceous materials filling the embayment thicken markedly southward and reach maximum thickness of about 3,500 feet in Mississippi, but farther south, as part of the Gulf Coast geosyncline, they thicken to more than 11,000 feet. The stratigraphic section is slightly compressed at the shallow end of the embayment and along the western flank, and expands toward the axis and toward the south.

### Depositional Environments

Paleontological evidence establishes the presence and, in part, the distribution of marine and non-marine environments during late Cretaceous deposition in the Upper Mississippi Embayment. Marine megafauna and microfauna occur in the fine clastics and carbonates of the glauconitic Selma and Coon Creek formations. The latter formation is characterized by an abundance of inner-neritic forms and the former by the abundance of outer-neritic forms. The southern part of the Coffee formation contains an inner-neritic marine fauna and the Owl Creek formation fauna is typified by outer-neritic forms. The sediments associated with the marine environments are characteristically calcareous, glauconitic, and fossiliferous silts, clays, and sands, and glauconitic and fossiliferous chalks. The fine-grained clastics grade laterally into the chalks and limestones.

Non-marine plant remains and lignite beds are abundant in the sands, silts, and clays of the McNairy formation and northern part of the Coffee formation. The textural aspects and sedimentary structures of the McNairy formation suggest fluviatile and deltaic depositional environments. The McNairy and Coffee formations grade laterally into the marine sediments, and in subsurface the McNairy formation sands are progressively more marine toward the southwest, becoming glauconitic and containing a marine fauna.

### Patterns of Sedimentation

The late Cretaceous sediments in the Upper Mississippi Embayment have in the northern part of the area sand-clay ratios of >2:1, which grade southward into sand-clay ratios of >1:2 (Fig. 4). The area of high clastic ratios coincides with that having low percentages of marine beds and occurs in the area of thinner sediments. The southward decrease in clastic ratios and increase in percentages of marine beds suggest sediment dispersion from north to south. This is supported by the textural analysis, which indicates the sands of coarser median diameter and of fluvial and upper delta type occur in the northeastern part of the area and changing slightly southwestward to sands of finer median diameters and lower delta type (Fig. 6). Directional properties of the McNairy formation also indicate a southwestward distribution of the sediments and the dispersal pattern is similar to that of deltaic deposits.

In general, therefore, the depositional picture is one of fluviatile and deltaic sediments in the northern and eastern end of the embayment, represented by the non-marine Coffee and McNairy formations, grading southwestward into the shallow-water, near-shore, marine Coon Creek formation. These near-shore marine sediments grade southward into the deeper-water, offshore, marine Selma formation. In Figure 17 three depositional environments are depicted: (1) the fluvial and upper delta, (2) the delta-front platform, and

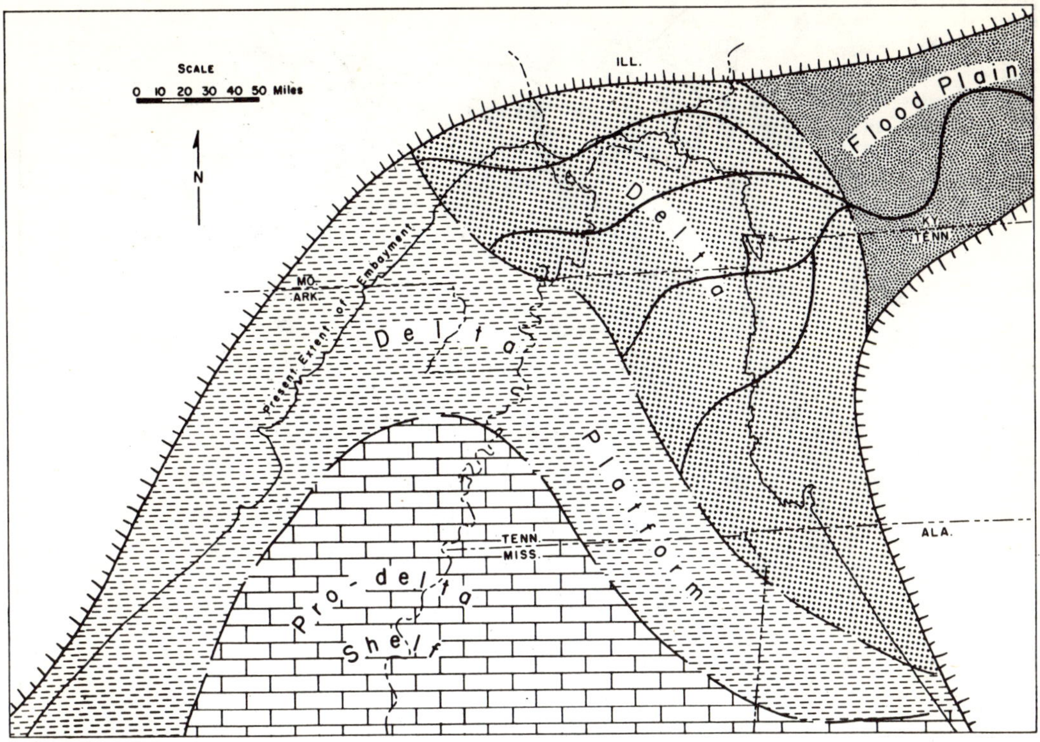


FIG. 17.—Interpretations of paleogeography and depositional environments of late Cretaceous Mississippi Embayment at time of deposition of McNairy formation.

(3) and pro-delta shelf. The fluvial-upper delta environment is represented by the McNairy and Coffee formations; the delta-front platform environment is represented by the Coon Creek formation and the calcareous part of the McNairy formation; and the pro-delta shelf environment is represented by the Selma and Owl Creek formations. The north-south cross section (Fig. 2) shows these facies relationships.

The fluvial and deltaic sediments are in general well sorted, highly cross-stratified sands, with subordinate amounts of lignitic, laminated silts and clays, and thin lignite beds. The delta-front platform sediments consist of glauconitic, fossiliferous (principally benthonic fauna), and slightly calcareous sands, silts, and clays. These sediments generally are lacking in lignitic materials and well developed cross-stratification. The pro-delta shelf sediments are represented by slightly glauconitic, fossiliferous (principally planktonic microfauna), very calcareous clays, chalks, and limestones. Small lenticular bodies of calcareous sand occur within the pro-delta shelf sediments.

The fluvial and deltaic sediments (Fig. 17) occupy the northeastern part of the embayment; the delta platform sediments occupy the south and west; and the pro-delta shelf sediments the south and center.

Fisk *et al.* (1954, p. 92) showed a delta distributory, bar-finger complex advancing seaward over the finer marine sediments of the delta platform in the recent Mississippi delta. Seaward from the distributary mouths, the sediments gradually become finer, and Shepard (1956, p. 2600) showed the deposition of carbonates gradually increased from the delta platform to the pro-delta shelf. Kruit (1955) showed the coarse clastic facies of the Rhone Delta transgressing seaward over the finer pro-delta materials in the Golfe du Lion. Although the sediments of the Mississippi Delta and the Rhone Delta are dissimilar from those of the late Cretaceous delta in the Upper Mississippi Embayment, the patterns of sediment transport and deposition are generally the same.

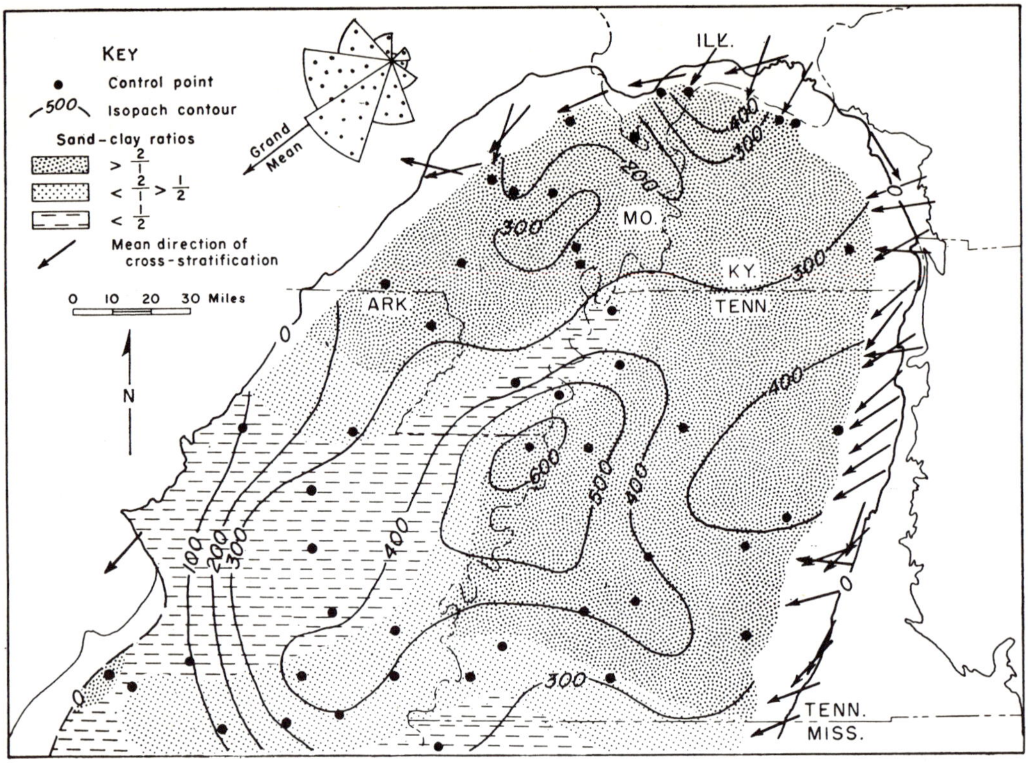


FIG. 18.—Isopach, lithofacies, and directional properties of McNairy formation, showing parallel trends.

Longitudinal filling, as suggested by Kuenen (1957), p. 194) for oblong basins, is the dominant pattern for these sediments.

### DEPOSITIONAL STRIKE AND PALEOSLOPE

The original depositional strike is deduced from the patterns of sediment dispersion and depositional environments (Fig. 17). It is roughly parallel with the isopach lines, with the equal clastic ratios, and with the percentage of marine beds; hence, it is parallel with the delta front. On these bases, the depositional strike is generally northwest-southeast.

The paleoslope is parallel with the plunge of the basin axis, with the direction of the paleocurrents (derived from cross-stratification), and with the gradient of the clastic ratios and percentage of marine beds, and is roughly perpendicular to the delta front. The paleoslope is toward the southwest, normal to depositional strike.

### SAND TRENDS

Establishing the general relations between sediment transport direction and depositional strike is of fundamental importance in predicting trends of sand bodies within a basin. Potter *et al.* (1958, p. 1041) showed a close parallelism of regional trends of the Mississippian Chester sands of the Illinois basin, the cross-bedding directions, and the regional slope.

To establish these regional relations in the descriptive depositional model of the late Cretaceous Mississippi Embayment, the trends of the McNairy formation sands are compared with the depositional strike and with the paleoslope.

Figure 18 is an isopachous and clastic-ratio map of the McNairy formation that shows the regional directional properties. The McNairy formation is depicted as an elongate body of sediments trending southwest perpendicular to the depositional strike and parallel with the paleoslope. The sand-clay ratios generally decrease southwest, with two elongate tongues of high sand-clay ratios extending southwest. The cross-stratification trends closely parallel the trend of both the thickness and clastic ratios.

Regionally, the sand trends of the McNairy formation in the Mississippi Embayment are

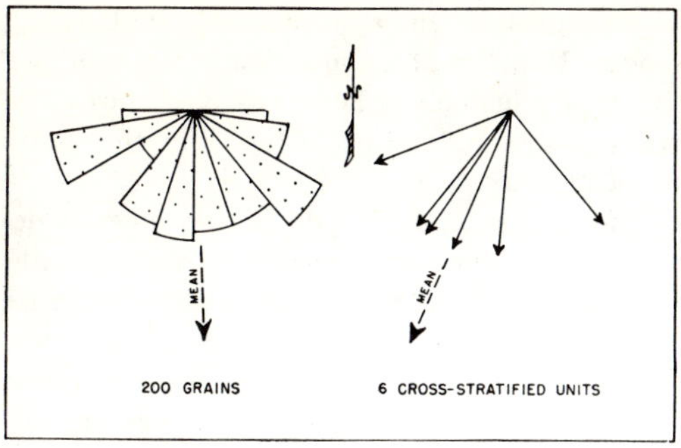


FIG. 19.—Relation of grain orientation and cross-stratification at outcrop level. (Sample No. 156)

toward the southwest, parallel with the paleoslope and sediment transport direction and normal to the depositional strike.

Cross-stratification was compared with the fabric of the sands at the outcrop level. Figure 19 shows the distribution of sand-grain orientation of a sample from a top-set bed and the vector mean of the cross-stratification observed in adjacent beds. Measurement of the orientation of 200 elongate grains on an oriented thin section indicated the fabric is oriented northwest-southeast, the vector mean azimuth being 177°. Directional measurements of six adjacent cross-stratified units gave a vectoral mean azimuth of 203° for the outcrop.

In the Cretaceous Mississippi Embayment the direction of sediment transport is essentially normal to the depositional strike and roughly parallel with the paleoslope. Sand trends, both regional and local, are parallel with the transport direction. Cross-stratification appears to be useful for predicting sand trends on both a regional and a local scale.

## Provenance

### source area

As indicated by the heavy-mineral suite, the principal contributing source area for the Mississippi Embayment Cretaceous sediments was a terrane composed largely of high-rank metamorphic rocks, including mica schists, gneisses, contact metamorphic rocks, and some acid and intermediate igneous rocks. The almost complete absence of rounded, overgrown, second-cycle quartz, tourmaline, or zircon grains precludes the contribution of any significant amounts of material from pre-existing sedimentary rocks and implies that most of the embayment sediments have had a relatively short history of abrasion.

Interpretations of the paleoslope and directional properties of the sediments suggest a source area on either the east or north. Source areas in these directions having the necessary compositions are the Blue Ridge and Piedmont Plateau of the southern Appalachians and, less possibly, the Pre-cambrian area of the southern Canadian Shield. The Blue Ridge and Piedmont Plateau most closely meet the compositional requirements. Sediments known to have been derived directly from the southern Canadian Shield (Ockerman, 1930; Ernst, 1954; Hamblin, 1958, pp. 21, 81, 101; and Driscoll, 1959, p. 13) have heavy-mineral suites different from those of the Mississippi Embayment sediments, whereas the heavy-mineral suites of the James River (Stow, 1939, p. 90) in Virginia and the New River (MacIntosh *et al.*, 1955, p. 286) in western Virginia and West Virginia that in part drain the Blue Ridge and Piedmont Plateau are closely comparable with those of the Cretaceous sediments in the embayment.

Although the heavy-mineral suite of the Eutaw formation (Needham, 1934) differs somewhat from that of the McNairy formation, it also is dominated by high-rank metamorphic minerals and probably was derived directly from the extreme southern part of the Piedmont Plateau and transported into the area by longshore currents from the southeast, as suggested by Tanner (1955, p. 2479) for the Cretaceous sediments of the East Gulf Coastal Plain.

The close gradational relations between the non-marine Coffee and McNairy formations and the marine Selma, Coon Creek, and Owl Creek formations are shown, both vertically and laterally, by the persistence of the heavy-mineral suite.

Evidence from paleoslope interpretations, from directional properties, and especially from the heavy-mineral suite, strongly favors the Blue Ridge and Piedmont Plateau of the southern Appalachians as the chief source area of the late Cretaceous sediments in the Upper Mississippi Embayment.

### paleoclimatology

Paleobotanic studies led Berry (1919, p. 30) to conclude that warm temperate to subtropical climatic conditions prevailed during late Cretaceous time in the Lower Mississippi Embayment and Eastern Gulf Coast. More recently, Lowenstam and Epstein (1954), on the basis of oxygen isotope studies on Coon Creek formation mollusks, arrived at mean paleotemperatures of 68°F.–82°F. for the late Cretaceous seas, suggesting a subtropical climate in the embayment similar to that of the Gulf of Mexico and the Caribbean Sea.

The absence of significant amounts of garnet and hornblende in the Upper Mississippi Embayment in sediments apparently derived from rocks bearing these minerals is analogous with the situation described by Dryden and Dryden (1946, p. 95), who show the destruction of garnet in place in the Wissahickon schist by weathering. This is in direct contrast to the findings of Goldich (1938, p. 53) whose studies of the weathering of the Black Hills crystalline rocks showed garnet resistant and concentrated in the weathering products. This may be the result of climatic differences, suggesting that the climate of the southern Appalachians during late Cretaceous time was humid, similar to the present climate of the area, and not semi-humid as is the present climate of the Black Hills.

The lack of significant quantities of feldspar in these sediments, which were derived in part from feldspathic rocks, suggests its destruction in place. Sand (1956, pp. 28–40) showed that the feldspathic rocks of the southern Piedmont Plateau and Blue Ridge weather readily and deeply, yielding clay minerals as weathering products, the destruction of feldspar being facilitated in a humid climate.

Paleontologic and petrologic evidence indicates humid, temperate to subtropical conditions for the southern Appalachians and Mississippi Embayment during the late Cretaceous.

## Paleogeography

The Upper Mississippi Embayment probably did not exist, as such, during early Cretaceous time. The topographic expression and the well developed soil horizon on the Paleozoic basement rock of the area (Mellen, 1937; Pryor, in preparation) suggest that it was mature and low-lying just prior to deposition of the Gulfian sediments. There was probably a well developed stream pattern on this surface that carried sediments southward during early Cretaceous time.

The lithologic character of the Tuscaloosa formation suggests local derivation, and, as the basal unit of the Gulf series, probably represents the results of slight uplift of the area north of and adjacent to the areas of deposition. Stearns (1958, p. 1094), however, suggested that the central part of the present embayment was uplifted and eroded. Uplift and erosion were accompanied by a subsidence in the southern part of the area and transgression of marine waters into the newly formed embayment.

The pattern of Eutaw deposition suggests development of the embayment into only the southeastern part of the area, followed by quiescence and slight erosion. The streams developed on the Paleozoic surface evidently did not contribute large amounts of sediment to the present, outcropping Eutaw formation, as the heavy-mineral suite in it is similar to that of the Gulf series in Alabama and Georgia (Needham, 1934) rather than to the sediments of the embayment.

The erosional interval was followed by general subsidence of the embayment and northward transgression of marine waters. Accompanying this transgression was the introduction into the northern part of the embayment of a new suite of clastic materials, represented by the Coffee formation, a deltaic deposit associated with renewed erosion in the source area.

Subsidence continued and the transgressive seas moved northward into the southeastern part of Missouri and southwestern Kentucky where they reached their early maximum extent. This major late Cretaceous transgression resulted in the deposition of transitional marine upper Coffee formation sediments and marine Selma formation chalks and calcareous clays, overlapping the deltaic lower Coffee sediments. Subsidence probably continued at a regular rate, but renewed erosion in the source area resulted in the dumping of large amounts of coarse clastics into the northeastern part of the embayment. These clastics (McNairy formation sands) formed a southward-building delta that rapidly filled the north end of the embayment, overcoming the subsidence and causing a general regression of the seas.

While the McNairy formation materials were being deposited in a floodplain and deltaic environment, the Coon Creek formation was being deposited at the delta front in shallow marine waters and the Selma formation was being deposited in the deeper waters on the south. As the "McNairy Delta" spread southward and westward it overlapped Selma and Coon Creek sediments previously deposited at the delta front. This period of regression reached its maximum extent when the seas withdrew to northern Mississippi and eastern Arkansas.

Subsidence then proceeded at a slower rate,

with the Coon Creek formation materials being deposited at the delta front and Selma deposition continuing in deeper waters. At the same time a minor contribution of sediment came from the source area. Subsidence apparently ceased for a time and slight uplift followed in the northern part of the embayment. This resulted in slight tilting and erosion of the McNairy sediments on the northern periphery of the embayment.

Following this hiatus in deposition was a pulse of general subsidence that resulted in the rapid northward transgression of the seas at least as far as southern Illinois and probably to southern Indiana. Reworking of the pre-existing "McNairy Delta" sediments plus a small amount of materials from the source area yielded the Owl Creek formation. During this maximum Gulfian transgression, the Prairie Bluff chalks were being deposited in deeper water at the south, contemporaneous with the Owl Creek clays in the north.

The axis of the embayment may have shifted slightly toward the west at this time, as was inferred by Stearns (1958, pp. 1094–95).

The last transgression during Cretaceous time was followed by a rapid regression in which the seas withdrew far toward the south, at least as far as southern Mississippi and Arkansas, and exposed the late Cretaceous sediments to subaerial erosion and soil development. Subsidence apparently halted for a time and did not resume until the great Midway transgression in early Paleocene time.

Preceding and accompanying the downwarp of the Mississippi Embayment, there was a limited amount of intrusive and volcanic activity. This is evidenced by the occurrence of igneous rocks in deep wells in the embayment and in outcrops in areas adjacent to the embayment. Moody (1949, p. 1427) summarized the present knowledge of these rocks and concluded that the northern Coastal Plain was invaded by magma in the Mesozoic era as early as the Triassic period and as late as the late Cretaceous. The later magma included both alkaline and basic types and resulted in both intrusive and extrusive rocks. Water-lain volcanic debris is reported in the Eutaw and Tuscaloosa formations of Mississippi and the equivalent Tokio formation of southeastern Arkansas. Mellen (1958, p. 53) reported "bentonite" in Mississippi in Cretaceous materials as young as the Ripley formation. A search of the literature fails to reveal mention of any bentonite in the Gulfian sediments in the northern part of the embayment. No bentonite or volcanic ash deposits were observed in the field. Although there undoubtedly was some igneous activity in the area during late Cretaceous time, it apparently resulted in no significant contribution of rock materials to the sediment fill of the Upper Mississippi Embayment.

Indications from the lithofacies maps and isopach maps (Figs. 3, 4, 18) suggest that the present-day extent of the Gulfian sediments closely approximates the original extent of the sediments. Stearns (1958, p. 1095) showed the original outline of the embayment extending only a short distance west of the present limits in southwest Arkansas. The Cretaceous sediments may have extended into Illinois 20–30 miles farther than the present limit. The near-shore and fluvial facies in the northern and western parts of the embayment strongly support the view that the Cretaceous sediments are present now to esentially the same extent as when they were deposited.

Farther southeast, however, in Tennessee and Mississippi, the eastern shoreward facies of the Selma and Coon Creek formations are not preserved. These facies probably were eroded and contributed to Tertiary sedimentation. Prior to Tertiary erosion the shoreward facies probably extended eastward to the vicinity of the Nashville dome.

The oldest erosional surfaces in the area (Lexingon-Lancaster) are believed to be of post-Cretaceous age, the result of late Tertiary erosion.

## Depositional Model

The descriptive model for the type of deposition illustrated by the late Cretaceous sediments in the Upper Mississippi Embayment embodies eight important and diagnostic features: basin geometry, depositional pattern, paleoslope, depositional strike, transport direction, sand trends, sedimentary associations, and tectonic activity. These features include several of the factors suggested by Potter (1959) in a discussion of "facies models." Table VIII summarizes the descriptive model.

## Summary and Conclusions

A deltaic system occupied the northeastern end of the Mississippi Embayment during late Cretaceous time. This system, the "McNairy Delta," dispersed sediments toward the south and west. The fluvial and upper delta coarse clastic facies

Table VIII. Depositional Model of Late Cretaceous Mississippi Embayment

| | |
|---|---|
| Basin geometry | Elongate, symmetrical or slightly asymmetrical, broad trough, widening down plunge |
| Depositional pattern | Longitudinal filling; deltaic pattern on shelf, becoming increasingly marine down paleoslope; stratigraphic section expands in direction of paleoslope and toward basinal axis |
| Paleoslope | Parallel or subparallel with axis of basin |
| Depositional strike | Normal to paleoslope |
| Transport direction | Normal to depositional strike and parallel with paleoslope |
| Sand trends | Parallel with transport direction and paleoslope, and normal to depositional strike |
| Sedimentary associations | Lithic sandstones (protoquartzite), generally lenticular and elongate, highly cross-stratified; dominantly non-marine, and associated with lignites, coals, and carbonaceous clays and shales. Impure carbonates, grading from calcareous clays and shales to argillaceous chalks and limestones; lutites rather than arenites |
| Tectonic activity | Mildly unstable shelf conditions; activity generally confined to subsidence and minor folding and normal faulting |

are represented by the Coffee and McNairy formations; the pro-delta shelf fine clastic-carbonate facies, are represented by the Selma and Owl Creek formations; and the transitional delta-front platform facies are represented by the Coon Creek formation and southern parts of the Coffee and McNairy formations. The distribution of the facies is controlled largely by the depositional strike of the deltaic system.

The following southwesterly trends in the late Cretaceous sediments were observed: the thickness increases, the clastic ratio decreases, the percentage of marine beds increases, and the sand types grade from fluviatile-upper deltaic to lower deltaic. These relations imply a northwest to southeast depositional strike.

Persistent, southwesterly paleocurrents are inferred from cross-stratification and indicate that the paleoslope was toward the southwest along the structural axis of the Mississippi Embayment.

The McNairy formation sands are well sorted protoquartzites, consisting chiefly of highly angular, first-cycle quartz grains with a matrix of muscovite, silt, and clay. The McNairy formation is fluviatile and grades southwestward into marine sands and clays. The mean orientation of cross-stratification in McNairy formation sands is 236° with a variance of 5,018 and a standard deviation of 70.8°, comparable with that of other fluvial-deltaic sandstones.

The Gulfian sediments of the Upper Mississippi Embayment were derived from an area composed principally of high-rank metamorphic rocks, the Blue Ridge and Piedmont Plateau of the southern Appalachians. These sediments were transported to the embayment by a single large stream system. Paleontological and petrographical evidences imply humid, temperate to subtropical climatic conditions for the source area and depositional area during the late Cretaceous.

The deposition of the Coffee formation sediments began with the general subsidence of the Mississippi Embayment and northward transgression of the late Cretaceous seas. Subsidence continued, but a large influx of coarse clastics from the northeast caused a general regression of the seas as the McNairy Delta built seaward. Subsidence halted briefly, followed by rapid regression of the seas in response to slight uplift in the northern part of the embayment. Resumption of subsidence resulted in a rapid transgression of marine waters deep into the continent, at least as far as southern Illinois and western Kentucky and possibly into southern Indiana. This maximum late Cretaceous transgression resulted in the deposition of the Owl Creek formation. Subsidence again ceased, followed by a general regression, which ended Mesozoic deposition in the Mississippi Embayment.

The descriptive model devised for deposition in intracratonic basins, applied here to the late Cretaceous Upper Mississippi Embayment, includes basin geometry, depositional pattern, paleoslope, depositional strike, transport direction, sand trends, sedimentary associations, and tectonic activity. In the depositional model of the Cretaceous Upper Mississsippi Embayment the major sand bodies are oriented parallel with the basin axis, paleoslope, and sediment transport direction and are normal to the depositional strike. The basin is a plunging open-end trough, the paleoslope is roughly parallel with the basin axis, and the depositional pattern is deltaic. Immature sands grading to impure clastics down the paleoslope constitute the sedimentary associations, and the tectonic activity is generally mild. The practical objective of constructing a descriptive model is to make it possible to predict where and how economically valuable deposits can be developed. Successful prediction requires adequate knowledge of the distribution of major sedimentary elements within a given association or facies, and a depositional model classifies such information to make it readily applicable.

## REFERENCES


ANDEL, TJ. H., VAN, 1950, *Provenance, Transport, and Deposition of Rhine Sediments*, pp. 5–43. H. Veenman & Zonen, Wageringen.

———, AND POSTMA, H., 1954, "Recent Sediments of the Gulf of Paria," *Verh. Kon. Nederlandse Akad. Wetensch.*, Afd. Nat., 1st reeks, deel 20, nr. 5, pp. 1–246.

BAILEY, S. W., BELL, R. A., AND PENG, C. J., 1958, "Plastic Deformation of Quartz in Nature," *Bull. Geol. Soc. America*, Vol. 69, No. 11, pp. 1443–66.

BERRY, E. W., 1917, "The Delta Character of the Tuscaloosa Formation," *Johns Hopkins Univ. Cir.*, N.S., No. 3, pp. 216–22.

———, 1919, "Upper Cretaceous Floras of the Eastern Gulf Region in Tennessee, Mississippi, Alabama, and Georgia," *U. S. Geol. Survey Prof. Paper 112*, pp. 1–173.

BRUSH, L. M., 1958, "Study of Cross-Stratification in a Large Laboratory Flume" (abs.), *Bull. Geol. Soc. America*, Vol. 69, No. 12, Pt. 2, p. 1542.

CUSHMAN, J. A., 1946, "Upper Cretaceous Foraminifera of the Gulf Coastal Region of the United States and Adjacent Areas," *U. S. Geol. Survey Prof. Paper 206*, pp. 1–241.

DAPPLES, E. C., KRUMBEIN, W. C., AND SLOSS, L. L., 1948, "Tectonic Control of Lithologic Associations," *Bull. Amer. Assoc. Petrol. Geol.*, Vol. 32, No. 10, pp. 1924–47.

DIXON, W. J., AND MASSEY, F. J., 1951, *Introduction to Statistical Analysis*. 370 pp. McGraw-Hill Book Company, Inc., New York.

DOEGLAS, D. J., 1946, "Interpretation of the Results of Mechanical Analyses," *Jour. Sed. Petrology*, Vol. 16, No. 1, pp. 19–40.

———, 1950, "De interpretatie van Korrelgrootee-analysen," *Verh. Kon. Nederlandse Geol. Mijnb.*, Gen. Geol. Ser., Deel 15, pp. 247–328.

DRISCOLL, EGBERT G., 1959, "Evidence of Transgressive-Regressive Cambrian Sandstones Bordering Lake Superior," *Jour. Sed. Petrology*, Vol. 29, No. 1, pp. 5–15.

DRYDEN, A. L., AND DRYDEN, CLARISSA, 1946, "Comparative Rates of Weathering of Some Common Heavy Minerals," *ibid.*, Vol. 16, No. 3, pp. 91–96.

DUTTON, C. E., 1928, "A Sedimentation Study of Cretaceous and Tertiary Sands of Southern Illinois," unpub. M. A. thesis, Univ. of Illinois.

EMRICH, G. E., ms., "Cross-Bedding and Textural Variation of the Miocene Hawthorne Formation in Northern Florida."

ERNST, W. G., 1954, "The St. Peter Sandstone-Glenwood Shale Contact in Minnesota," *Amer. Mineralogist*, Vol. 39, Nos. 11–12, pp. 1025–31.

FARKUS, S. E., 1958, "Analysis of Cross-Lamination in the Upper Cambrian Franconia Formation of Southwestern Wisconsin," unpub. M. S. thesis, Univ. Wisconsin.

FISK, H. N., 1944, "Geological Investigation of the Alluvial Valley of the Lower Mississippi River," pp. 1–78. Mississippi River Commission, Vicksburg.

———, MCFARLAN, E., JR., KOLB, C. R., AND WILBERT, L. J., JR., 1954, "Sedimentary Framework of the Modern Mississippi Delta;" *Jour. Sed. Petrology*, Vol. 24, No. 2, pp. 76–99.

GLENN, L. C., 1906, "Underground Waters of Tennessee and Kentucky West of the Tennessee River and of an Adjacent Area in Illinois," *U. S. Geol. Survey Water-Supply Paper 165*, pp. 22–29.

GOLDICH, S. S., 1938, "A Study in Rock Weathering," *Jour. Geol.*, Vol. 46, pp. 17–58.

GRIM, R. E., 1936, "The Eocene Sediments of Mississippi," *Mississippi Geol. Survey Bull. 30*.

HAMBLIN, W. K., 1958, "Cambrian Sandstones of Northern Michigan," *Michigan Dept. Cons., Geol. Survey Div. Pub. 51*. 146 pp.

HARPER, L., 1857, "Preliminary Report on the Geology and Agriculture of the State of Mississippi," pp. 72–115.

HARRISON, P. W., 1957, "New Technique for Three-Dimensional Fabric Analyses of Till and Englacial Debris Containing Particles from 3 to 40 mm. in Size," *Jour. Geol.*, Vol. 65, No. 1, pp. 98–105.

HAYES, W. E., 1951, "The Smaller Fossils from the Upper Cretaceous of the Starkville Area," unpub. M.S. thesis, Mississippi State College. 147 pp.

HILGARD, E. W., 1860, "Report on the Geology and Agriculture of the State of Mississippi," pp. 60–106.

KRUIT, C., 1955, "Sediments of the Rhone Delta," *Verh. Kon. Nederlandse Geol. Mijnb.*, Gen. Geol. Ser., deel 15, nr. 3, pp. 1–141.

KRUMBEIN, W. C., SLOSS, L. L., AND DAPPLES, E. C., 1949, "Sedimentary Tectonics and Sedimentary Environments," *Bull. Amer. Assoc. Petrol. Geol.*, Vol. 33, No. 11, pp. 1859–91.

KUENEN, PH. H., 1957, "Longitudinal Filling of Oblong Sedimentary Basins," *Verh. van net Koninkl. Nederlandse Geol. Mijnb. Genoot.*, Geol. Ser., Vol. XVII, pp. 189–95.

LOEBLICH, A. R., AND TAPPAN, HELEN, 1957, "Planktonic Foraminifera of Paleocene and Eocene Age from the Gulf and Atlantic Coastal Plains," *U. S. Nat. Mus. Bull. 215*, pp. 173–200.

LOWENSTAM, H. A., AND EPSTEIN, S., 1954, "Paleotemperatures of the Post-Aptian Cretaceous as Determined by the Oxygen Isotope Method," *Jour. Geol.*, Vol. 62, No. 3, pp. 207–48.

MCDOWELL, J. P., 1957, "The Sedimentary Petrology of the Mississagi Quartzite in the Blind River Area," *Ontario Dept. Mines Geol. Cir. 6*. 31 pp.

MACINTOSH, C. A., MIXSON, A. R., AND WILLARD, D. K., 1955, "Mineralogical Studies of Sediments from New River, Virginia" (abs.), *Virginia Jour. Sci.*, Vol. 6, No. 4, p. 286.

MCKEE, E. D., AND WEIR, G. W., 1953, "Terminology for Stratification and Cross-Stratification in Sedimentary Rocks," *Bull. Geol. Soc. America*, Vol. 64, No. 4, pp. 381–90.

MELLEN, F. F., 1937, "The Little Bear Residuum," *Mississippi Geol. Survey Bull. 34.*

———, 1958. "Cretaceous Shelf Sediments of Mississippi," *ibid., Bull. 85.* 112 pp.

MOODY, C. L., 1949, "Mesozoic Igneous Rocks of Northern Gulf Coastal Plain," *Bull. Amer. Assoc. Petrol. Geol.*, Vol. 33, No. 8, pp. 1410–28.

NEEDHAM, C. E., 1934, "The Petrology of the Tombigbee Sands of Eastern Mississippi," *Jour. Sed. Petrology*, Vol. 4, No. 2, pp. 55–59.

NIEHOFF, WOLFGANG, 1958, "Die primar gerichten sedimentstrukturen, insbesondere die schragschichtung im Koblezquarit am Mittellhein," *Geol. Rundschau*, Bd. 47, Heft 1, pp. 252–321.

OCKERMAN, J. C., 1930, "A Petrographic Study of the Madison and Jordan Sandstones of Southern Wisconsin," *Jour. Geol.*, Vol. 38, No. 4, pp. 346–53.

PELLETIER, B. R., 1958, "Pocono Paleocurrents in Pennsylvania and Maryland," *Bull. Geol. Soc. America*, Vol. 69, No. 8, pp. 1033–64.

PETTIJOHN, F. J., 1957, *Sedimentary Rocks.* 2d ed. 718 pp. Harper and Brothers, New York.

PHLEGER, F. B, AND PARKER, F. L., 1951, "Ecology of Foraminifera, Northwest Gulf of Mexico," *Geol. Soc. America Mem. 46*, pp. 1–152.

PIERCE, G. R., 1942, "Foraminiferal Zones of the Upper Cretaceous of Mississippi," unpub. Ph.D. thesis, Iowa Univ.

POTTER, P. E., 1959, "Facies Model Conference," *Science*, Vol. 129, No. 3358, pp. 1292–94.

———, AND GLASS, H. D., 1958, "Petrology and Sedimentation of the Pennsylvanian Sediments in Southern Illinois—a Vertical Profile," *Illinois Geol. Survey Rept. Inv. 204.* 59 pp.

———, AND OLSON, J. S., 1954, "Variance Components in Some Basal Pennsylvanian Sandstones in the Eastern Interior Basin—Geological Application," *Jour. Geol.*, Vol. 62, No. 1, pp. 50–73.

———, AND PRYOR, W. A., ms., "Dispersal Centers of Paleozoic and Later Clastics of the Upper Mississippi Valley and Adjacent Areas."

———, AND SIEVER, RAYMOND, 1956, "Sources of Basal Pennsylvanian Sediments in the Eastern Interior Basin, Part 1. Cross-Bedding," *Jour. Geol.*, Vol. 64, No. 3, pp. 225–44.

———, NOSOW, EDMUND, SMITH, N. M., SWANN, D. H., AND WALKER, F. H., 1958, "Chester Cross-Bedding and Sandstone Trends in Illinois Basin," *Bull. Amer. Assoc. Petrol. Geol.*, Vol. 42, No. 5, pp. 1013–46.

POWERS, M. C., 1953, "A New Roundness Scale for Sedimentary Particles," *Jour. Sed. Petrology*, Vol. 23, No. 2, pp. 117–19.

PRYOR, W. A., "Dip Direction Indicator," *ibid.*, Vol. 28, No. 2, p. 230.

———, 1959, "Cretaceous Geology and Petrology of the Upper Mississippi Embayment," unpub. Ph.D. thesis, Rutgers Univ.

———, ms., "Geology of the Cairo Quadrangle and Parts of the Adjacent LaCenter and Thebes Quadrangles," *Illinois Geol. Survey Cir.*

———, AND GLASS, H. D., ms., "Cretaceous-Tertiary Clay Mineralogy of the Upper Mississippi Embayment."

RUSNAK, G. A., 1957, "A Fabric and Petrologic Study of the Pleasantview Sandstone," *Jour. Sed. Petrology*, Vol. 27, No. 1, pp. 41–55.

SAFFORD, J. M., 1864, "Cretaceous and Superior Formations of West Tennessee," *Jour. Sci. and Arts*, Vol. 27, 2d Ser., pp. 360–72.

SAND, L. B., 1956, "On the Genesis of Residual Kaolins," *Amer. Mineralogist*, Vol. 41, pp. 28–40.

SCHWARZACHER, W., 1953, "Cross-Bedding and Grain Size in the Lower Cretaceous Sands of East Anglia," *Geol. Mag.*, Vol. 90, pp. 322–30.

SHEPARD, F. P., 1956, "Marginal Sediments of Mississippi Delta," *Bull. Amer. Assoc. Petrol. Geol.*, Vol. 40, No. 11, pp. 2537–2634.

SHRODE, R. S., 1954, "Graphite in the Cretaceous of Southern Illinois," *Jour. Sed. Petrology*, Vol. 24, No. 3, pp. 219–20.

STEARNS, R. G., 1958, "Cretaceous, Paleocene, and Lower Eocene Geologic History of the Northern Mississippi Embayment," *Bull. Geol. Soc. America*, Vol. 68, No. 9, pp. 1077–1100.

———, AND ARMSTRONG, C. A., 1955, "Post-Paleozoic Stratigraphy of Western Tennessee and Adjacent Portions of the Upper Mississippi Embayment," *Tennessee Div. Geol. Rept. Inv. 2*, pp. 1–29.

STEPHENSON, L. W., 1914, "Cretaceous Deposits of the Eastern Gulf Region and Species of Exogyra from the Eastern Gulf Regions and the Carolinas," *U. S. Geol. Survey Prof. Paper 81.*

———, 1955, "Owl Creek (Upper Cretaceous) Fossils from Crowleys Ridge, Southeastern Missouri," *ibid., Prof. Paper 274-E*, pp. 97–139.

———, AND MONROE, W. H., 1938, "Stratigraphy of Upper Cretaceous Series in Mississippi and Alabama," *Bull. Amer. Assoc. Petrol. Geol.*, Vol. 22, No. 11, pp. 1639–57.

———, KING, F. B., MONROE, W. H., AND IMLAY, R. W., 1942, "Correlation of Outcropping Cretaceous Formations of the Atlantic and Gulf Coastal Plains and Trans-Pecos Texas," *Bull. Geol. Soc. America*, Vol. 53, No. 3, pp. 435–36.

STOW, M. H., 1939, "Reflection of Provenance in Heavy Minerals of James River, Virginia," *Jour. Sed. Petrology*, Vol. 9, No. 2, pp. 86–91.

TANNER, W. F., 1955, "Paleogeographic Reconstruction from Cross-Bedding Studies," *Bull. Amer. Assoc. Petrol. Geol.*, Vol. 39, No. 12, pp. 2471–83.

TUKEY, J. W., 1954, "Chi-Square Test of Orientation," Comment No. 1-A, Earth Sciences Panel Review Group CSPS-ASA, unpub. communication.

WADE, BRUCE, 1926, "The Fauna of the Ripley Formation on Coon Creek, Tennessee," *U. S. Geol. Survey Prof. Paper 137*, pp. 1–272.

WILLMAN, H. B., 1942, "Feldspar in Illinois Sands," *Illinois Geol. Survey Rept. Inv. 79*, pp. 10–11.

WURSTER, PAUL, 1958, "Geometrie und geologie von kreuzschichtungkorpen," *Geol. Rundschau*, Bd. 47, Heft 1, pp. 322–59.

**Reprinted from:**
BULLETIN OF THE AMERICAN ASSOCIATION OF PETROLEUM GEOLOGISTS
VOL. 48, NO. 5 (MAY, 1964), PP. 637-658, 14 FIGS.

# LATE MISSISSIPPIAN RHYTHMIC SEDIMENTS OF MISSISSIPPI VALLEY[1]

DAVID H. SWANN[2]
Urbana, Illinois

## ABSTRACT

The upper part of the Mississippian column in the type region is the Pope Megagroup, a cyclic alternation of limestone-dominated and clastic-dominated units. Its lower boundary transects time, so that in different districts the Pope equals part of the Chesterian, all of the Chesterian, or includes the Chesterian plus some late Valmeyeran. Within the type region these rocks are essentially confined to the Illinois basin, which as it gradually sank was repeatedly filled by sediments carried from eastern Canada by the Michigan River system, and by locally precipitated limestone. The fill averaged half mud, a quarter sand, and a quarter limestone.

The shoreline fluctuated several hundred miles landward (northeastward) and seaward, with 70 or more minor reversals in direction superimposed on about 15 major cycles of advance and retreat. Sea-level is inferred to have been relatively static, and variation in rainfall was the primary factor controlling sedimentary cycles. A clastic unit was deposited when the Michigan River delta prograded scores or hundreds of miles across the subsiding basin in response to increased sediment yield that was triggered by a drier climate in the source area. Lunate sand bars at the mouths of advancing distributaries were lengthened into bar fingers. Unweathered grooves, later filled with more sand, were gouged far below sea-level behind the advancing mouth bars. The delta surface supported vegetation, and in some cycles became a great sea-level swamp when increased humidity and more effective plant cover in the source region so reduced sediment yield that clastic sedimentation could no longer keep pace with basin sinking. When plant growth could not keep up with continued subsidence, the sea again covered the area and limestone was deposited until return to a drier climate at the sediment source brought the cycle full turn.

A few cycles seem correlated with changes in sea-level, as were many Pennsylvanian cycles. The alternation between tropical pluvial and subtropical interpluvial stages that caused the cyclic sedimentation in what is now the northern hemisphere corresponded in some manner to fluctuations of the boundary of continental ice in the southern hemisphere. However, the direct control of the sedimentary rhythms was rainfall, not glaciation.

## INTRODUCTION

The earliest geologists in the Mississippi Valley recognized a sequence of alternating limestone-dominated and sandstone-dominated units above the Subcarboniferous "Mountain Limestone" and beneath the Carboniferous "Millstone Grit." The term Chester had been informally applied to this succession for several years before it first appeared in print in an article by Worthen (1860) that specifically called attention to the alternation of rocks containing terrestrial floras and those containing marine faunas. The term is now used in time stratigraphy in the form Chesterian Series to designate the uppermost of three (or four) series in the Mississippian. In rock stratigraphy the cyclic succession is called the Pope Megagroup[3] (Swann and Willman, 1961). Because the clastic parts of several of the early cycles extend over only part of the region, the boundary between the clearly cyclic Pope Megagroup and the underlying Mammoth Cave Limestone Megagroup transgresses time. The Mammoth Cave includes some early Chesterian rocks in areas of low clasticity, whereas the Pope extends down into the Valmeyeran (Middle Mississippian) in strongly clastic parts of the region. However, the lower boundary of the Pope lies within a hundred feet of the base of the Chesterian in all except a few small areas. The top of the Pope corresponds with the top of the Mississippian, that is, the top of the Chesterian Series.

Within the type Mississippian region, the cyclic

[1] Manuscript received, October 28, 1963. Presented as part of the Society of Economic Paleontologists and Mineralogists Research Committee symposium on cyclic sedimentation at Houston, March 25, 1963.

[2] Illinois State Geological Survey, Urbana, Illinois. The writer is indebted to Elwood Atherton for the use of maps and manuscripts in preparation. Elwood Atherton, T. C. Buschbach, R. H. Howard, C. A. Ross, and H. B. Willman of the Illinois State Geological Survey, P. E. Potter of Indiana University, J. E. Sanders of Yale University, H. R. Wanless of the University of Illinois, and J. M. Weller of the University of Chicago have read the manuscript. D. M. Sullivan and C. B. Rexroad of the Indiana Geological Survey have given advice on specific problems. C. A. Bays has read and discussed pertinent literature. Mrs. L. L. Whiting, Mrs. D. H. Swann, and Mrs. W. M. Swann have read to me.

[3] This paper follows the stratigraphic policy of the Illinois State Geological Survey (Willman *et al.*, 1958), which differs in minor detail from the code of the American Commission on Stratigraphic Nomenclature (1961). Megagroup is preferred to Supergroup, the synonym used in the code, because of the use of the prefix "super" in stratigraphy to imply spatial relation as well as rank, a conflict that does not occur in biologic nomenclature. Trinomials are used where appropriate in the names of groups and megagroups (Mammoth Cave Limestone Megagroup) as well as in units of infraformational rank.

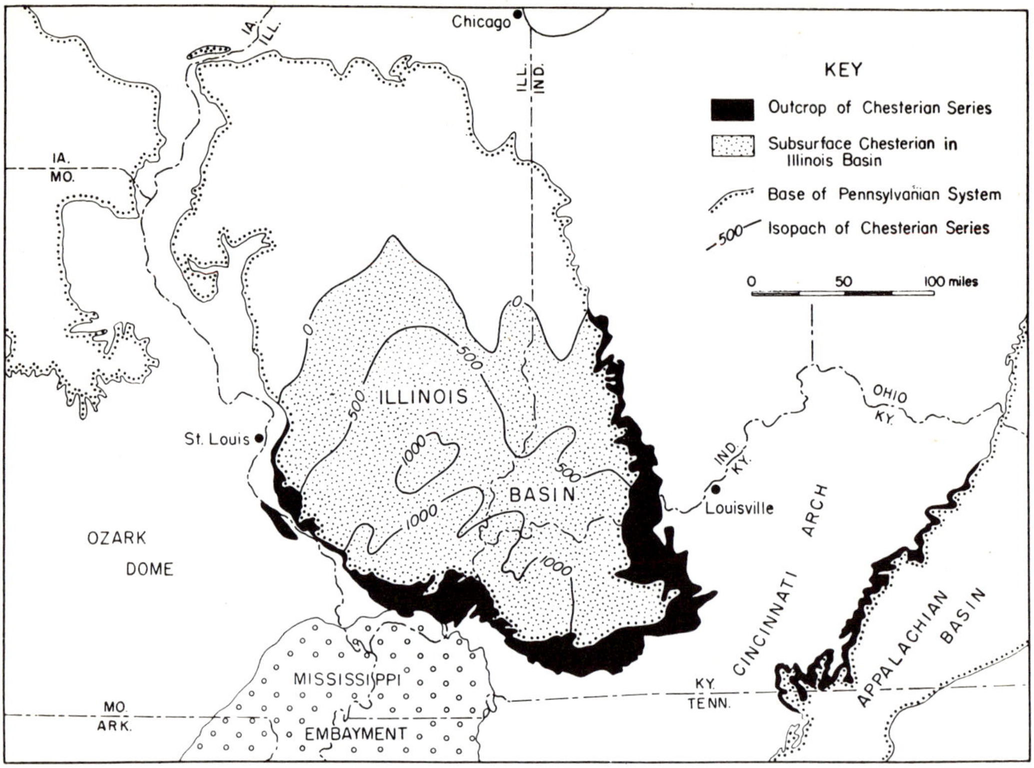


FIG. 1.—Regional tectonic setting and thickness of late Mississippian cyclic sediments in Illinois basin.

Pope sediments are essentially confined to the Illinois basin (Fig. 1). The rocks crop out in a horseshoe around the southern rim of the basin but are overlapped by Pennsylvanian strata at the north. They are nearly 1,500 feet thick near the center of the outcrop belt. Originally, they thinned moderately toward the neighboring Ozark and Cincinnati Arch regions (Fig. 4) but their present configuration is the result of pre-Pennsylvanian truncation.

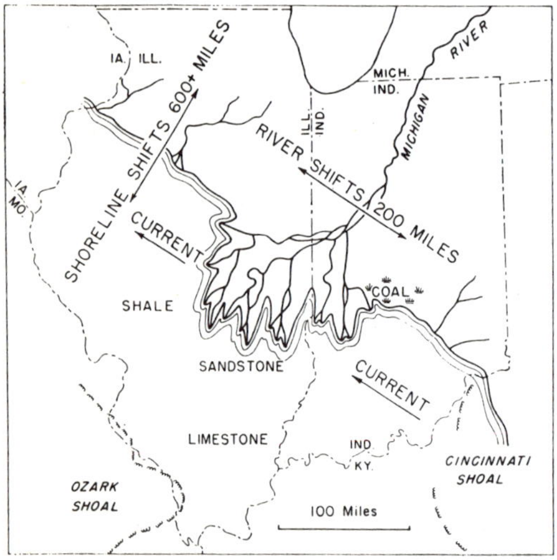


FIG. 2.—Typical paleogeography during deposition of late Mississippian cyclic succession.

The limestone that forms a quarter of the late Mississippian fill of the Illinois basin was locally derived. However, the sandstone forming a second quarter and the shale making up the remainder of the fill are composed of grains derived from sources farther northeast, largely from eastern Canada. They were carried by a river system (Fig. 2) that has been named the Michigan River (Swann, 1963). The Michigan River discharged into the Illinois basin and fed sediment to it (and at times to the Michigan and Black Warrior basins) much as the contemporaneous Ontario River (Pepper *et al.*, 1954) fed sediment to the northern end of the Appalachian basin.

The shore that divided land on the northeast from sea on the southwest was not static. Outcrops of coal and of rocks with *Stigmaria* roots *in situ* prove that many times land extended south to the outcrop belt, whereas open-sea limestones show that at other times the shore lay well north of the basin.

Clastic units resulted when the Michigan River prograded its delta across the gradually sinking Illinois basin, pushing the shoreline southwestward. Limestone was deposited when the Michigan River could not supply enough sediment to offset continued sinking and the sea transgressed northeastward over the land. The position of the shore is a delicate indicator of the balance between sedimentation and subsidence. A plot of relative shore position (Fig. 3) forms a continuous curve that reflects the dominance of carbonate or clastic sedimentation. At least 70 reversals in the direction of shore movement are indicated in the stratigraphy of the Genevievian Stage (latest Valmeyeran) and the three Chesterian stages, although some of the minor fluctuations can not be shown at the scale of Figure 3. These reversals are superimposed on 12–15 major cycles. It is difficult to define "major cycle" precisely, because cycles that appear important in one region seem insignificant or can not be recognized elsewhere.

The position of the mouth of the Michigan River is marked by northeast-southwest belts of sandstone concentration (Swann and Bell, 1958, Figs. 10, 12, 14, 16; this paper, Figs. 5–8). In most of the clastic units these belts cross the basin near its center, but in a few they lie near the western or eastern edges (Swann and Bell, 1958, Figs. 10, 14). Thus the position of the lower course and mouth of the Michigan River shifted laterally through a range approaching 200 miles, but returned again and again to the axis of the developing basin. The lateral shifts are indicated on the curve at the right in Figure 3.

These late Mississippian rocks differ in many respects from those of the Tertiary of the central and western Gulf Coast, but the entire sedimentary process presented here is remarkably similar to that envisioned by Rainwater (1963) for the Gulf Coast.

The two curves of Figure 3 summarize the stratigraphy of the Pope Megagroup. Although some subjective elements entered into their construction, the curves are based on comparison of well logs or outcrop sections from different parts of the basin.

The details on which the generalities of this paper are based have been or will be presented elsewhere. The stratigraphy is from Swann (1963). Lithofacies and oil production were reported by Swann and Bell (1958), cross-bedding by Potter *et al.* (1958), sandstone distribution by Potter (1962, 1963), general lithology by Atherton *et al.* (1960), clay mineralogy by Atherton *et al.* and by Smoot (1960), grain size by Swann *et al.* (1959) and by Potter (1963), limestone distribution by Atherton *et al.* (*in prep.*), and provenance by several of these authors and by Potter and Pryor (1961). The illustrations for this paper have largely been modified from these publications and simplified by omitting datum points, well identification, and similar items that can be referred to in the originals.

## Clastic Units

In cross sections several miles long, limestone generally appears as continuous bands but sandstone occurs as discrete bodies within clastic units that also include much shale and some thin coal beds. Figure 4, a section across the eastern third of the basin, is typical, although sandstone bodies are more numerous and larger in the central third. A small amount of shale and sandstone associated with certain of the limestones does not obscure the basic rhythmic alternation of limestone and clastic units. Maps of sandstone within individual clastic units (Figs. 5–8) show that some sandstone bodies are truly isolated, but most are integrated into a network resembling the veining of a leaf whose blade is the shale lying between and extending beyond the sandstone ribs.

The maps of sandstone thickness (Figs. 5–9, 12) are based on maps by Potter (1962, 1963). Each shows the total thickness of sandstone within a single clastic formation, as determined from electric logs (*cf.* Fig. 10). The shale component of the formation is not indicated. Some fine detail was omitted from the regional maps (Figs. 5–8), which were based on originals using 6,000–11,000 wells.

## Large Sandstone Bodies

The most striking features of the clastic units are large, massive sandstone bodies. They are many miles long, a few miles wide, several tens of feet thick, and generally extend northeast to

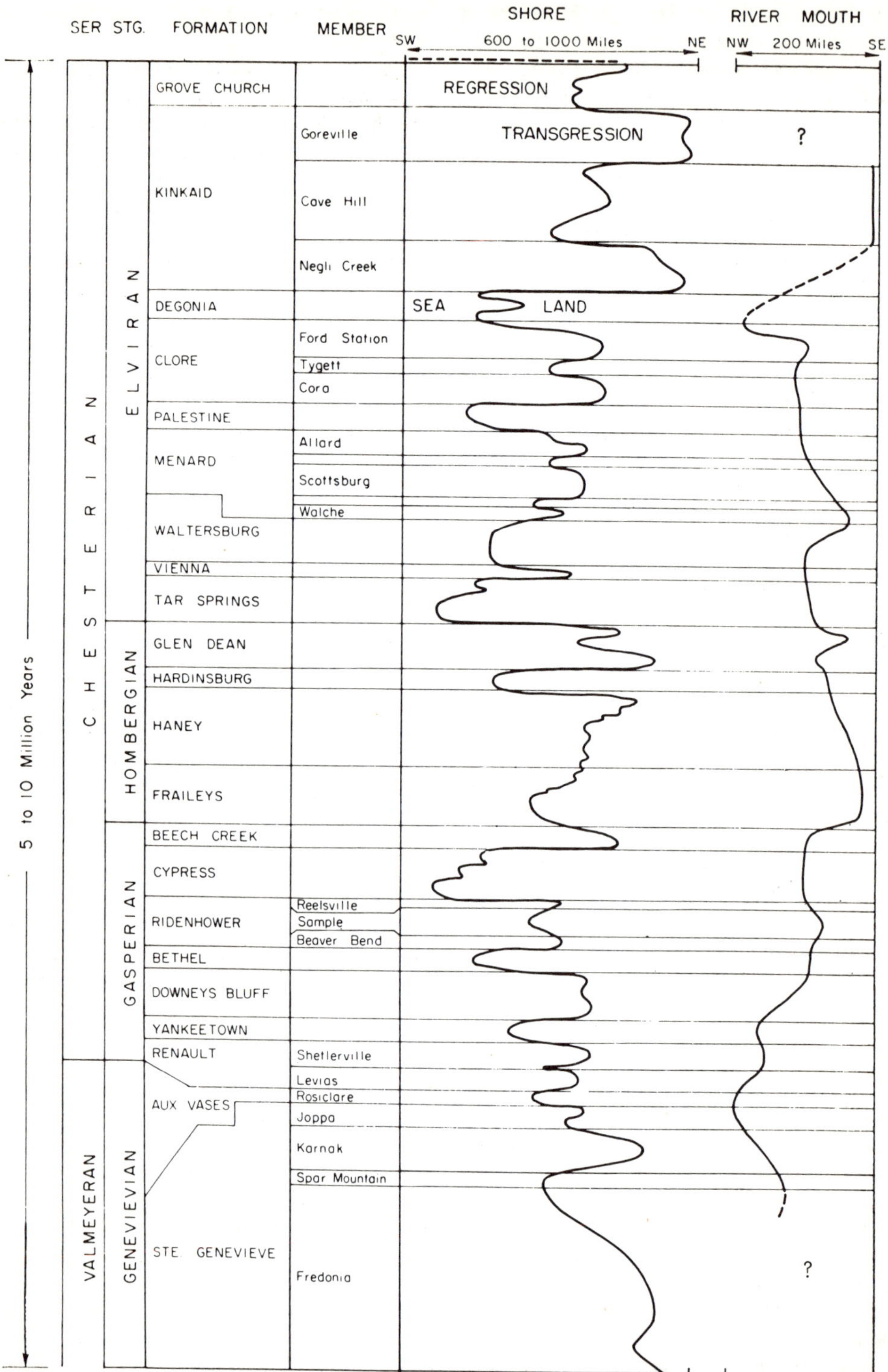


FIG. 3.—Graphs showing the position of shore and lateral shift in course of Michigan River during late Mississippian.

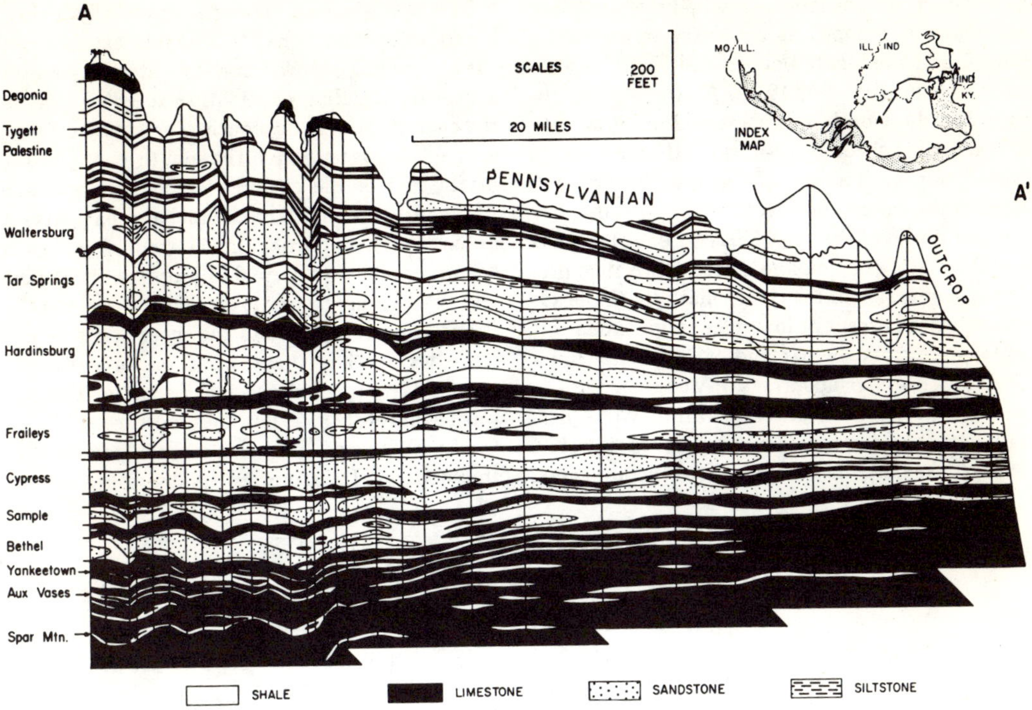


FIG. 4.—Cross section of late Mississippian strata across eastern third of Illinois basin from subsurface in western Kentucky oil fields to outcrop at east edge of basin along Indiana-Kentucky boundary. (Modified from Swann and Atherton, 1948, Fig. 6, by permission of University of Chicago Press, publishers of Journal of Geology.)

southwest. They are well exposed as prominent scarps, in contrast to the meager outcrops of the shaly phases of the same units. The type sections of nearly all of the clastic units lie within these larger sandstone bodies.

The large sandstone bodies are fine-grained to very fine-grained, well sorted, porous, and permeable. They are generally barren of oil, except on the largest structural features in the basin. The bedding is most commonly flat, even, or parallel and the beds are several inches to a few feet thick. However, each body includes some cross-bedded sandstone, and in a few it is the major constituent. The average dip direction of cross-bedding parallels the length of individual sandstone bodies, dipping toward the south or west end. Most of the cross-bedding data reported by Potter *et al.* (1958) came from massive sandstone bodies and much of it from these large first-order ones.

Outcrops of both the cross-bedded and flat-bedded phases can be seen in most well exposed large sandstone bodies. Outcrops with the base of the sandstone exposed are not common, but in those the writer has seen the flat-bedded phase is conformable on lower strata, whereas the cross-bedded phase is unconformable. A large sandstone body extends north and south for at least 15 miles and dominates the outcrop of the Aux Vases Sandstone in Monroe and Randolph Counties, southwestern Illinois. The massive flat-bedded phase of this body is well exposed at the waterfall in a branch of Prairie du Rocher Creek in SE. $\frac{1}{4}$, SE. $\frac{1}{4}$, SW. $\frac{1}{4}$, sec. 3, T. 5 S., R. 9 W., at the Monroe-Randolph county line, Renault Quadrangle. The cross-bedded phase is exposed in the Mississippi River bluff at the mouth of Barbeau Hollow 2 miles southeast of Prairie du Rocher, in the SW. $\frac{1}{4}$, SE. $\frac{1}{4}$, sec. 26, T. 5 S., R. 9 W., Randolph County, Renault Quadrangle. Cross-bedding in this latter outcrop was mapped in detail by Potter (1963, Fig. 48), and at least 15 feet of relief at the base of the Aux Vases can be seen in the Mississippi bluff exposures northwest of the hollow.

Deltaic patterns (Figs. 5–8) show that the large sandstone bodies mark distributary channels. However, in a strict sense most are bar fingers (Fisk, 1952, 1961; Fisk *et al.*, 1954) rather than channel sandstones. The fine sand bedload of a fresh-water stream entering a body of standing salt water is dumped as a transverse crescentic mouth bar, the lunate bar, several times as wide as the channel (Bates, 1953). As the stream approaches the mouth, it widens and shallows into an expanding sheet of fresh water that retains its momentum as it rides up the broadening back slope of the bar and out over the denser salt water beyond the lip of the bar. The crest line of the bar stabilizes at a position and water depth that prevent the invasion of the river mouth by the heavier sea water, except at low river stages when salt water may cross the bar and extend many miles upstream as a temporary salt-water wedge beneath the fresh river water (Scruton, 1960). The salt water and any mud deposited in it are flushed beyond the bar when the next river flood begins.

The back slope of the bar is essentially a shallowed and proportionately widened extension of the stream bed over which river water passes with a velocity equaling or slightly exceeding that in the channel above the mouth. The cross-sectional area of the water above the mouth bar approximates that of the channel farther upstream, so that width and depth of channel and bar passage are inversely proportional. The front slope lies in sea water of nearly normal salinity that is moving very slowly landward to replace water lost by friction and turbulence to the sheet of overriding fresh water. The upstream slope is extremely gentle, averaging only 10 feet per mile. The seaward slope is steeper, but does not exceed 2°.

The Michigan River was a little smaller than the modern Mississippi, so its distributaries and

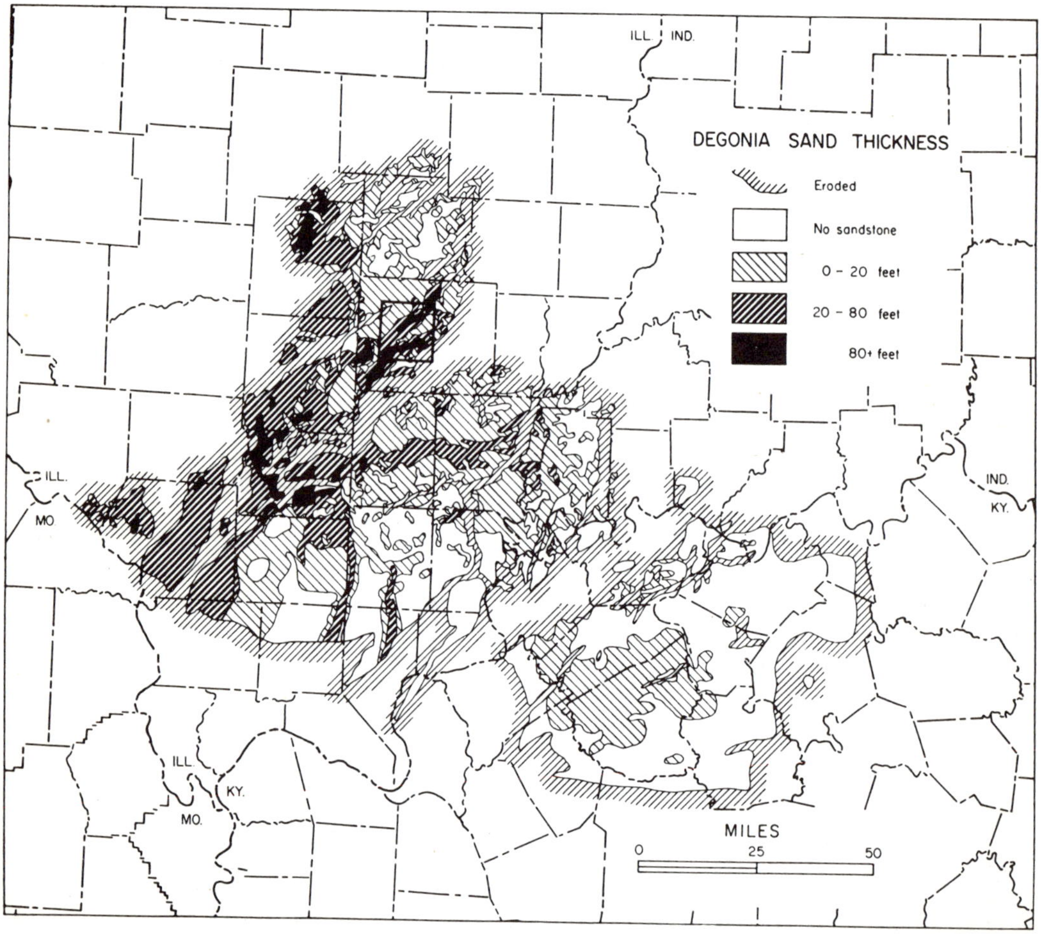


FIG. 5.—Total thickness of sandstone in Degonia Formation as determined from electric logs. Small rectangle indicates area shown in detail in Figure 9. (Simplified from Potter, 1962, Pl. 1; 1963, Pl. 1D.)

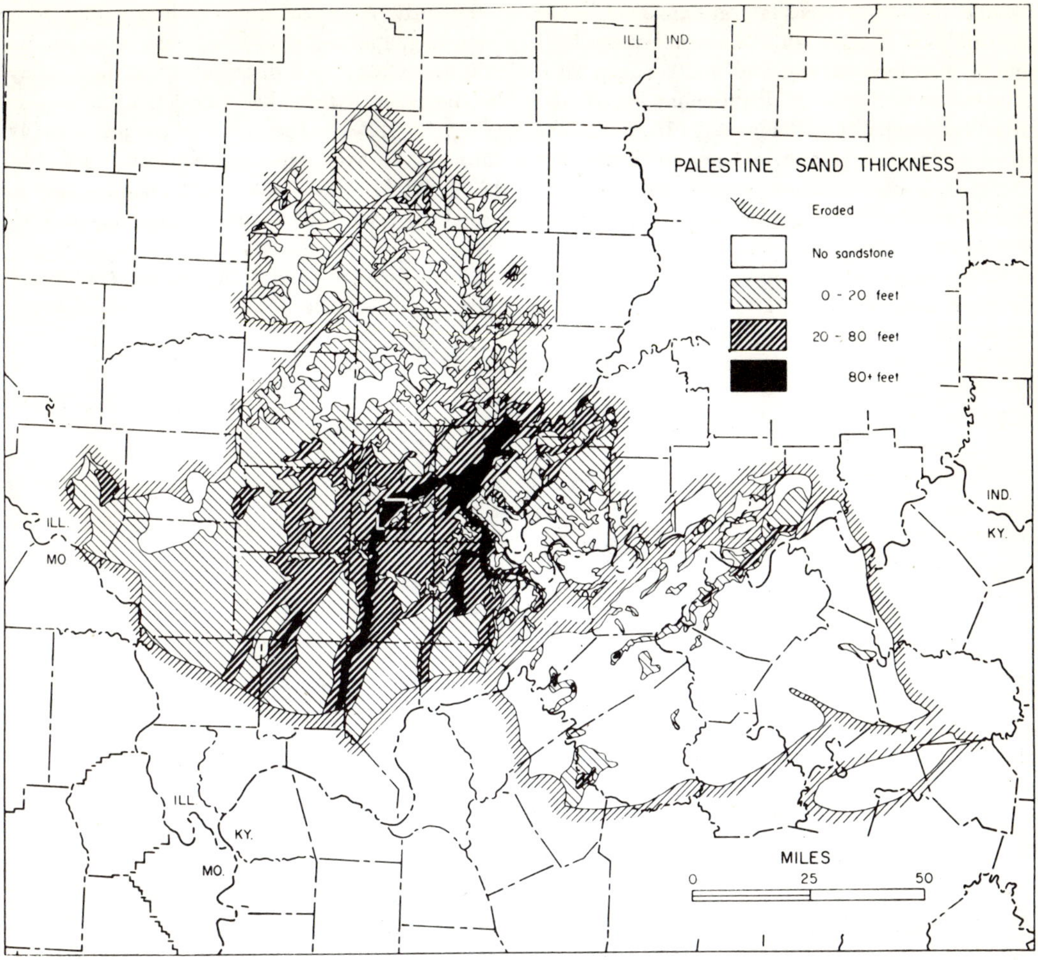


FIG. 6.—Total thickness of sandstone in Palestine Formation as determined from electric logs. Small rectangle indicates area shown in detail in Figure 12A. (Simplified from Potter, 1962, Pl. 2; 1963, Pl. 1E.)

mouth bars were not so large. Most distributary channels were $\frac{1}{4}$–$\frac{1}{2}$ mile wide and probably 30–60 feet deep, though they ranged to more than 200 feet. Typical mouth bars were 2–5 miles wide and their crests must have been about 10 feet below sea-level. The thickness of the bar, largely determined by the depth of water in which it is constructed, was between 30 and 100 feet for the Michigan River. The details on which the depth estimates are based are discussed in a later section. Most of the Michigan River distributary mouth bars rested on relatively firm bases, whereas those of the modern Mississippi are precariously perched in gravitational unbalance atop mounds of less dense pro-delta soupy slush. Density features comparable to Mississippi delta mud lumps and overthickened bar fingers are very rare in the Michigan River sediments.

A crescentic river-mouth bar is not static. It is subject to wave and tidal attack that tend to destroy it and redistribute the sand as beaches, delta-front sheets, or tidal-current ridges. Where these agents are unusually active or where the stream load is small, all of the sediment may be redistributed and no deltaic deposits remain.

A major river system provides more sediment to its lunate mouth bars than waves and tide can redistribute. Each mouth bar advances seaward as sand is pushed up its back slope and dumped over on the seaward slope. As the forward edge of the bar advances, the central part of the older portion is excavated to form the distributary channel, while the broad lateral flanks are left as the sandy bases upon which natural levees of silt and mud are constructed. The lunate mouth bar is thus the leading edge of a lengthening broad sand prism called a bar finger (Figs. 11, 12B, 12C). The channel forms a groove in the top of the bar finger, generally a sixth or an eighth of its width but apparently varying from a twentieth to as much as a quarter. The thickness of the bar finger and the depth of the channel are essentially independent, the former being largely a function of the depth of water into which the stream is emptying, and the latter a function of stream hydraulics. Grooves of distributary channels may be contained within their respective bar fingers (Figs. 11, 12B), or they may be cut more deeply through the bar finger into older deposits (Fig. 12C). The Michigan River deltas, like the several post-Pleistocene Mississippi deltas, include both types.

Channel cutting and filling processes are intermittent but simultaneous. The channel groove is

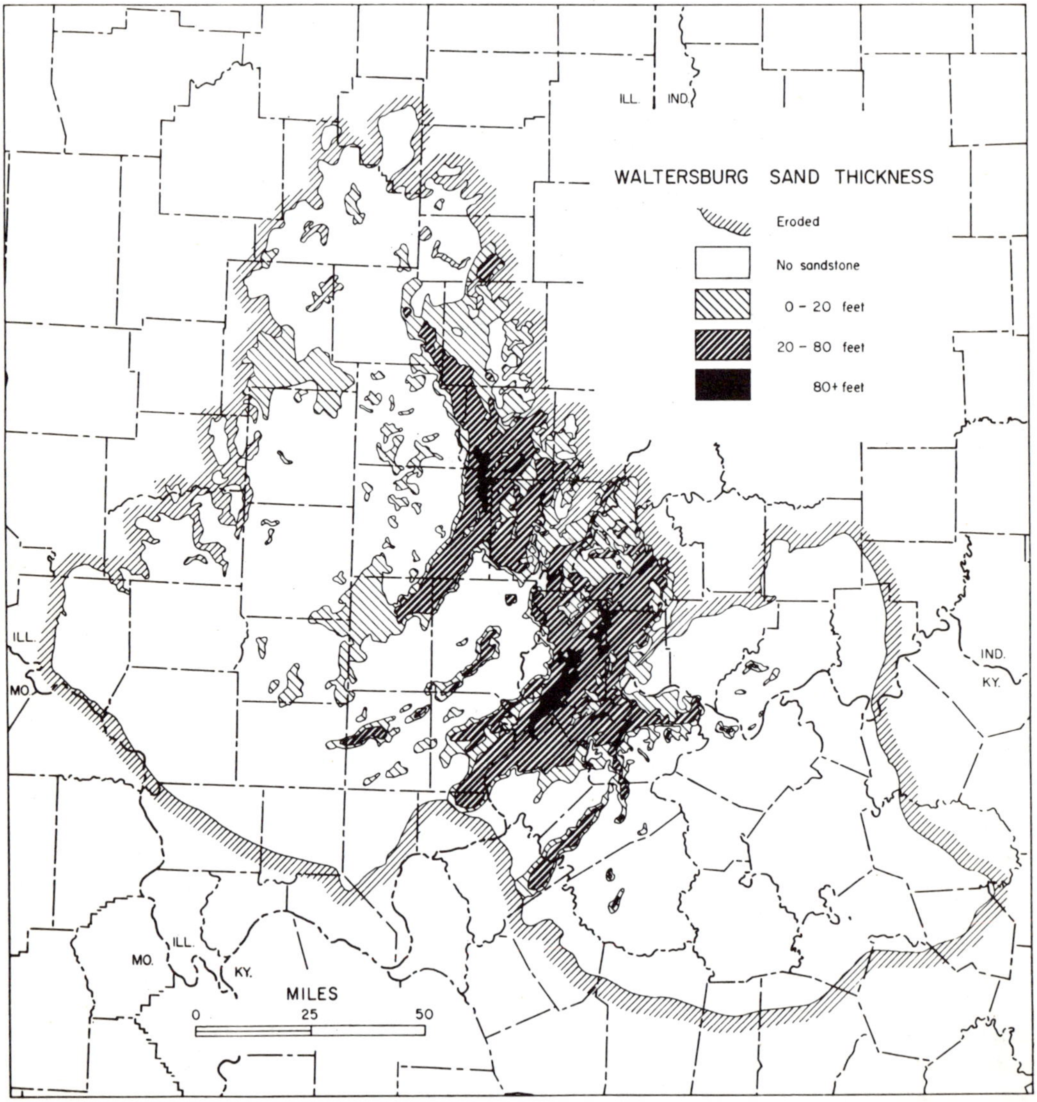


FIG. 7.—Total thickness of sandstone in Waltersburg Formation as determined from electric logs. (Simplified from Potter, 1962, Pl. 3; 1963, Pl. 1F.)

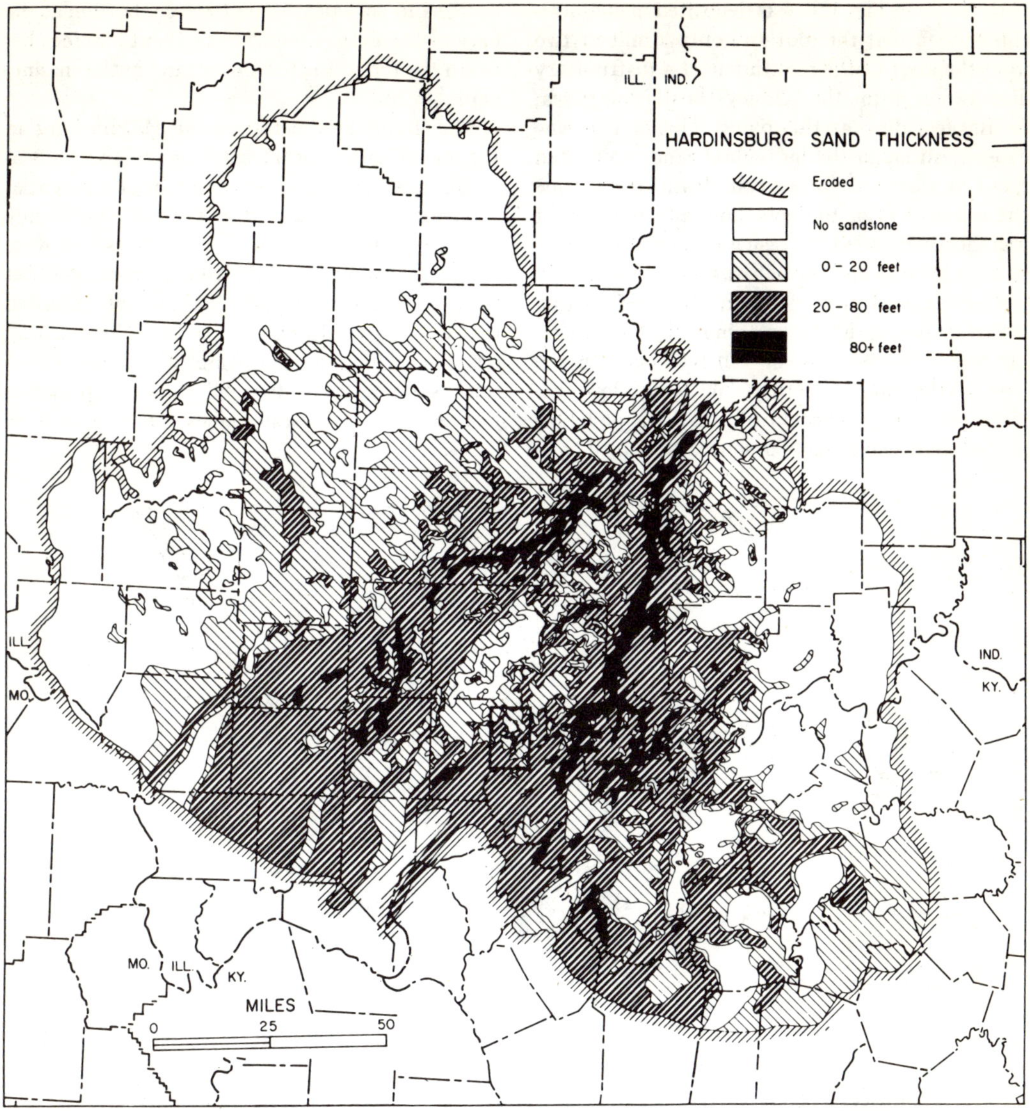


FIG. 8.—Total thickness of sandstone in Hardinsburg Formation as determined from electric logs. Small rectangle indicates area shown in detail in Figure 12D. (Simplified from Potter, 1962, Pl. 4; 1963, Pl. 1G.)

partially filled most of the time with cross-bedded bottom deposits. These deposits represent bed-load that has been halted, generally temporarily, on its passage downstream from headwaters to mouth bar. Sand may remain in a channel-bottom deposit for minutes, months, or millennia before it is again picked up and shifted farther downstream. Because much of this bottom material moves as sand waves of varying dimensions, it is largely cross-bedded and in this respect differs from the material in the main body of the bar finger that is generally massive and has widely spaced parallel bedding. In an active distributary, the channel-bottom deposits are re-excavated. However, as flow through a waning distributary diminishes, the channel is not re-excavated so deeply, and it gradually fills with cross-bedded sand. After the distributary ceases to function, the upper few feet of the channel become filled with mud and marsh deposits indistinguishable from interdistributary fill and levee material.

The sand at the base of the flat-bottomed part

of the body in Fig. 12C was deposited in conformable sequence at the foot and only a mile or two from the crest of the mouth bar of a distributary discharging into the Haney-Hardinsburg sea, 80–100 feet deep at this point. This is a marine deposit, although the individual sand grains had spent at most only a few minutes in the marine environment prior to deposition, and the deposit was shortly invaded by fresh ground water. Deposition was rapid and the bar was built to within 10 feet of sea-level in a few years. Channel cutting began as soon as the bar crest moved past the site, but the extreme depth of 120 feet beneath the base of the bar, more than 200 feet below sea-level, was not reached until the delta had prograded many miles farther south. Not until then could sand be deposited on the river bottom at the base of the groove, which was about 200 feet beneath the top of the bar-finger sand in the channel bank.

The hump in the top of the Hardinsburg in Figure 12C is a typical feature on cross sections of such sandstone bodies. It results from differential compaction. Limestone and shale in the Haney were coherent enough to form pebbles but were still so porous during Hardinsburg deposition that their subsequent compaction has far exceeded that of the sand filling the channel. The resulting regional settling has left compactional highs over thick sand sections that fill incised depressions just as it has over sand bodies built up on level bases.

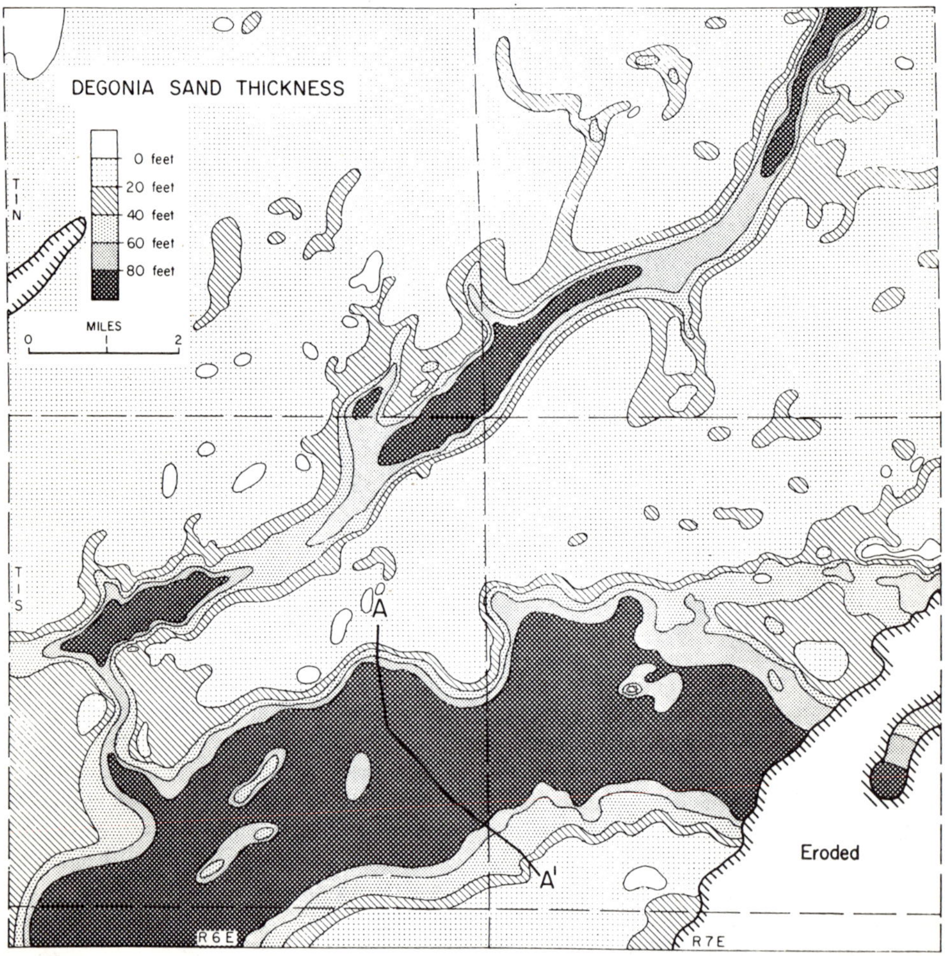


FIG. 9.—Thickness of sandstone in Degonia Formation in part of Wayne County, Illinois, based on 1150 electric logs. Line of cross section in Figure 10 is shown. (Simplified from Potter, 1962, Fig. 11.)

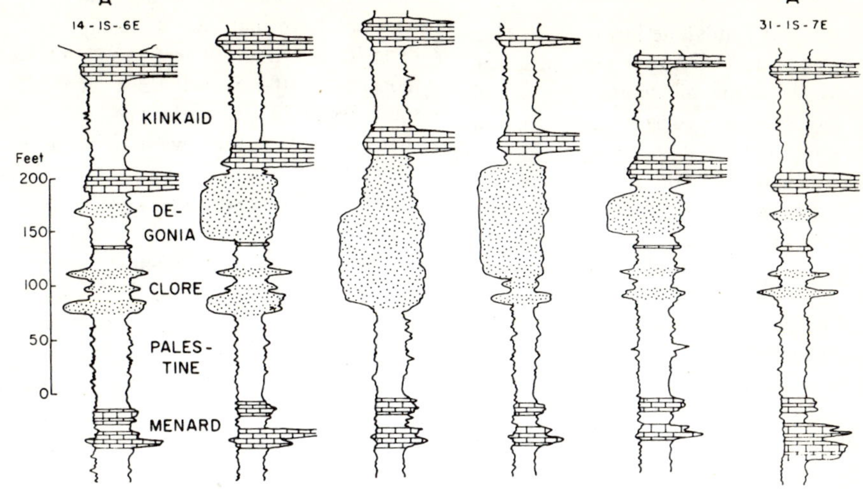


FIG. 10.—Electric log cross section of Degonia and adjacent formations along line A-A′ shown on Figure 9. (Modified from Potter, 1962, Fig. 12.)

Although most distributary channels in an advancing delta deteriorate, one or more are destined to become main river channels. On the old part of the delta the main channel may coincide with a minor bar finger or perhaps may cut across what were originally interdistributary areas. In still older portions, meandering and lateral cut-and-fill may so widen channel deposits that the bar fingers are obliterated. The relatively wide excavated groove containing the Degonia sand body of Figures 9 and 10 is an example.

The major sandstone bodies—those many miles long, a few miles wide, and several tens of feet thick—are largely of composite bar-finger and channel origin. Many contain only a thin thread of channel material running through the flat-bedded bar finger, whereas others are largely of cross-bedded channel origin. However, the relative abundance of massive sandstone compared with cross-bedded sandstone in the outcrops of these major bodies suggests that sandstone of bar-finger origin is several times as common as channel sandstone. In these and in most other Carboniferous cyclic deposits there should be no sharp differentiation on regional maps between deltaic and channel sandstones. The channel system penetrates and permeates the delta; deltaic sandstones (bar fingers) were the precursors of the channels and of the channel-filling sandstones.

The picture presented here of the origin of the major sandstone bodies is consistent with the details of these bodies in outcrops and the subsurface. It contains some subjective judgments. Modern point bars and other river deposits have bedding like that in the ancient cross-bedded phase. The writer knows of no description of bedding in modern lunate mouth bars. However, a lunate bar is composed of grains which have been carried as bed load at the base of the rapidly moving fresh-water stream, but which fall freely through a wedge of standing or slowly moving sea water (Bates, 1953) and come to rest on a surface that slopes less than 2°. The elements needed to produce cross-bedding are absent in this environment. The advance of a mouth bar during a single flood stage appears consistent with the deposition of a bed several inches to a few feet thick on the front slope of the bar.

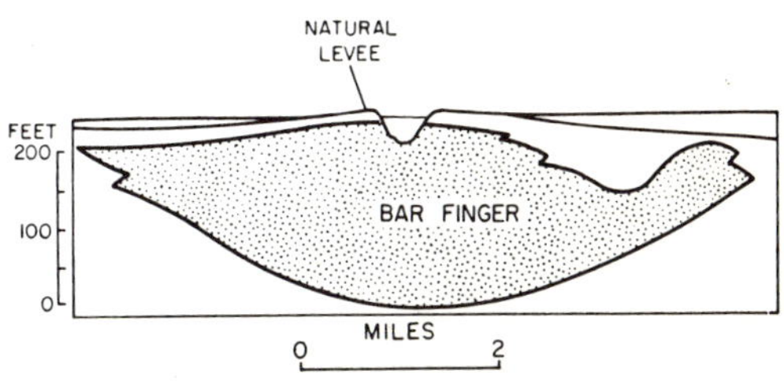


FIG. 11.—Cross section about 5 miles above mouth of Southwest Pass, Mississippi River delta. (After Fisk, 1961, Fig. 2, by permission of author.)

### Small Sandstone Bodies

At least half the sandstone in the Pope Megagroup is in the large bodies previously discussed, but most of the Pope oil comes from smaller sandstone bodies of a variety of origins and shapes. Some are sheets, like those in the Clore Formation in Figure 10. Others are thicker, elongate, and either straight or curved, like smaller versions of the major bodies. Some are attached and form part of the main sandstone framework of their respective formations. Others are isolated, surrounded by shale or by shale and limestone. Most occur in and near the major sandstone belts, but a few are found 30 to 50 miles from the nearest deltaic complex.

Many of the small sandstone bodies are composed of sand that was eroded from mouth bars and redistributed by waves or tidal currents. The processes of delta destruction by wave and current attack (Scruton, 1960) begin concurrently with delta construction, but the destructive processes appear to have been relatively minor in the case of the Michigan River deltas. Few attempts have been made to map late Mississippian sandstone deposits less than 10 feet thick. Axes of bodies more than 10–15 feet thick generally extend southwestward down the regional slope rather than along the northwest-trending shorelines. Therefore, beaches, cheniers, offshore bars, and similar bodies that are elongate parallel with shorelines must be rare. Many of the thin bodies may be delta-front sheet sands.

The trend of many of the thicker bodies at right angles to the shoreline, their parallelism, and their spacing suggest tidal-current ridges. Theodore Off (1963) listed sandstone bodies of the Palestine as a possible example of ancient tidal-current ridges, basing his conclusion on a partial map of sand distribution (Potter *et al.*, 1958, Fig. 13). The complete distribution of the Palestine sandstone bodies (Fig. 6) is clearly deltaic. Sandstone bodies south and east of the largest one in the Waltersburg (Fig. 7) and others in the Big Clifty and Aux Vases are more likely examples of tidal action.

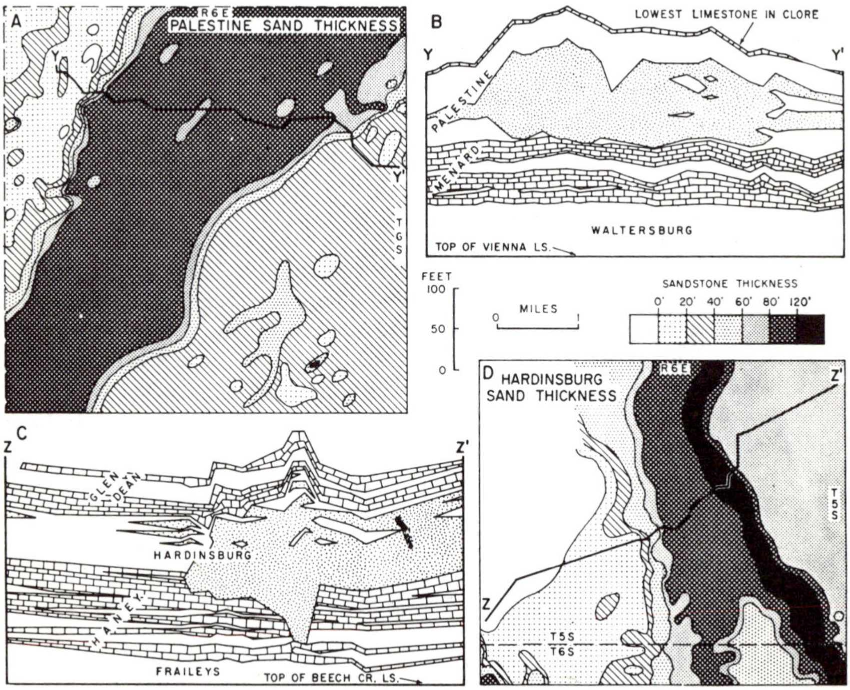


FIG. 12.—A. Thickness of sandstone in Palestine Formation in part of Hamilton County, Illinois, based on 500 electric logs. B. Cross section of Palestine and adjacent formations along line Y–Y′ indicated on A, based on 29 electric logs. C. Cross section of Hardinsburg and adjacent formations along line Z–Z′ indicated on D, based on 14 electric logs. D. Thickness of sandstone in Hardinsburg Formation in part of Hamilton County, Illinois, based on 170 electric logs. (Based on Potter, 1963, Figs. 38 and 42.)

Many minor sandstone bodies are simply smaller versions of bar fingers and filled channels. Some crevasses or breaks in channel walls develop into main channels or major distributaries; others less favorably situated in time or space carry only a small proportion of the total discharge and eventually heal. Because the latter are smaller, their mouth bars are narrower and have shallower sills. Their deposits are likely to lie in interdistributary basins that are protected by the more active distributaries and may already be partly filled with mud. The channels may be narrow, no more than a few hundred feet across.

Although minor sandstone bodies are of many origins, most are closely related to delta growth. Like the major bodies, most smaller ones were deposited below sea-level although the tops of spits and beaches can reach sea-level, and upstream channel fills may even extend above it.

### Shale

Whereas the bedload of a fresh-water stream entering a body of standing salt water is dropped on the distributary-mouth bars, the suspended load may be carried far. If the sandstone that was deposited at the river mouth is thought of as forming the vein network of a leaf-like clastic unit, the extensive blade of the leaf is shale derived from the river's suspended load. Shale surrounds many of the sand bodies, fills the interdistributary areas, and extends beyond the delta sand complex as a thinning sheet that generally reaches the basin margins. These shale sheets were originally more extensive. The central location of the delta complexes has resulted in preservation of an abnormally large proportion of sand, as suggested by the relatively high sand:shale ratio of 1:2 for these rocks.

Shale is distributed asymmetrically. There is more shale northwest than southeast of the sand belts. This is shown even on the generalized facies map for the entire section above the base of the Aux Vases (Swann and Bell, 1958, Figs. 5, 6) but is more striking on maps of shorter intervals (Swann and Bell, 1958, Figs. 9–16) or on maps of individual units. There are exceptions, but typically the shale in a clastic unit is twice as thick west of a delta complex as it is east of the complex. This suggests a slow residual current to the northwest (Fig. 2), ineffective at shifting sand and coarse silt but capable of carrying finer material.

As a rule, the shale is a trifle coarser or siltier and somewhat more variable in the regions dominated by sand. It is finer, more homogenous and smoother, and has slightly lower electrical resistivity in areas remote from sandstone bodies. It is generally dark gray, olive-gray, or greenish gray, centered at the Munsell designation of GY 4/1. More than nine-tenths of the shale has a hue of N, G, GY, or Y, a value of 3 to 5, and a saturation of 0 (N) to 2. About 5 per cent of the shale is red (5YR to 10 RP 3/6). Individual red shale units may be traced widely, largely in areas remote from contemporaneous sand and most commonly in the upper part of clastic units. Green shale and black or brownish black shale are much rarer, the green generally associated with limestone and the black occurring toward the top or bottom of clastic units in regions where sand is fairly common. The occurrence of red and green shales generally can be correlated with greater distance from an active delta. There is a vague suggestion that the gray shales are somewhat darker, and thus more carbonaceous, near the delta complex.

Marine fossils are rare or apparently absent in much of the shale. However, in many localities they are common to abundant in the lower part of a clastic unit, in some places at the top, and commonly in the distal areas where a clastic tongue wedges out between carbonate units. Plant fossils occur sporadically and are locally abundant in both shale and sandstone in the upper part of several of the clastic units.

Most of the suspended load of a major stream is carried out to sea through the distributary channels and then transported by marine currents. Some of it drifts far to sea, some is carried considerable distances parallel with shore, but much is swept back toward the shore and deposited in interdistributary bays and marshes. During unusually heavy storms, such mud can be piled upon the shore above the normal sea-level (Ball *et al.*, 1963). Mud also is deposited in these areas directly from river water that overflows the channel banks during flood times. Flood-deposited marsh sediments are continuous with upstream flood plains. They grade and intertongue into marine mud. Marsh and floodplain deposits rise to slightly higher elevations and are, at least theoretically, coarser next to the channels, where they form the natural levees. However, the differences in shale type produced in different parts of the delta are subtle, and their significance in interpreting these ancient sediments has not been recognized.

### Coal and Associated Rocks

Although volumetrically insignificant, the non-clastic constituent, coal, is important in interpreting the clastic units.

Thin coal seams, less than a foot and generally less than an inch thick, are found in some localities in the upper part of nearly every clastic unit. Coal has been noted at three or more levels in the upper halves of the Cypress, of the Hardinsburg, and of the Tar Springs, and at two levels in the upper part of some other units. There are only one or two examples of coal toward the bottom of a clastic unit.

Even thin coals are underlain by modified sediments (underclays and other seat-rocks), generally only a few inches thick, which grade downward into normal sediments but which lack bedding, contain *Stigmaria* and plant rootlets, are slickensided (clays) or powdery (silts and sandy silts), and are relatively light-colored. Coals are overlain by relatively dark, exceptionally well bedded or fissile shales. The associated rocks extend beyond the coals and mark the potential position of the absent coal. Although less obvious, the associated rocks are commoner in the late Mississippian section than are the actual coals. The coals and associated rocks indicate establishment of a stand of vegetation upon the deltaic surface.

### Limestone Units

The Haney Limestone, which underlies the Hardinsburg Sandstone and overlies the Fraileys Shale, is typical of the limestones that lie between the clastic units. The thickness of the Haney and the percentage of limestone in it are shown on Figure 13. A network of about 550 control points was used for this figure. The contours were smoothed by using the running averages at the midpoint of each block of four adjacent wells rather than the individual values.

The Haney is about 90 feet thick at several places along the southern outcrop and thins northward. It can not be recognized at the northern corner of the basin where shales of the Hardinsburg and Fraileys are in contact and can not be distinguished from each other.

Other limestone units of the cyclic succession thin northward or northeastward with much variation in individual configuration and rate of thinning. In many units the northern edge of limestone deposition lies within the basin, but the thickness and character of other limestones suggest that they originally extended northward, far beyond the limit now imposed by pre-Pennsylvanian erosion.

The section in Figure 10 is near the northern limit of the Kinkaid Formation. Its upper and lower limestone members are as massive as they are at the outcrop 100 miles south, and each is about half as thick as at the outcrop, suggesting that they originally extended farther north. On the other hand, the upper limestone member of the Clore Formation is represented only by a single thin limestone bed. The lower limestone member of the Clore can not be recognized in these logs, although it contains good but thin limestone beds only a few miles south. The boundary between the Clore and the Palestine can not be recognized reliably on these electric logs. Detailed examination of samples might show that plant debris, coal, or underclay mark the upper part of the Palestine and that calcareous fossiliferous shale indicates the lower part of the Clore, but such indications would also disappear a few miles farther north.

Except for the northward thinning, variations in thickness of the Haney seem unsystematic. In some other limestone units, thinning over unusually thick individual sandstone bodies and over deltaic complexes is more marked. There is less thinning of limestone units than of clastic units toward the east and west flanks of the basin (Fig. 4) and several limestone units actually thicken toward the flanks. Limestone units are thinned, or even cut out, in small areas by channel erosion related to the overlying clastic unit. Such occurrences are so uncommon that they affected only four or five of the nearly 550 wells used in constructing Figure 13. In the Beech Creek and the Ridenhower, thick limestone bodies were deposited in channels left unfilled at the end of the preceding clastic episode.

The limestone units contain some shale and even sand. The proportion of limestone in the Haney is mapped on Figure 13 at equal intervals of the arc sine or theta transformation of the percentage values (Krumbein, 1957). The non-limestone material in the Haney is shale, distributed in one to five or six irregular beds (Fig. 12C).

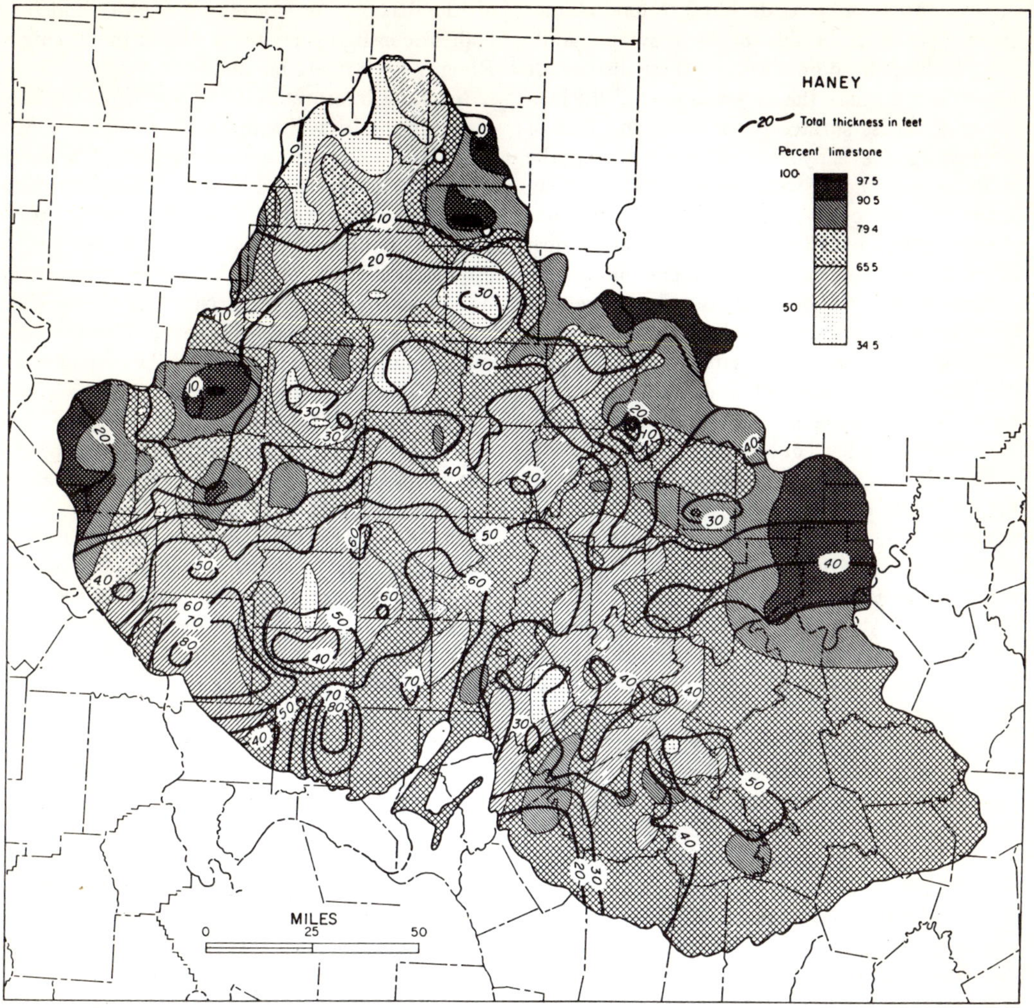


FIG. 13.—Thickness of Haney Formation and percentage of limestone, based on about 550 control wells. Proportion of limestone is indicated at equal increments of arc sine transformation. (After Atherton *et al.*, *in prep.*)

Some of the shale bodies are isolated, but most extend northward to merge with the Hardinsburg or Fraileys Formations. Like the formation as a whole, the individual limestone beds within the Haney thin and disappear northward. The final disappearance of the limestone bed at the top or the bottom of the formation necessitates a step in the formational boundary to the next limestone bed and the consequent assignment of the intervening shale bed to the adjacent clastic formation. The northward-thinning wedge of the Haney, therefore, does not taper smoothly because there are abrupt steps from limestone bed to limestone bed in both its upper and lower surfaces. The thinning wedge is not only physical; it also represents a wedge in time. The early phases of Haney deposition in the south were contemporaneous with the later phases of Fraileys deposition in the north. The later part of Haney deposition corresponded with the beginning of Hardinsburg deposition in the north.

The limestones tend to be coarser, lighter in color, and more oölitic on the east and west flanks than near the central axis of the basin.

## Discordant Surfaces

Each clastic unit contains some sandstone bodies that lie entirely or in part in unweathered grooves gouged in earlier sediments of the same or underlying units (Figs. 10, 12C). Laterally the surfaces of these grooves die out into conformable bedding planes at the sides of the individual sandstone bodies. At no one level do these discordant surfaces cover more than a small part of the basin's area. These surfaces were cut many tens of feet below sea-level by river-bottom erosion in the main channel and distributaries of the Michigan River. In some cycles, stretches of channel a few miles long are now filled with as much as 200 feet of sandstone that represents a pre-compaction channel depth of about 250 feet. Such stretches alternate with others eroded much less deeply. Although the cross-bedded sandstone at the bottom of these grooves was laid down in fresh river water, it represents deposition at the greatest depth below sea-level of any part of the cycle.

The discordant surfaces appear mainly in the major deltaic complexes, *i.e.*, in the belts dominated by sandstone. Away from these regions, the contacts between limestone and clastic units are conformable and normally gradational. Still farther from the deltaic complexes, the clastic units fade out entirely and even their position is lost within a conformable sequence of shallow-water limestones.

In western Illinois there are prominent discordant surfaces beneath sandstone bodies in the Aux Vases, Yankeetown, and even the Renault, for the mouth of the Michigan River lay quite far west during the deposition of these units (Fig. 3). On the east flank of the basin, in Kentucky and Indiana (Fig. 4), the same clastic units are represented only intermittently by shale or by shaly limestone that is conformable with the overlying and underlying limestone beds. Later, during the deposition of the Bethel-Cypress interval, the mouth of the Michigan River was located centrally. On Figure 4, the clastic units in this interval are largely continuous shales conformable with the adjacent limestone. Were a cross section drawn paralleling Figure 4 but a few miles farther south, the Bethel clastic unit would fail to reach the eastern outcrop and the Cypress and Sample clastic zones would be represented only intermittently. The entire section beneath the Fraileys (Big Clifty) is represented at the southeast corner of the basin by a continuous sequence of shallow-water limestones (McFarlan *et al.*, 1955), and similar conditions prevail over the top and along the east edge of the Cincinnati arch (McFarlan and Walker, 1956; Peterson, 1962). Discordant relations appear in the eastern part of the basin (Fig. 4) only beneath sandstone bodies deposited when the mouth of the Michigan River lay relatively far east.

In many interpretations of the late Mississippian rhythmic sediments and of similar cyclic units, the discordant surfaces beneath the sandstone bodies have been cited as evidence for subaerial erosion, which demanded considerable regional uplift. The continuous deposition of sequences of conformable shallow-water limestone on the nearby Cincinnati arch during the development of several cycles in the basin makes this interpretation untenable for these particular cycles. The importance of the discordant surfaces present beneath some sandstone bodies, in both the Mississippian and the Pennsylvanian, has been exaggerated in classifications of cyclic deposits in which the base of the sandstone is chosen as the base for the cyclic unit. In such a classification system, the products of a single pulse of clastic deposition are unnaturally split between two adjacent units. Pro-delta shale is assigned to one cyclic unit whereas the deltaic sand from which it was winnowed is placed in the overlying "younger" one. This puts the primary division of cyclic units at the heart of the delta-building process whose continuity and importance are being emphasized here.

## Michigan River Delta

The immediate cause of the late Mississippian rhythmic sedimentation was the repeated advance of the delta of the Michigan River system across the gradually subsiding Illinois basin. The intervals of deltaic prograding were separated by other periods when the basin was covered by a clear shallow sea in which limestone was deposited.

The depth of the sea into which the river emptied can be determined with some precision. Only a few inches, or at most a few feet, of compactible pro-delta clay was laid down on the underlying limestone ahead of the distributary mouth. The crest of the distributary-mouth bar, now the top of the bar finger, rose to about 10 feet below sea-level. Pro-delta clays may have compacted more than 50 per cent, but the clean bar sand has compacted less than 20 per cent since deposition. Therefore, the water depth can be calculated by taking two or three times the thickness of fossiliferous shale separating a limestone from an overlying major sandstone body with thick, even beds, plus the thickness of the sandstone increased by

15–25 per cent, plus 10 feet. This generally was 50–75 feet, but for different advances ranged from 30 feet or less to a little more than 100 feet.

The sea was generally shallower over the adjacent positive features and on the east and west flanks of the basin than near the basin axis. Sediments along the crest of the Cincinnati arch from Ohio to Alabama are thin and composed almost entirely of limestone of shallow-water origin. Outliers near the crest of the Ozarks and on its south flank, are thin, non-cyclic, and include much shallow-water limestone. Despite their origin in shallow water, sequences in both areas seem to represent essentially continuous deposition. Several times as the Michigan River advanced across the center of the Illinois basin it converted the shallow water there into a delta plain without affecting the still shallower water at the basin margins.

During certain of its advances, the Michigan River prograded several hundred miles. However, the Illinois basin provides a record of less than 200 miles for the earlier advances and only 100 miles for the later. The record lies toward the upstream end of some advances (Figs. 5, 9, 10) and the terminus of others (Fig. 7). In some, the delta pattern was simple (Fig. 6), but in others the system included two subdeltas (Fig. 8) or was very complex (Cypress and Tar Springs Formations).

That the basin was remote from the source of sediment is attested by the lack of obvious sedimentary gradient in the direction of movement. Gradients of grain size (Swann *et al.*, 1959) are directed away at right angles to the stream path rather than downstream.

The clastic rocks in the late Mississippian cyclic sequence in the western part of the Black Warrior basin in northeastern Mississippi must have been transported across the Illinois basin. They are isolated from potential sources in the southern Appalachians by the contemporaneous limestone sequence of Alabama and from Llanoria by deep-water deposits of the Ouachitas. This sequence of Michigan River deposits does not differ significantly from the Illinois basin sequence despite its location 300 miles farther downstream. The overall pattern of sedimentation is similar. Drilling in the southern basin is sparse, so that correlation of individual units with their counterparts in Illinois and with equivalent strata in the limestone sequence outcropping immediately to the east in northwestern Alabama is not yet secure.

The Black Warrior section includes the only other late Mississippian sediments deposited directly by the Michigan River. The Mississippian of Michigan is not young enough. Fossiliferous marine shale in the Pella Formation in central Iowa represents very fine material carried leeward from the delta during latest Valmeyeran and earliest Chesterian time.

In its more extensive advances the Michigan River delta was probably halted near the southeast corner of Arkansas at the northeast edge of the rapidly subsiding Ouachita trough. Here sediments accumulated until, with further sinking and overloading, they slid down the steep northeast side and pushed westward down the axis of the trough as turbidity currents (Cline and Shelburne, 1959). The relatively clean and mature character of the sandstone in the Chesterian Jackfork Formation, compared with the less mature aspect of the underlying Stanley Formation, can be attributed to the incorporation in the Jackfork of large amounts of sand brought to the northeast edge of the trough by the Michigan River. Only material from a southern source was available during Stanley deposition. Bokman (1953) assigned about 90 per cent of the heavy minerals of the Jackfork to his suite 1, whereas he assigned 80–85 per cent of the heavy minerals in the Stanley to suite 2. His suite 1 is identical with the heavy-mineral suite of the Illinois basin Chesterian, but his suite 2 contains elements not found in the basin. During some of the cycles the river could not have reached the Ouachita trough, but enlarged channels suggest that at least during Bethel, Cypress, Tar Springs, and Degonia time the river could have emptied directly into the geosyncline.

The area drained by the Michigan River was in general eastern Canada (Potter and Pryor, 1961). Much of the sediment was derived from older super-mature sedimentary rocks, but there is a minor immature component that later dominated Pennsylvanian sediments. This component can be identified most readily by angular tourmaline fragments. Part of it may have been supplied by the basement rocks of the Canadian Shield, but more probably it came from the northeastern extension of the Appalachians in the Maritime Provinces. Streams from the southern and central Appalachians emptied directly into the Appalachian basin from the southeast. Those farther north in the New York-New England region were the eastern tributaries of the Ontario River, which flowed southward to empty into the northern end of the Appalachian basin in northern West Virginia. Streams heading in the Appalachian system still farther northeast must have swung in a wider arc to join the Michigan River and provide the relatively immature component for the Illinois basin.

### Periodic Rate

The late Mississippian rocks were deposited within a few million years, perhaps more than 5 but no more than 10. During this time about 70 reversals took place in the direction of shoreline movement. That is, there were about 35 cyclical pulses of which 12–15 could be considered major. Advances of the delta recurred a few times per million years.

This rate differs from that of the cycles of transgression and regression in such geologic times as the Cambrian, Cretaceous, and Tertiary when a cycle lasted about 10–30 million years. The entire late Mississippian might be considered the recessional phase of such a cycle. The periodic rate of the late Mississippian cycles discussed here was comparable with that of Pennsylvanian cyclothems or Pleistocene glaciations.

### Origin of Rhythmic Sedimentation

This paper is primarily concerned with the origin of late Mississippian rhythmic sediments, but other similar late Paleozoic cyclic deposits, or cyclothems, probably have a similar origin. The causes suggested for such rhythmic sedimentation have been reviewed exhaustively by Weller (1956) and Beerbower (1961). Many of the proposed origins could not apply to the cycles considered here. For instance, the near absence of coal and the dearth of compactible clay or shale in many sections rule out compaction as an important factor, and such shoreline processes as the building and breaching of a succession of bay-mouth bars do not fit the paleogeography.

The rhythmic sediments represent a periodic imbalance in the rates of basin sinking and of supply of sediment to the basin. A change in either rate could affect the balance.

### Rate of Basin Subsidence

Apparent changes in the rate of basin sinking due to regional tectonic or world-wide eustatic changes in sea-level can be ruled out, at least for some of the earlier cycles, by the non-rhythmic character of contemporaneous shallow-water limestones on adjacent positive features. The possibility of sea-level changes during some of the later of the late Mississippian cycles can not be eliminated, and the evidence for such changes during some Pennsylvanian cycles is undeniable. However, base level changes can not be invoked for the non-marine but otherwise typical cycles in the latest Pennsylvanian or early Permian Dunkard Group of West Virginia, Ohio, and Pennsylvania, which were deposited in a land-locked basin (Beerbower, 1961). Moreover, the importance of the rise and the fall of sea-level in cyclic sedimentation has been magnified out of proportion by the interpretation that "channel sandstones" imply subaerial erosion.

Local tectonic causes for oscillations in the rate at which the basin sank seem improbable. Cyclic sedimentation was essentially contemporaneous in a wide variety of geologic settings. It began at about the same time during the Mississippian in the Illinois and Appalachian basins in North America, the Yoredale deposits of northern England, and the Donets basin of Russia. The simultaneous appearance of cycles with similar characters in many basins that vary greatly in size and structural type demands more than a purely local cause.

### Rate of Sediment Supply

Tectonic control of sediment supply faces even greater difficulties than tectonic control of the rate of basin subsidence. Source regions for contemporaneous rhythmic sediments are more varied geologically than are the depositional basins. The change in gradient necessary to alter significantly the carrying power of a system the size of the Michigan River demands a staggering amount of periodic uplift in the upstream region. Several authors have noted that difficulties are compounded for tectonic theories that include simultaneous uplift of the source region and downwarping of the depositional area. Tectonic movement is of course responsible for the broad pattern of source areas and basins in which cyclic sedimentation developed, but the cyclicity is superimposed upon this pattern rather than originating in it.

Interruption in the supply of sediment by changes in the drainage pattern—"delta shifting"

—has interested many authors. It is particularly attractive if only a small region can be studied, as is true for the English Yoredale succession. Its weakness is apparent where the entire deltaic region is known, as in the Donets and Illinois basins. Here there is no *terra incognita* to hide the clastic delta during deposition of the carbonate half of the cycle.

The possibility that the Michigan and Ontario Rivers were alternate outlets for the same drainage system is a modification of the delta-shifting theory. The idea was mentioned by Pepper *et al.* (1954), who suggested that the early Mississippian "Michigan Berea" may have been deposited during a temporary westward diversion of the Ontario River, which was the source of the type Berea Sandstone of the Appalachian basin. This explanation seems reasonable for a single episode, or pair of episodes, as tenuously correlated as deposition of the Berea of the Michigan and Appalachian basins. It seems less logical when the entire Mississippian or the Mississippian-Pennsylvanian record is examined. Were the episodes of clastic and carbonate deposition in the Michigan, Illinois, Black Warrior basins and in the Appalachian basin in phase or did they alternate? Most correlations suggest that they were in phase. Correlations of many units can be disputed, but those of the Big Clifty, the Hardinsburg or "Pencil Cave," the Glen Dean, the Menard, and several Pennsylvanian cycles seem firm, both on lithologic and paleontologic grounds.

Moreover, the Michigan River appears to have been larger and to have extended farther from its source than the Ontario River. The deposits of the Ontario River did not extend far south into the Appalachian basin, most of whose clastics came from the southeast. Comparison of published subsurface and outcrop sections in the northern Appalachian region shows a sharp southwestern (downstream) decrease in grain size is noteworthy in the deposits of the Ontario River both during the early and late Mississippian, but a similar decrease can hardly be discerned in those of the Michigan River. It therefore seems unlikely that the two could have been alternate outlets for the same stream. Nor is any mechanism known that could have caused the numerous repetitive diversions necessary if the streams were alternate outlets. The known parts of the courses of the two streams were 300–500 miles apart and essentially parallel, both with each other and with the persistent Cincinnati-Findlay-Algonquin positive structural axis that separated them. Any diversion point must have been several hundred miles above the mouths of the streams, athwart the extension of the structural axis on the Canadian Shield. Single cases of stream piracy in a region undergoing erosion occur for a variety of causes. Repeated diversions take place only on an aggrading region such as the North China Plain, where the Hwang Ho empties first north, then south of the Shantung Peninsula. There is no evidence for a depositional basin high on the eastern Canadian Shield during the Mississippian, and such an area aligned with a positive structural axis seems particularly unlikely.

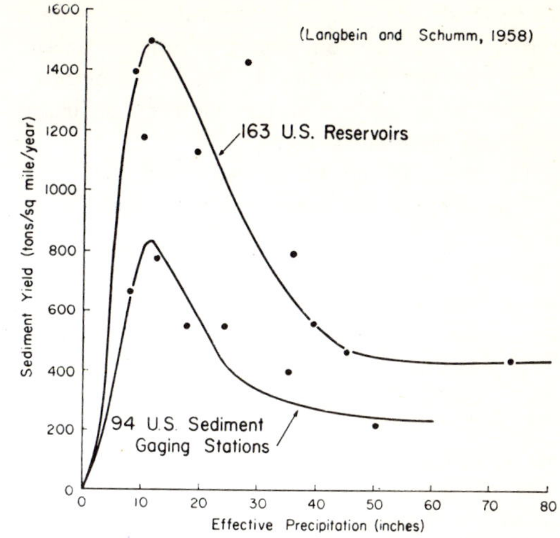


FIG. 14.—Relation of sediment yield to precipitation. (After Langbein and Schumm, 1958, Figs. 2 and 3, by permission of authors and American Geophysical Union.)

Neither structural nor sedimentary processes, then, seem adequate to produce an intermittent supply of sediment. There remain climatic influences.

## RAINFALL CONTROL OF CYCLIC SEDIMENTATION

Rainfall greatly influences yield of sediment per unit of source area, as is shown by the data of Langbein and Schumm (1958) reproduced in Figure 14. In an extremely dry climate, the amount of sediment produced increases with increased rainfall. However, at a somewhat higher rate of precipitation sediment yield drops rapidly, then remains nearly constant. Precipitation exerts its greatest influence where the abrupt decrease marks the change from desert to humid climate. Here a modest change in climatic conditions has a disproportionate effect on the amount of sediment delivered into a river system.

The effect of rainfall on sediment yield is of course modified by many other factors. Most, however, average or cancel when total yield is considered for a large basin over successive periods of time. For instance, the relief and the average composition of the surficial rocks affect the rate of sediment yield greatly, but these factors change gradually and secularly rather than periodically. Temperature must be considered, for precipitation is more effective at low temperatures than at high temperatures where evaporation is rapid. The curves of Figure 14 were adjusted to an annual mean temperature of 50°F., and would be displaced to the right for higher temperatures and to the left for lower. Seasonal or monsoonal rainfall is much more effective in producing sediment than is evenly distributed rainfall. The results attributed here to drier climate may instead have been produced by a change from evenly distributed to markedly seasonal rainfall with little or no change in the annual total.

Carboniferous plants undoubtedly differed greatly from modern ones in their ability to halt erosion. The transition from desert to well vegetated landscape during the Mississippian and Pennsylvanian must have been different from the present. One can theorize that because xerophytic adaptions were less prevalent, the transition was sharper and took place at a higher level of precipitation. Such factors would modify the exact form of the graphs of Figure 14 without altering their basic shape. As today, there were deserts and regions with effective plant cover. The erosive power of a raindrop falling on bare rock is greater than that of another striking vegetation. Therefore, precipitation:erosion curves have been sigmoid since the beginning of land vegetation, although the sharpness and coordinates of the flexure may have varied.

It is here suggested that the cause of the late Mississippian sedimentary cycles was an alternation of relatively wet and dry climates in eastern Canada—the source region. This alternation produced an intermittent or periodically varying supply of sediment to the Michigan River and other rivers of the region. With a drier climate and consequently reduced plant cover, an increased yield of sediment converted the Michigan River into a prograding stream that rapidly pushed its delta scores or hundreds of miles across the gradually subsiding Illinois basin.

The fine sand bedload of the river was deposited in tributary-mouth bars, which rapidly elongated into bar fingers in whose upper surfaces channel grooves were excavated. Because the water was shallow, little of the suspended load was carried far out to sea in pro-delta deposits. Most was swept back toward shore and deposited in interdistributary troughs and along the shore in the lee of the distributary network. Some suspended load was deposited during flood stages directly from river water as marsh and floodplain deposits and natural levees. Because of the rapidity of the advance, modification of the network of sandy bar fingers was only moderate. The record of the dry spell is preserved as the leaf-like clastic unit previously described whose sandstone vein network is dominated by bar fingers. The blade of the leaf is the shale that lies between and extends far beyond the sandstone network.

With a return to a more humid climate and more effective plant cover in the source region, the sediment load decreased and deposits of the river could no longer compensate for the continuous downwarping of the basin. The plant cover of the delta during these late Mississippian cycles is only sketchily preserved, but during comparable Pennsylvanian cycles a great sea-level swamp developed over the whole delta surface and considerable thicknesses of peat were deposited and later converted to coal. These major cyclic coals covering an entire delta were much more extensive than the local coals developed in interdistributary swamps during the delta's advance.

Eventually, as the basin continued to sink, plant growth, too, was unable to keep pace and the entire delta region was again inundated. Marine deposition dominated by limestone then continued until another sedimentary cycle was initiated by a return to the dry climate phase.

Paleobotanical evidence suggests that eastern North America had a tropical or subtropical climate during the Mississippian and Pennsylvanian. The alternation of humid or pluvial and arid or interpluvial climates in the drainage basin of the Michigan River was caused by shifts in climatic boundaries that placed eastern Canada first largely in the equatorial humid zone, then in one of the subtropical arid zones.

The validity of rainfall control of rhythmic sedimentation is not dependent upon a particular paleogeographic reconstruction. However, the reconstructions based on recent paleomagnetic work place the Carboniferous north pole in eastern Asia or the western Pacific relative to the present position of North America. This puts eastern Canada along the boundary between the equatorial and the northern subtropic zones and fits many other facets of the distribution of climatically significant rock types.

The general atmospheric circulation links the climates throughout the world into a single system. As Brückner (1953) has pointed out, climatic variation may thus eventually provide a tool for very precise interregional correlation. Evaporation must equal precipitation for the system as a whole. Therefore, an increase in precipitation in one region must be compensated by corresponding decreases elsewhere, and a shift in a single climatic boundary implies correlative shifts in other boundaries.

To keep the precipitation and evaporation balanced for the world system, there must be regions with opposing climatic cycles, the pluvial stages in one corresponding to interpluvial stages in the other. During much of the late Paleozoic, the fluctuations in climatic zones responsible for sedimentary cyclic development in the present northern hemisphere must have corresponded in some manner to fluctuations in the boundary of the continental ice sheet, or sheets, of the southern hemisphere. The changes in the total load of ice in turn caused a rise and fall of sea-level that modified cyclic deposition, particularly during the middle Pennsylvanian. However, these sea-level changes can not have been the primary factor in cyclic deposition, because the late Mississippian and the Dunkard Group cycles did not involve sea-level fluctuation yet are in general comparable with the middle Pennsylvanian cycles.

The primary cause of the sedimentary pattern was the change in the rainfall. The climatic fluctuations that began during the Valmeyeran were at first too weak to cause widespread glaciation, although they initiated rhythmic sedimentation. Later fluctuations were stronger and did cause repeated glaciation.

The ultimate cause of the climatic rhythms is still obscure. As the same cause presumably operated during the Pleistocene, the search for the ultimate bases of Pleistocene multiple glaciation should take into account the much longer but less complete record of the late Paleozoic climatic cycles.

## References


American Commission on Stratigraphic Nomenclature, 1961, Code of stratigraphic nomenclature: Am. Assoc. Petroleum Geologists Bull., v. 45, p. 645–665.

Atherton, Elwood, Eldridge, W. F., and Swann, D. H., *in prep.*, Facies in the Golconda Group: Ill. Geol. Survey Circ. (number not assigned).

Atherton, Elwood, Emrich, G. H., Glass, H. D., Potter, P. E., and Swann, D. H., 1960, Differentiation of Caseyville (Pennsylvanian) and Chester (Mississippian) sediments in the Illinois Basin: Ill. Geol. Survey Circ. 306.

Ball, M. M., Shinn, E. A., and Stockman, K. W., 1963, Geologic record of hurricanes (abs.): Am. Assoc. Petroleum Geologists Bull., v. 47, p. 349.

Bates, C. C., 1953, Rational theory of delta formation: Am. Assoc. Petroleum Geologists Bull., v. 37, p. 2119–2162.

Beerbower, J. W., 1961, Origin of cyclothems of the Dunkard Group (Upper Pennsylvanian–Lower Permian) in Pennsylvania, West Virginia, and Ohio: Geol. Soc. America Bull., v. 72, p. 1029–1050.

Bokman, John, 1953, Lithology and petrology of the Stanley and Jackfork Formations: Jour. Geology, v. 61, p. 152–170.

Brückner, W. D., 1953, Cyclic calcareous sedimentation as an index of climatic variations in the past: Jour. Sed. Petrology, v. 23, p. 235–237.

Cline, L. M., and Shelburne, O. B., 1959, Late Mississippian—Early Pennsylvanian stratigraphy of the Ouachita Mountains, Oklahoma, *in* The geology of the Ouachita Mountains, a symposium: Dallas Geol. Soc. and Ardmore Geol. Soc., p. 175–208.

Fisk, H. N., 1952, Geological investigation of the Atchafalaya Basin and the problem of Mississippi River diversion: U. S. Corps Engineers, Waterways Expt. Sta., Vicksburg, Miss., 145 p.

——— 1961, Bar-finger sands of Mississippi, *in* Geometry of sandstone bodies: Am. Assoc. Petroleum Geologists, p. 29–52.

——— McFarlan, E., Jr., Kolb, C. R., and Wilbert, L. J., Jr., 1954, Sedimentary framework of the modern Mississippi delta: Jour. Sed. Petrology, v. 24, p. 76–99.

Krumbein, W. C., 1957, Comparison of percentage and ratio data in facies mapping: Jour. Sed. Petrology, v 27, p. 293–297.

Langbein, W. B., and Schumm, S. A., 1958, Yield o sediment in relation to mean annual precipitation: Trans. Am. Geophys. Union, v. 39, p. 1076–1084.

McFarlan, A. C., Swann, D. H., Walker, F. H., and Nosow, Edmund, 1955, Some old Chester problems—correlations of lower and middle Chester formations of western Kentucky: Ky. Geol. Survey, ser. 9, Bull. 16.

McFarlan, A. C., and Walker, F. H., 1956, Some old Chester problems—correlations along the eastern belt of outcrop: Ky. Geol. Survey, ser. 9, Bull. 20.

Off, Theodore, 1963, Rhythmic linear sand bodies caused by tidal currents: Am. Assoc. Petroleum Geologists Bull., v. 47, p. 324–341.

Pepper, J. F., de Witt, Wallace, Jr., and Demarest, D. F., 1954, Geology of the Bedford Shale and Berea Sandstone in the Appalachian Basin: U. S. Geol. Survey Prof. Paper 259.

Peterson, M. N. A., 1962, The mineralogy and petrology of Upper Mississippian carbonate rocks of the Cumberland Plateau in Tennessee: Jour. Geology, v. 70, p. 1–31.

Potter, P. E., 1962, Late Mississippian sandstones of Illinois: Ill. Geol. Survey Circ. 340.

——— 1963, Late Paleozoic sandstones of the Illinois Basin: Ill. Geol. Survey Rept. Inv. 217.

——— Nosow, Edmund, Smith, N. M., Swann, D. H., and Walker, F. H., 1958, Chester cross-bedding and sandstone trends in Illinois Basin: Am. Assoc. Petroleum Geologists Bull., v. 42, p. 1013–1048.

Potter, P. E., and Pryor, W. A., 1961, Dispersal centers of Paleozoic and later clastics of the upper Mississippi Valley and adjacent areas: Geol. Soc. America Bull., v. 72, p. 1195–1250.

Rainwater, E. H., 1963, Transgressions and regressions in the Gulf Coast Tertiary (abs.): Am. Assoc. Petroleum Geologists Bull., v. 47, p. 367.

Scruton, P. C., 1960, Delta building and the deltaic sequence, *in* Recent sediments, northwest Gulf of Mexico: Am. Assoc. Petroleum Geologists, p. 82–102.

Smoot, T. W., 1960, Clay mineralogy of pre-Pennsylvanian sandstones and shales of the Illinois Basin. Part III.—Clay minerals of various facies of some Chester formations: Ill. Geol. Survey Circ. 293.

Swann, D. H., 1963, Classification of Genevievian and Chesterian (late Mississippian) rocks of Illinois: Ill. Geol. Survey Rept. Inv. 216.

Swann, D. H., and Atherton, Elwood, 1948, Subsurface correlation of lower Chester strata of the Eastern Interior Basin: Jour. Geology, v. 56, p. 269–287; reprinted as Ill. Geol. Survey Rept. Inv. 135.

Swann, D. H., and Bell, A. H., 1958, The habitat of oil in the Illinois Basin, *in* The habitat of oil: Am. Assoc. Petroleum Geologists, p. 447–472.

Swann, D. H., Fisher, R. W., and Walters, M. J., 1959, Visual estimates of grain size distribution in some Chester sandstones: Ill. Geol. Survey Circ. 280.

Swann, D. H., and Willman, H. B., 1961, Megagroups in Illinois: Am. Assoc. Petroleum Geologists Bull., v. 45, p. 471–483.

Weller, J. M., 1956, Argument for diastrophic control of late Paleozoic cyclothems: Am. Assoc. Petroleum Geologists Bull., v. 40, p. 17–50.

Willman, H. B., Swann, D. H., and Frye, J. C., 1958, Stratigraphic policy of the Illinois State Geological Survey: Ill. Geol. Survey Circ. 249.

Worthen, A. H., 1860, Remarks on the discovery of a terrestrial flora in the Mountain Limestone of Illinois (abs.): Am. Assoc. Adv. Sci. Proc., v. 13, p. 312–313.

**Reprinted from:**
BULLETIN OF THE AMERICAN ASSOCIATION OF PETROLEUM GEOLOGISTS
VOL. 51, NO. 5 (MAY, 1967), PP. 761-779, 8 FIGS.

# OUTLINE OF GEOLOGY OF NIGER DELTA[1]

K. C. SHORT[2] AND A. J. STÄUBLE[2]
Port Harcourt, Nigeria

ABSTRACT

The coastal sedimentary basin of Nigeria has been the scene of three depositional cycles. The first began with a marine incursion in the middle Cretaceous and was terminated by a mild folding phase in Santonian time. The second included the growth of a proto-Niger delta during the Late Cretaceous and ended in a major Paleocene marine transgression. The third cycle, from Eocene to Recent, marked the continuous growth of the main Niger delta. A new threefold lithostratigraphic subdivision is introduced for the Niger delta subsurface, comprising an upper sandy Benin Formation, an intervening unit of alternating sandstone and shale named the Agbada Formation, and a lower shaly Akata Formation. These three units extend across the whole delta and each ranges in age from early Tertiary to Recent. They are related to the present outcrops and environments of deposition. A separate member of the Benin Formation is recognized in the Port Harcourt area. This is the Afam Clay Member, which is interpreted to be an ancient valley fill formed in Miocene sediments. Subsurface structures are described as resulting from movement under the influence of gravity and their distribution is related to growth stages of the delta. Rollover anticlines in front of growth faults form the main objectives of oil exploration, the hydrocarbons being found in sandstone reservoirs of the Agbada Formation.

## Introduction

The search for oil in Nigeria has led to a great increase in the knowledge of the geology of the coastal sedimentary basin. Details of the surface geology of the area already have been published by the Geological Survey of Nigeria, in the form of 1:250,000-scale maps which include the results of geologists of the Shell-BP Petroleum Development Company of Nigeria. More recently the formation nomenclature has been improved and formations have been defined by Reyment (1965). The contents of this paper are based mainly on the work of past and present Shell-BP staff whose numbers unfortunately preclude their being mentioned by name. However, this does not diminish the writers' indebtedness to them, for without their efforts this paper could not have been written. Acknowledgment is also made to the Shell-BP Petroleum Development Company of Nigeria Limited for permission to publish this information.

Since its inception in Early Cretaceous time, the southern basin of Nigeria has been the scene of at least three major depositional cycles. The first, possibly a double cycle, commenced in pre-Albian time and involved mainly marine deposition within the whole basin. It terminated with a brief phase of folding of Santonian age. The second cycle began with a Campanian marine transgression and included the growth of a proto-Niger delta in the northern part of the basin. This ended with renewed widespread marine transgression in Paleocene time. These two cycles are considered briefly because, during them, the basic framework was formed on which the present delta was to build. The main part of the paper deals with the third cycle, from the time when the Niger delta began to grow southward in Eocene time. The principal rock-stratigraphic units found in the basin and their age ranges are shown diagrammatically in Figure 1.

[1] Read before the Association, April 26, 1966. Published with permission of Shell-BP Petroleum Development Co. of Nigeria Ltd. Manuscript received, July 5, 1966; accepted, July 27, 1966.

[2] Shell-BP Development Co. of Nigeria Ltd. 1976 addresses: Shell Internat. Pet. Mij., The Hague, Netherlands; and Nishi Nihon Sekiyu Kaihatsu K. K., Tokyo, Japan, respectively.

## Stratigraphy of Basin Frame

### APTIAN TO SANTONIAN SEDIMENTARY PHASE

The oldest sediments found in southern Nigeria are non-fossiliferous, arkosic, gravelly, and in general ill-sorted, commonly cross-bedded sand and quartzitic sandstone. They form the weathering products of the nearby and underlying crystalline and metamorphic Basement complex. The thickness of these sandstones probably is related to the paleomorphology of the basement surface, being thickest in old river valleys and in front of fault scarps.

Albian deposits are known to occur in eastern and northern Nigeria (Fig. 2) where they are exposed at various localities in the Abakaliki and Calabar areas. The main rock-stratigraphic unit is

**NIGER DELTA**

| SUBSURFACE | | | SURFACE OUTCROPS | | |
|---|---|---|---|---|---|
| YOUNGEST KNOWN AGE | | OLDEST KNOWN AGE | YOUNGEST KNOWN AGE | | OLDEST KNOWN AGE |
| RECENT | BENIN FORMATION (Afam Shale Member) | OLIGOCENE | PLIO/ PLEISTOCENE | BENIN FORMATION | MIOCENE? |
| RECENT | AGBADA FORMATION | EOCENE | MIOCENE | OGWASHI-ASABA FORMATION | OLIGOCENE |
| | | | EOCENE | AMEKI FORMATION | EOCENE |
| RECENT | AKATA FORMATION | EOCENE | L. EOCENE | IMO SHALE FORMATION | PALEOCENE |
| EQUIVALENTS NOT KNOWN | | | PALEOCENE | NSUKKA FM | MAESTRICHTIAN |
| | | | MAESTRICHTIAN | AJALI FORMATION | MAESTRICHTIAN |
| | | | CAMPANIAN | MAMU FORMATION | CAMPANIAN |
| | | | CAMP./MAEST. | NKPORO SHALE | SANTONIAN |
| | | | CONIACIAN/ SANTONIAN | AWGU SHALE | TURONIAN |
| | | | TURONIAN | EZE AKU SHALE | TURONIAN |
| | | | ALBIAN | ASU RIVER GROUP | ALBIAN |

FIG. 1.—Table of formations, Niger delta area, Nigeria. Note added in proof: Afam Shale Member should be Afam *Clay* Member.

the Asu River Group which consists of dark, micaceous, and sandy shale and fine-grained micaceous sandstone. In the Calabar area as well as in the Benue trough, shale alternates with thick shelly, fossiliferous limestone, the latter being associated with lead-zinc mineralization.

The Asu River Group has yielded a well-developed ammonite fauna which has been described by Reyment (1955). This indicates a middle to late Albian age.

Sedimentary rocks of undoubted Cenomanian age have been established only in the Odukpani Formation cropping out north of Calabar in eastern Nigeria. They consist of a very heterogeneous sequence of sandstone, alternating with sandy shale, shale, and fossiliferous limestone, deposited in a very shallow marine environment close to a shoreline along the bordering Oban massif. Based on ammonite evidence, early and late Cenomanian are represented (Reyment, 1954, 1955).

The fact that sediments of undoubted Cenomanian age are known only from the Calabar flank led Buser (1965) to suggest a major regression for this time. However, no positive evidence is yet available of any break in the pre-Santonian succession of the Abakaliki area that might confirm this possibility.

The important Turonian transgression is characterized by the development of the thick Eze Aku Shale composed of gray, flaggy calcareous shale with intercalated limestone lenses. These lenses have numerous ammonites and other mollusks. Species of the vascoceratids and hoplitoids have been used to form a zonal subdivision of the Turonian and Coniacian (Barber, 1957; Reyment, 1955). The Eze Aku Shale grades laterally into sandy shale and calcareous sandstone (Amaseri Sandstone). They are overlain by the sandy Makurdi Formation and the Awgu Shale, the latter ranging into the Santonian.

With the onset of crustal movements in Coniacian time, the facies development is more heterogeneous. Sandstone extends farther basinward, alternating with limestone and sandy shale (Awgu Shale, Amaseri Sandstone, Agbani Sandstone).

The Aptian to Santonian sedimentary phase ended with the onset of a phase of folding, faulting, and uplift in early Santonian time, resulting in the erosion of Coniacian, Turonian, and even Albian deposits of the uplifted Abakaliki anticlinorium.

### SANTONIAN TO EOCENE SEDIMENTARY PHASE

The Santonian folding phase was followed

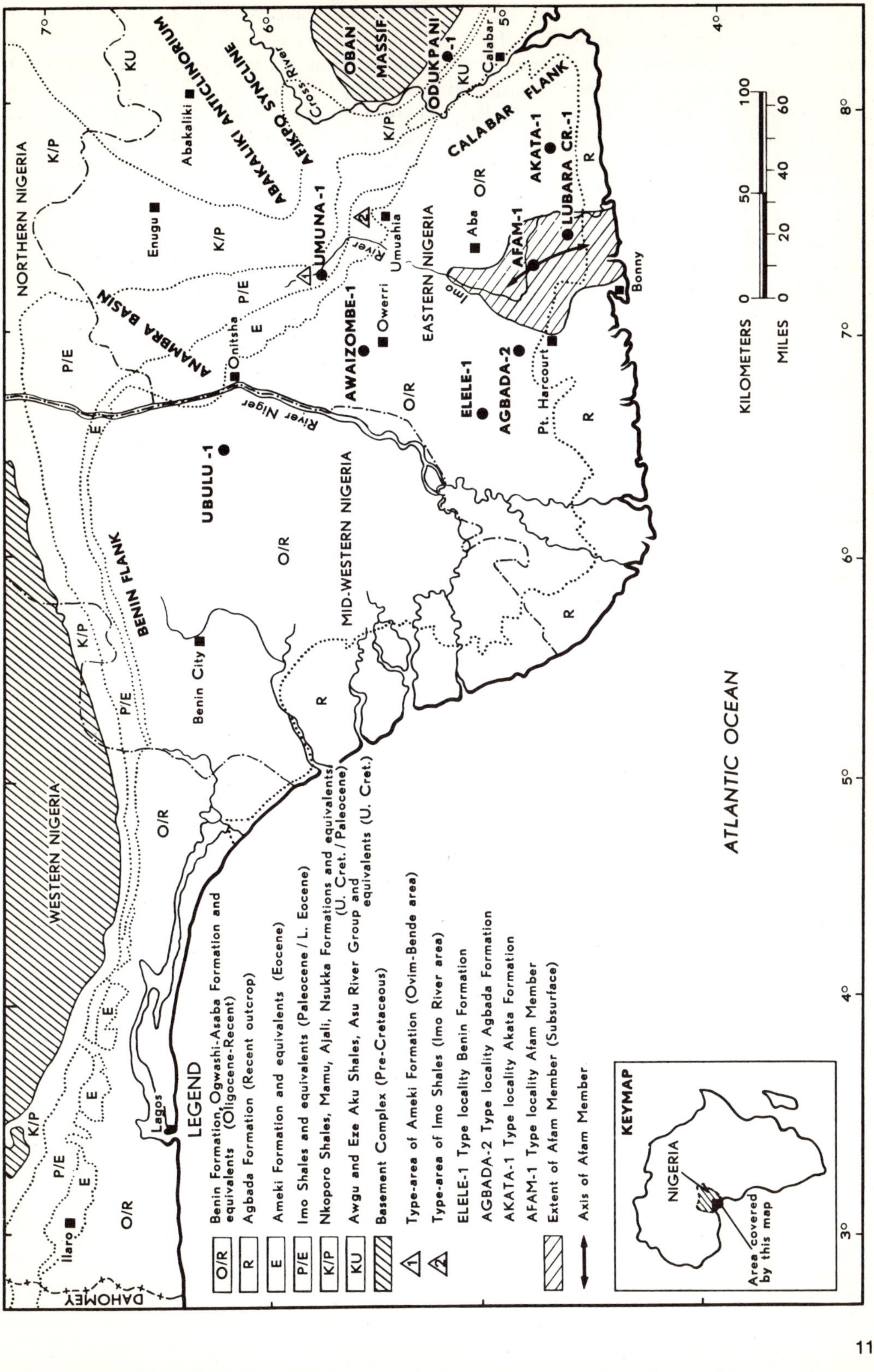


FIG. 2.—Map of southern Nigeria showing outcrops of Cretaceous and Tertiary formations and type localities of subsurface stratigraphic units.

by subsidence which initiated a new marine transgression, resulting in the deposition of the Nkporo Shale of Campanian-Maestrichtian age, and its lateral equivalents, the Owelli Sandstone and the Enugu Shale. West of the Niger the marine Nkporo Shale ranges from Campanian to late Maestrichtian. In the east the Maestrichtian is represented by deltaic deposits—the Mamu Formation, the Ajali Formation, and the Nsukka Formation—containing coal seams at several levels which are mined in the Enugu area. The marine influence in these deposits is restricted to the lowermost part of the Mamu Formation and becomes insignificant in higher levels (Bain *et al.*, 1924; Reyment, 1965).

This regressive period, with the formation of a proto-Niger delta, continued throughout the latest Cretaceous and deltaic conditions may, especially in eastern Nigeria, have lasted locally into the Paleocene.

Beginning with the Tertiary, the sea transgressed the whole of southern Nigeria, terminating the advance of the Upper Cretaceous Niger delta and separating it stratigraphically from the modern Niger delta which began forming in the Eocene.

The main rock-stratigraphic unit of Paleocene age is the Imo Shale Formation, which crops out in an arcuate belt from western to eastern Nigeria. The Imo Shale is blue-gray, in a few places sandy, and commonly very fossiliferous. In western Nigeria the shale of the Imo grades in part into thick shelly limestones of the Ewekoro Formation, which are mined for cement manufacture. In eastern Nigeria the shale in some places is sandy to silty, or grades into shaly sandstone. The Imo Shale Formation ranges into the early Eocene (Stolk, 1963) and is overlain by the sandy Ameki Formation which marks the onset of a regression and the formation of the modern Niger delta.

The microfauna of the Maestrichtian-Oligocene sedimentary rocks has been studied in detail by Stolk (1963) who divides it into several faunal zones based on planktonic Foraminifera.

## Stratigraphy of Basin Center

### OUTCROPS OF TERTIARY NIGER DELTA

During the early Eocene a regression began, resulting in the deposition of the more sandy Ameki Formation. West of the River Niger the Ameki still shows a predominantly shaly development as far as the area north of Lagos, where it passes laterally into the sandy Ilaro Formation and into the lagoonal clay of the Oshosun Formation. East of the Niger the Ameki Formation is very heterogeneous, consisting of alternating sandstone and shale, sandy or calcareous shale, marl, and a few fossiliferous shale and limestone beds. These abrupt, irregular alternations indicate deposition in a shallow marine environment with sediment supply from a nearby coast. During the middle and late Eocene, the sedimentary rocks become increasingly sandy, marking the onset of a general regression and of deltaic deposition.

Surface evidence of Oligocene and Miocene deposits is limited and much of the evidence for the age determinations is inferred. The main rock-stratigraphic unit is the continental Ogwashi-Asaba Formation and its equivalent, the Ijebu Formation in the Lagos area, which contains some sparse marine faunas. Both formations are predominantly sandy, the sand alternating with lignite seams and a few beds of clay as in the Ogwashi-Asaba Formation, or with a few thin clay beds with scarce marine faunas as in the Ijebu Formation (Reyment, 1965, p. 102).

The youngest rock-stratigraphic unit is the Benin Formation of possible Miocene to Recent age. This unit consists predominantly of yellow and white continental sand, alternating with pebbly layers and a few clay beds (Reyment, 1965, p. 104).

### RECENT NIGER DELTA

In the arcuate outcrop belt of Tertiary deposits which extends from western to eastern Nigeria (Fig. 2), biostratigraphic dating is possible only in the more marine sequences such as the Paleocene-Eocene Imo Shale and the Eocene Ameki Formation. South of these dated formations the quality of the outcrops becomes gradually poorer and the younger Ogwashi-Asaba and Benin Formations are commonly masked by thick, partly weathered layers of Recent Niger delta sediments and by dense vegetation, ranging from secondary bush to fresh-water swamp flora, and, at the coast, mangrove swamp. In the subsurface sedimentary sequences, separate rock-stratigraphic units have been established. An understanding of this subsurface sequence can be derived by comparison with sediments of the Recent Niger delta. The modern delta appears from sedimentological, faunal, and floral evidence to have a configuration today similar to those of the past.

The physiography of the Recent Niger delta is governed by several factors which influence transport, ultimate deposition of the sediment load, and shape and growth of the delta. A detailed account of these factors is given by Nedeco (1954, 1961) and Allen (1963, 1965); therefore only the more pertinent features are reviewed briefly in this paper

The climate of the Niger delta area is dependent on the movements of the intertropical front which marks the boundary between the humid air masses from the south and the dry air masses from the north. The intertropical convergence zone moves from south to north, depending on the season. In general the humid air masses and the associated rain-bearing southwesterly *monsoon* winds dominate the delta and the dry *harmattan* reaches the coast only occasionally, in the months of January and February. Temperatures over the delta are very uniform and range from 20°C. at night to more than 30°C. during daytime. Under the influence of southwesterly winds, the area receives a considerable amount of rain, ranging from 400 inches per year in the east to 150–170 inches over the delta, and decreasing markedly toward the north and west to 75 inches, as in the Onitsha and Lagos areas.

Because it faces the high-energy dynamic environment of the Atlantic Ocean, the shape of the Nigerian coast is influenced by the regular strong swell of the South Atlantic coming in from a general S. 24° W. direction. This swell reaches the coast at an oblique angle, creating strong longshore currents which flow eastward on the eastern flank of the delta and westward on the western flank. The strength of these currents diverts the sediments brought into the sea by the rivers, distributing them along the coast and preventing the formation of a bird-foot-type delta. Swell, wave, and longshore-current activity is accentuated by tidal currents created by tides with an amplitude ranging from 3 feet at the Benin River mouth to 10 feet on the Cross River. Apart from the scouring effect of the tidal currents along the coast and in estuaries and rivers, a physical-chemical effect is caused by the heavy salt-water wedge which, at high tide, pushes far inland.

The discharge of the River Niger ranges from 5,420 cubic meters per second ($171 \times 10^9$ cubic meters per year) to 9,400 cubic meters per second ($298 \times 10^9$ cubic meters per year). These figures show the discharge of the river to be about two-fifths that of the Mississippi. This water carries an estimated $2.62 \times 10^6$ cubic meters of sand of which, however, only about 35 per cent, the finer fractions, reaches the sea, the rest being trapped in the delta.

The Recent Niger delta area may be divided into three main sedimentary environments, the continental environment, the transitional environment, and the marine environment (Fig. 3).

The *continental environment* comprises the alluvial environments, including the braided-stream and meander-belt systems of the *upper deltaic plain*. The sediments deposited in this zone are predominantly sandy. Feldspar grains are fairly common and sand grains commonly are limonite-coated. Finer-grained sediments (silt and clay) are deposited in the adjacent fresh-water backswamps and oxbows, together with large quantities of plant remains.

The *transitional environment* comprises the brackish-water *lower deltaic plain* (mangrove swamps, flood-plain basin, marsh) and the coastal area with its beaches, barrier bars, and lagoons. The sediments in this environment are distinctly finer-grained than in the continental environment. Feldspar is scarce, and brackish-water faunas may occur.

The *marine environment* includes the submarine part of the delta, the delta fringe with its fine sand, silt, and clay, and the associated marine faunas. This environment grades laterally into the *holomarine environment* which is not affected by deltaic activity.

## SUBSURFACE SEDIMENTARY SEQUENCE

In an advancing delta such as that of the River Niger, sediments of the three environments described above become stratigraphically superimposed. The submarine delta fringe will encroach on holomarine sediments and will, in turn, be covered by a younger lower deltaic plain which subsequently will be overlain by the next-younger upper deltaic plain.

The same sequence is found in the Niger delta subsurface but is modified by the numerous transgressions which have occurred from time to

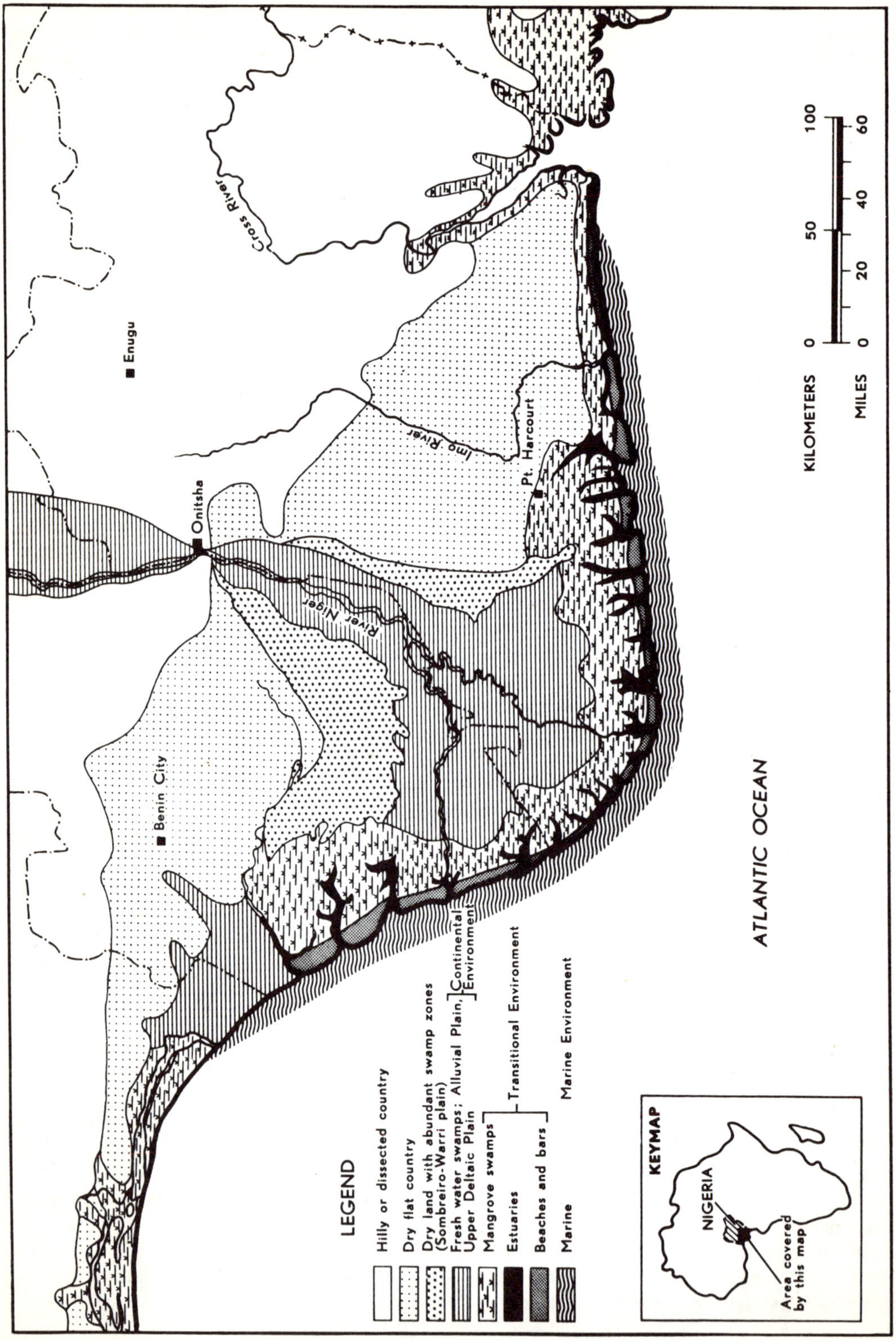


FIG. 3.—Morphological features and sedimentary environments of southern Nigeria.

time, breaking the continuity of the main over-all regression.

Based on this consistent picture the following rock-stratigraphic subdivision is proposed: (1) Benin Formation; (2) Agbada Formation; and (3) Akata Formation.

#### BENIN FORMATION

*Type section.*—Elele-1, new-field wildcat, drilled 24 miles NNW. of Port Harcourt.

*Interval.*—0–4,600 feet below derrick floor.

*Location.*—Geographical coordinates: Lat. 5°4′ 12″N.; Long. 6°50′4″E.

*Derrick floor elevation.*—98 feet above mean sea-level.

*Total depth.*—10,635 feet.

*Samples.*—Ditch-cutting samples every 30 feet. Stored with The Shell-BP Petroleum Development Company of Nigeria Limited and with the Geological Survey of Nigeria, Kaduna. Well drilled by The Shell-BP Petroleum Development Company of Nigeria Limited.

*Description of type section.*—The top of the type section is at the surface; the base of the type section is at a depth of 4,600 feet, and is the top of the highest shale bearing a marine fauna. This boundary is in fact an arbitrary one selected in a gradational sequence. On the log of type section (Fig. 4), a more pronounced break appears at 3,700 feet but this was found to have only local significance. Of much greater importance is the highest appearance within the shale of the first fully marine Foraminifera. This highest appearance can be recognized easily throughout the delta and has been selected to pinpoint the otherwise arbitrary boundary.

*Lateral extent.*—The Benin Formation occurs across the whole Niger delta from the Benin-Onitsha area in the north to beyond the present coast line.

*Lithology.*—The sequence penetrated by Elele-1 is predominantly (more than 90 per cent) sandy with a few shale intercalations; these become more abundant toward the base (Fig. 4). The sand and sandstone are coarse-grained, commonly very granular and pebbly to very fine-grained. In general they appear to be very poorly sorted. The grains are subangular to well-rounded. The sand and sandstone are white or, because of a limonite coating, yellowish brown. Lignite occurs in thin streaks or finely dispersed fragments; hematite grains and feldspars are common. The shale is grayish brown, sandy to silty, and contains some plant remains and dispersed lignite. Composition, structure, and grain size of the sequence indicate deposition in a continental, probably upper deltaic environment. The sand and sandstone may represent point-bar deposits, channel fills, or natural levees, whereas the shales may be interpreted as backswamp deposits and oxbow fills.

*Age of type section.*—The highest dated shale underlying this barren sandy sequence indicates an early Miocene age. The total age span is thus Miocene to Recent.

*Remarks.*—The Benin Formation as described here is that of Parkinson (1907, p. 311), for which no type section was given. The name was reinstated by Reyment (1965, p. 104) for the "Coastal Plain Sands" cropping out in Benin, Onitsha, and Owerri Provinces, where a Pliocene-Pleistocene age is given. The subsurface extent of the Benin Formation is delimited here for the first time. Genetically the Benin sands and sandstones are mainly deposits of the continental upper deltaic plain environment. In the subsurface, the age of the Benin Formation is Oligocene in the north, and becomes progressively younger southward. The formation is easily recognizable because of the high sand percentage (70–100 per cent), the few minor shale intercalations, and the lack of a characteristic marine or brackish-water microfauna. The thickness is variable and may be more than 6,000 feet. In the eastern part of the Niger delta, the normally continuous body of the Benin Formation is interrupted by the Afam Clay Member, which is discussed in a subsequent section. To date, very few hydrocarbon deposits have been found in this highly porous, generally fresh-water-bearing formation.

#### AGBADA FORMATION

*Type section.*—Agbada-2, new-field wildcat, drilled 7 miles NNW. of Port Harcourt.

*Interval.*—5,760–9,500 feet below derrick floor.

*Location.*—Geographical coordinates: Lat. 4°55′39.94″N.; Long. 7°1′50.92″E.

*Derrick floor elevation.*—80 feet above mean sea-level.

*Total depth.*—9,500 feet.

*Samples.*—Ditch-cutting samples every 15 feet. Stored with The Shell-BP Petroleum Develop-

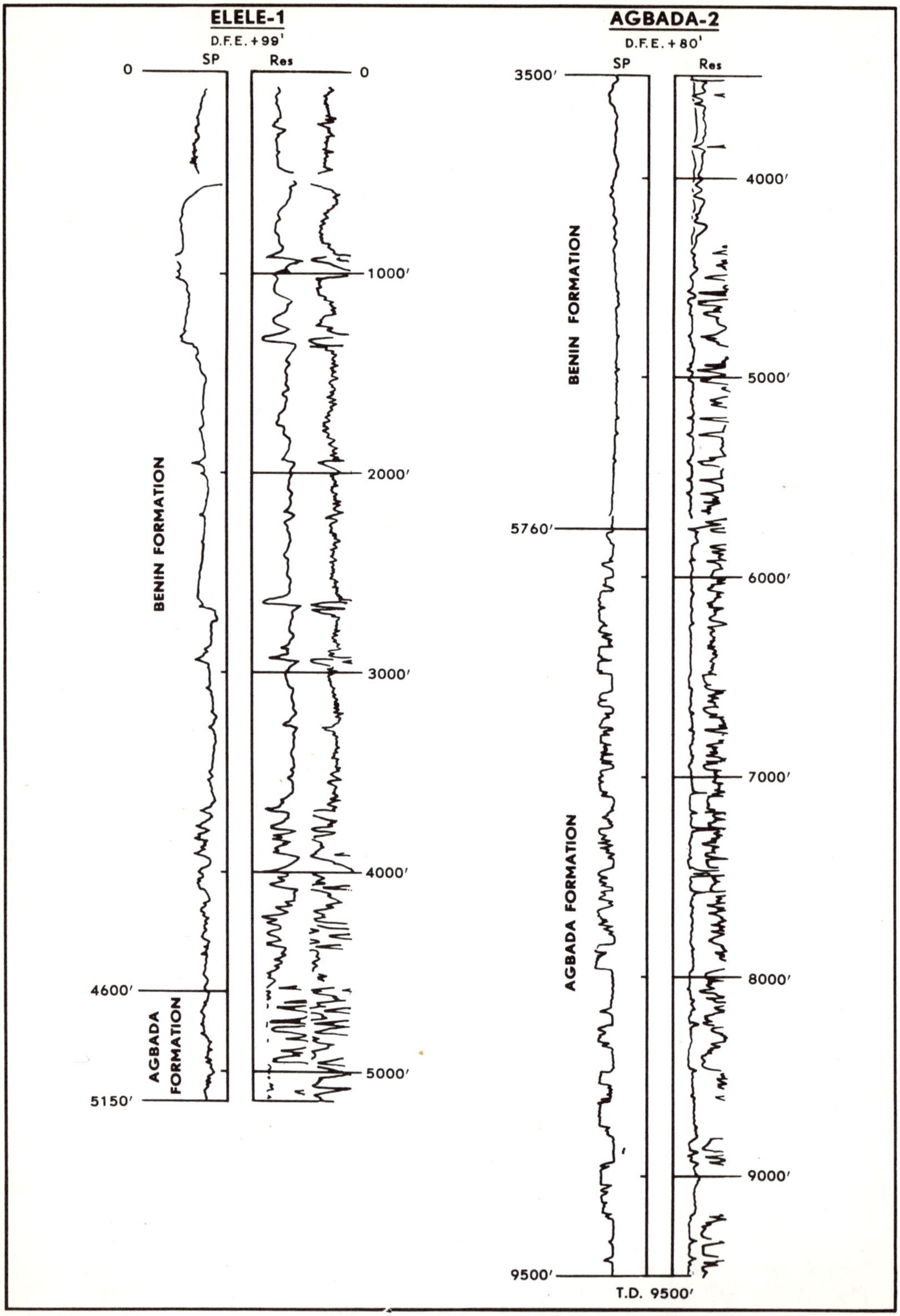


FIG. 4.—Type sections of Benin Formation and Agbada Formation. Depths in feet.

ment Company of Nigeria Limited and with the Geological Survey of Nigeria, Kaduna. Well drilled by The Shell-BP Petroleum Development Company of Nigeria Limited.

*Description of type section.*—The top is the highest occurrence of a shale with a brackish-water to marine fauna. The base was not reached at a total depth of 9,500 feet. The true base of the Agbada Formation is better shown in Akata-1 where the top of the Akata Formation was reached (Fig. 5). The base of the Agbada Formation is defined as the base of the deepest significant sandstone body in all vertical sections through the delta sediments; the base corresponds with the top of the marine Akata Shale.

*Lateral extent.*—The Agbada Formation occurs through the whole Niger delta area, south of the outcrop of the Ogwashi-Asaba Formation.

*Lithology.*—The Agbada Formation is characterized by the alternation of sandstone and sand bodies with shale layers. The sequence penetrated in Agbada-2 may be subdivided into two units: an upper one in which sandstone-shale alternations are abundant and the shale intercalations relatively thin; and a lower unit in which the shale units become more prominent and in some places are thicker than the intercalated sandstone or sand bodies. The sandstone percentage ranges from 75 in the upper unit to 50 in the lower unit (Fig. 4). The sandstone and sand are very coarse to very fine-grained, predominantly unconsolidated, or slightly consolidated with a calcareous matrix. Sorting generally is poor except where the sand or sandstone grades into shale. Lignite streaks and limonite grains are common. However, limonite-coated grains and feldspars which characterize the overlying Benin Formation are very scarce. Some fine-grained sand and sandstone units contain shell fragments and glauconite.

The shale beds are gray and dense at the base; they become markedly sandy and silty upward and grade into the overlying sand and sandstone bodies. Commonly they contain a microfauna which is best developed at the base of the shale units and becomes sparse or absent in the upper part. This indicates an increase in the rate of deposition in front of the approaching delta.

Compared with the foraminiferal development on the Recent Nigerian continental shelf, the depths of these fossil assemblages range from littoral-estuarine-marsh types to faunas developed at a water depth of approximately 300 feet. The occurrence of some larger Foraminifera at the base of some of the more prominent shale units suggests that, in a few places, non-deltaic conditions were present at the onset of a transgression. The coarseness and the poor sorting of many sand and sandstone bodies indicate that they are of fluviatile origin. The well-sorted, in general finer-grained, sand and sandstone beds with glauconite grains and shell fragments represent beach or coastal barrier sand deposits.

*Remarks.*—The Agbada Formation extends throughout the Niger delta subsurface. It is probably continuous with the outcrops of the Ogwashi-Asaba and Ameki Formations of Eocene-Oligocene age, which correspond approximately with the twofold subdivision described in the type section. However, this subdivision can not be recognized in all parts of the delta, and the Agbada Formation generally is more marine. Therefore a single formation name has been preferred for the subsurface extension. The Recent outcrops of the Agbada Formation are formed by the present lower delta plain and the marine sediments of the continental shelf fronting the delta (Fig. 2). The alternations of sand, sandstone, and shale are the result of differential subsidence, variation in the sediment supply, and shifts of the delta depositional axes which cause local transgressions and regressions. In general, however, the upper part of the formation has a higher sandstone percentage than the lower part. This demonstrates the progressive, if somewhat irregular, seaward advance of the Niger delta through geological time despite the numerous, at times regional, transgressive sequences found in the Agbada Formation. The thickest known section of the Agbada Formation is about 10,000 feet but the maximum thickness may well be much greater.

The age of the formation decreases from north to south from Eocene in Awaizombe-1 and Ubulu-1 to possibly Plio-Pleistocene far from the present coast line; the unit is Recent at the present delta surface. The Agbada Formation is the main objective in the exploration for oil in southern Nigeria. Hydrocarbons have been found in intervals from Eocene to probably Pliocene age.

### AKATA FORMATION

*Type section.*—Akata-1, new-field wildcat,

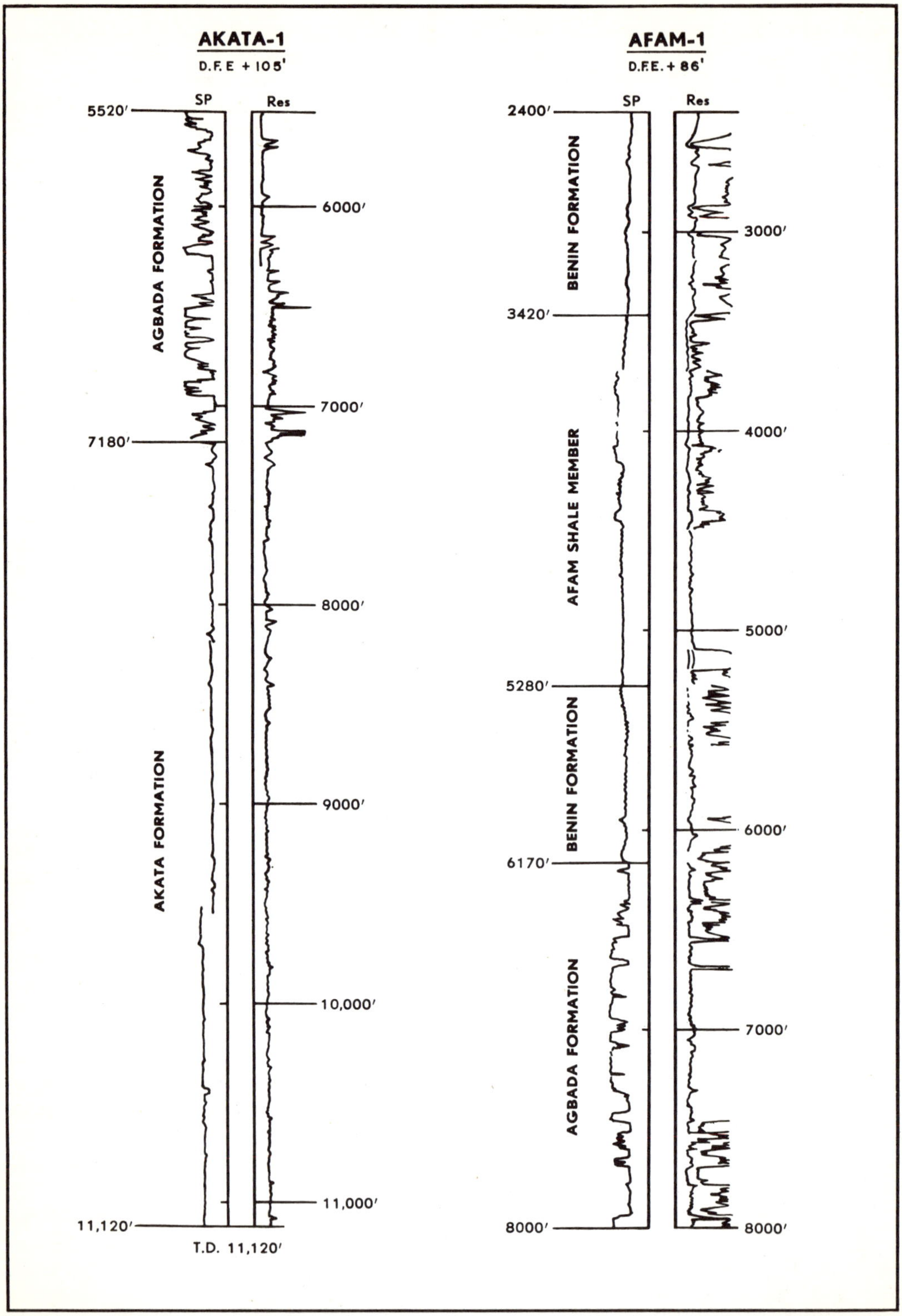


FIG. 5.—Type sections of Akata Formation and of Afam Clay Member of Benin Formation. Depths in feet Note added in proof: Afam Shale Member shown in Afam-1 well is an error; should read Afam *Clay* Member.

drilled 50 miles east of Port Harcourt.

*Interval.*—7,180–11,121 feet below derrick floor.

*Location.*—Geographical coordinates: Lat. 4° 41′50.5″N.; Long. 7°46′58.6″E.

*Derrick floor elevation.*—106 feet above mean sea-level.

*Total depth.*—11,121 feet.

*Samples.*—Ditch-cutting samples every 20 feet. Stored with The Shell-BP Petroleum Development Company of Nigeria Limited and with the Geological Survey of Nigeria, Kaduna. Well drilled by The Shell-BP Petroleum Development Company of Nigeria Limited.

*Description of type section.*—The top of the Akata Formtion is the deepest occurrence of deltaic sandstone beds at 7,180 feet. The base was not reached at the total depth of 11,121 feet.

*Lateral extent.*—The formation probably underlies the whole Niger delta south of the Imo Shale outcrop.

*Lithology.*—The Akata Formation is characterized by a uniform shale development (Fig. 5). The shale in general is dark gray, in some places sandy or silty, and contains, especially in the upper part of the formation, plant remains and some mica. Toward the top of the formation where it grades into the overlying Agbada Formation, some thin sandstone lenses may occur. The fauna generally is very rich. Planktonic Foraminifera may make up more than 50 per cent of the microfauna. The benthonic foraminiferal assemblages indicate deposition of the shale on a shallow-marine shelf. No environments of deposition indicating a bathyal or deeper-water environment have been found.

*Remarks.*—The Akata Formation is a marine sedimentary sequence laid down in front of the advancing delta. Intervals of variable thickness have been penetrated by several wells. The base has been reached only along the delta margin. From these data, the base of the Akata Formation can be defined only provisionally as the top of the highest major sandstone body or the first major unconformity found below the top of the shale. Probably the shale of the Akata extends across the whole delta area. At depth one may expect that the shale gradually assumes a bathyal or abyssal character, perhaps interrupted by a few turbidity-current deposits. The Akata Formation may be continuous with the outcrops of the Imo Shale Formation, but continuity between the two type sections, which are of very different age, is not yet proved, though theoretically is probable. The known age of the Akata Formation ranges from Eocene to Recent.

#### AFAM CLAY MEMBER OF BENIN FORMATION

*Type section.*—Afam-1, new-field wildcat, drilled 22 miles east of Port Harcourt.

*Interval.*—3,420–5,540 feet below derrick floor.

*Location.*—Geographical coordinates: Lat. 4° 48′56.95″N.; Long. 7°19′47″E.

*Derrick floor elevation.*—86 feet above mean sea-level.

*Total depth.*—10,207 feet.

*Samples.*—Ditch-cutting samples every 20 feet. Stored with The Shell-BP Petroleum Development Company of Nigeria Limited and with the Geological Survey of Nigeria, Kaduna. Well drilled by The Shell-BP Petroleum Development Company of Nigeria Limited.

*Description of type section.*—The top of the Afam Clay Member is the top of the uppermost clay intercalation in the gradational sequence between it and the overlying part of the Benin Formation. The base is the unconformity at the bottom of the main clay body overlying the lower part of the Benin Formation.

*Lithology.*—The member consists mainly of clay with a few intercalated sandstone bodies (Fig. 5). The clay is browish gray to dark gray, laminated, commonly silty and finely sandy and bearing siltstone stringers and lenses. The clay contains a moderate amount of pyritized plant remains, some lignite, and some finely dispersed pyrite. The upper part of the sequence is more sandy and barren of microfauna. Below 4,700 feet, a poorly preserved, limonite-stained, and pyritized fauna occurs, which in the basal part of the member is overlain by a better-preserved arenaceous foraminiferal assemblage. The intercalated sandstone is medium- to fine-grained, commonly clayey or with clay lenses and streaks, finely bedded, and well sorted. Limonite staining of the grains is scarce and feldspars occur in very few samples.

*Remarks.*—The Afam Clay Member has been recognized only in the subsurface of eastern Nigeria in an area extending westward from a line connecting the Imo River northwest of Aba with the estuary of the Imo River (Fig. 2). The member has the form of a canyon fill which strikes in a south-southeast direction, from slightly north of Afam-1 to west of the Imo River estuary. Along this axis a maximum thickness of more than 2,500 feet has been measured in the Lubara Creek-1 well. Everywhere the Afam Clay Member unconformably overlies older continental deposits. A basal sand or gravel bed commonly is developed just above the unconformity. The overlying clay contains a predominantly arenaceous fauna of marine origin. These clearly marine associations are confined to deposits in the thickest sections and have not been found in the upper part of the formation which occurs across a much wider area. This upper part of the Benin is either barren or contains a partly reworked, partly pyritized fauna in which very small, yellowish-stained *Bolivinas* predominate. The scarcity of the fauna, its arenaceous character, and poor preservation, together with the lamination of the clay, the silt content, and the siltstone and sandstone streaks and lenses, suggest moderately high rates of deposition in a very restricted marine to transitional environment with no known counterpart in the modern Niger delta.

### PALEOGEOGRAPHY OF TERTIARY NIGER DELTA

The main regression and the formation of the present Niger delta began in middle Eocene time with the deposition of the Ameki Formation, west and east of the Niger River. Sediment supply was derived from two drainage systems, the Niger-Benue system through the Anambra basin north of Onitsha, and the less important Cross River system through the Afikpo syncline (Fig. 2). In Eocene-Oligocene time the two systems appear to have been separated from each other by the Abakaliki anticlinorium and its southwestern extension.

The sands derived from the Eocene-Miocene Cross River system are characterized by a heavy-mineral association containing much garnet, kyanite-andalusite, and rutile-anatase-titanite, but little staurolite; in contrast, the sands of the Niger system have relatively little kyanite-andalusite and rutile, but generally have a high staurolite, garnet, and tourmaline content.

The Cross River system did not supply much sediment during all of middle and late Eocene time. The delta built up mainly in the Bende-Ameki area, although the increasing amount of sand toward the top of the Ameki Formation indicates a possible movement of the coast line toward the south.

At the same time the Niger delta advanced very rapidly over the shallow Anambra shelf. During the middle Eocene it straddled the Onitsha-Benin area and advanced farther toward the Kwale area (Ogume-1, Ethiope-1, Benin West-1). It reached the Aboh area in late Eocene time (Fig. 6). Because of the slow movement of the Cross River delta system, a marine interdeltaic area named the Ihuo embayment formed between the two deltas, extending beyond the Orlu area (Fig. 6).

This rapid advance of the Niger delta continued in post-Eocene time. In the Oligocene the delta front reached the Ubie-1 area east of the Niger and extended westward across the river to the Warri area. In late Oligocene-early Miocene time, it had advanced to a line extending from Elele-1, east of the Niger, to Kumbowei-1 and Ubefan-1 on the west.

In late Eocene-early Oligocene time, sedimentation in the Cross River delta area became more active and a small but important and rapidly advancing delta developed in the Olumbe-1 area. At this stage it was separated from the main Niger delta by the Ihuo embayment which, being a fairly shallow bay, was filled rapidly by the advancing Cross River delta and probably by local rivers. In late Oligocene-early Miocene time, the sediments of the Cross River merged with those of the Niger delta. Henceforth, extensive mixing of sediments by longshore and tidal currents makes the distinction of the two deltas difficult. However, the later growth of the Cross River delta can be traced southward from the Imo River field area to the Lubara Creek area.

Little is known either of the movement of the Cross River delta in Miocene-Pliocene time or of its shift from the Lubara Creek area to its present northwest-southeast course, reaching the sea southeast of Calabar.

In late Miocene time the subaerial Niger delta reached a position beyond the present coast line. In Pliocene-Pleistocene time it advanced farther seaward and reached its maximum extent very probably at approximately the edge of the present continental shelf. Coarse channel sands of fluviatile origin and shallow marine to estuarine-lagoonal faunas have been obtained from sea-floor samples taken from various localities up to 40 miles offshore (Allen, 1965).

The post-glacial rise in sea-level caused the sea

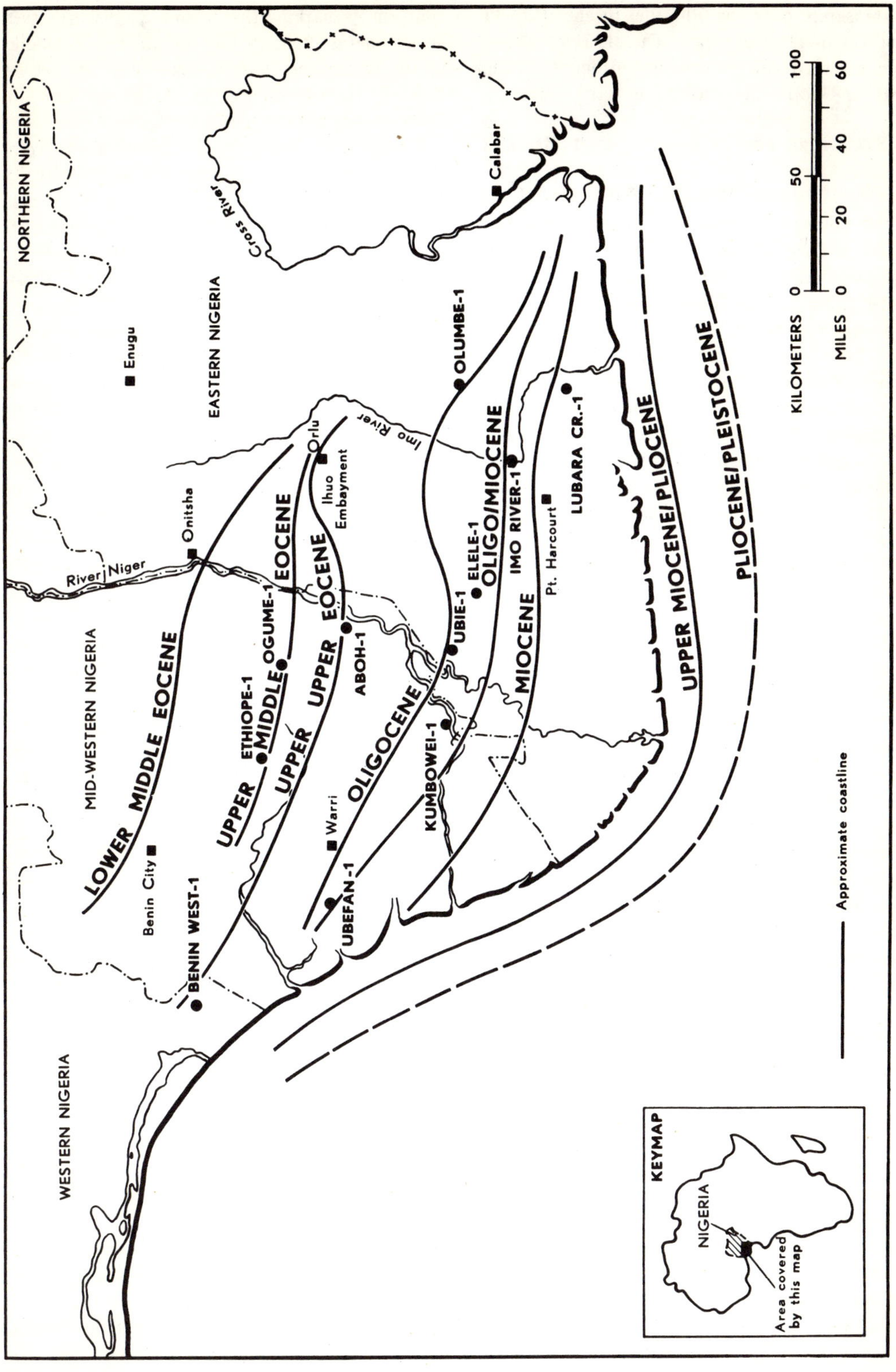


FIG. 6.—Paleogeography of Tertiary Niger delta—stages of delta growth.

to transgress over the Plio-Pleistocene delta beyond the present coast line. The old river systems were drowned and scoured by strong tidal currents and form the existing estuary system.

## Structure of Nigerian Coastal Basin

### Structural history of delta "frame"

The sedimentary basin of southern Nigeria originated in Early Cretaceous time as an X-shaped depression oriented northeast-southwest and northwest-southeast. This depression formed in the Basement complex of the African shield (Fig. 7). The form of this depression, and the rather straight course of long reaches of the Niger and Benue Rivers, have led to speculation that this depression is fault-controlled. Supporting evidence is forthcoming from recent gravity surveys and mapping, which suggest that the Benue Valley originated as a rift valley (Cratchley and Jones, 1965).

Thickest sedimentation occurred along the predominant northeast-southwest limb of the depression, in the Benue trough and its *en échelon* equivalent, the Abakaliki trough. The latter trough existed from Albian to Turonian time on the site of the present Abakaliki anticlinorium; the Anambra basin was an adjacent shelf area. In contrast, the northwest-southeast limb, containing the Middle Niger and Mamfe embayments, has only a thin sedimentary sequence.

Structural movement in the trough began in Coniacian time and culminated during the Santonian. Sediments of the Benue trough were folded in long gentle anticlines and synclines parallel with the trough margins. The Abakaliki trough containing less competent shaly sediments was folded more intensely into the Abakaliki anticlinorium, along the same northeast-southwest axes. This structural inversion of the Abakaliki trough was accompanied by the formation of depressions on either flank, the small Afikpo syncline on the southeast and the much wider Anambra basin on the northwest. The sediments affected by these movements were intruded by numerous small basic igneous bodies.

The succeeding period of erosion bared the core of the Abakaliki anticlinorium to the Asu River Group. Much of the sediment formed by erosion of the anticlinorium filled the Anambra basin, which had become the major site of deposition in Late Cretaceous-Eocene time. The unconformity marking this period of erosion is seen on the aerial photographs around the Abakaliki uplift, where the Nkporo Shale can be seen to overstep the outcrops of the Awgu and Eze-Aku Formations onto the Asu River Group.

In the field, however, the unconformity commonly is difficult to place because of the paucity of outcrops at the base of the escarpment formed by the overlying Mamu Formation. However, its existence is certain, based on three points.

1. The relatively flat-lying nature of the Nkporo Shale, compared with the folded older beds.
2. Although igneous intrusions are numerous in the Awgu Shale and older formations of the Abakaliki anticlinorium, none is known to penetrate the Nkporo Shale.
3. A marked paleontological break occurs between the Eze-Aku and Nkporo Shale formations.

In Paleocene-Eocene time, following the transgression of the Imo Shale Formation, further uplift took place. The Abakaliki anticlinorium was rejuvenated and, from time to time, emerged and was eroded. The evidence of this erosion may be seen today (1) in the form of sandstone bodies in the Imo Shale, which thicken toward the Abakaliki uplift, and (2) by a general thinning of the Imo Shale over the southwest nose of the uplift. The maximum extent of the Paleocene transgression in the Abakaliki area, and hence the amount of the succeeding uplift, are not clearly known.

### Formation of present delta surface

Subsequent to this period of mild early Tertiary uplift, the structural history was one of subsidence along the front of the growing delta accompanied by some southward tilting. Although the southern prolongation of the Abakaliki anticlinorium determined the main areas where the Niger and Cross Rivers deposited sediments, no clear evidence of further uplift is known from the basin.

Concomitant with the phases of subsidence and folding in the troughs just described, uplift of the surrounding basement massifs also occurred. This is evidenced by a series of erosion surfaces separated by scarps, forming a series of steps up to the summit of the Jos Plateau in northern Nigeria. These erosion surfaces have been described by several authors in various areas (*e.g.*, Pugh and King, 1952; Clayton, 1958), but they are not well enough known to make possible a clear cor-

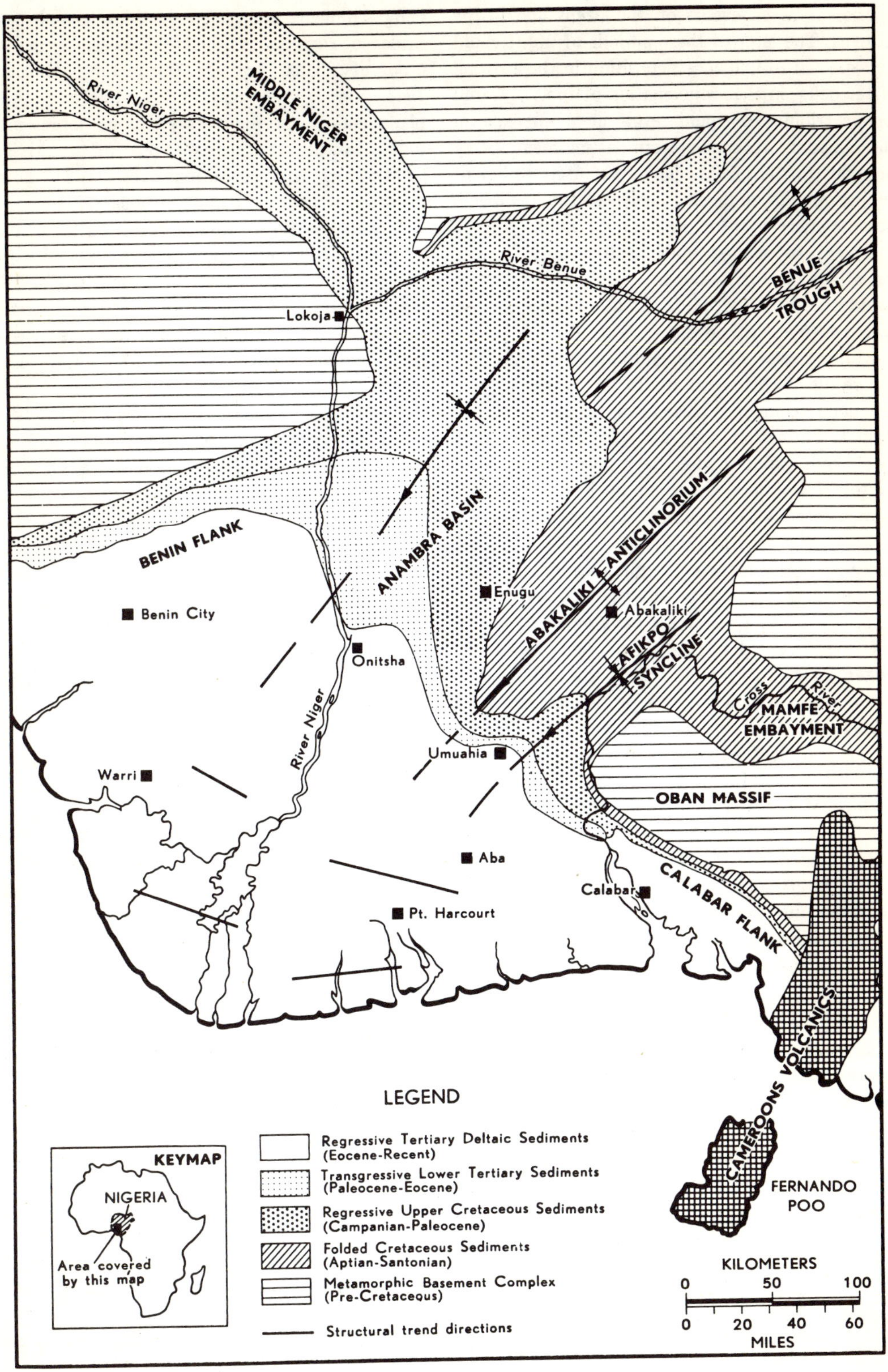


FIG. 7.—Structural units of Niger delta area.

relation between events on the swells and in the troughs. The later of these periods of uplift could be expected to influence deposition in the growing Niger delta, in that rejuvenation of the Niger and Benue tributaries would result in an increase in the amount of coarser-grained sediment being brought to the delta. This increased sediment supply would diminish slowly as the pediplane became more mature until a new influx marked the next succeeding uplift. Alternately, similar variations in the sediment supply to the delta could be caused by major climatic changes, about which even less is known.

Major cycles of sedimentation which may be the result of this cause are, in fact, discernible in the delta sediments, particularly in the sensitive[3] Agbada Formation. Two, and possibly three, relatively more sandy phases have been distinguished in this formation, separated by one or possibly two relatively less sandy phases. Dating is difficult for reasons given previously, but the three relatively more sandy phases may be of Oligocene, Miocene, and possibly late Miocene ages. A fourth sandy phase must have occurred during the Pleistocene lowering of sea-level when the Niger River cut a wide valley through its own delta. The remnants of this pre-Pleistocene delta surface are to be found in the dry land areas north and east of the Sombreiro-Warri Plain (Fig. 3) which forms part of the Niger erosion surface of Clayton (1958). The Pleistocene valley is now being filled again as sea-level returns to "normal."

The intervening Sombreiro-Warri Plain represents an intermediate flood-plain level, perhaps formed during some interglacial or postglacial interstadial stage.

### SUBSURFACE STRUCTURE OF DELTA

In middle to late Eocene time, the Niger delta began to build outward over the edge of the African continent. The actual form of the continental shelf at this time has been discussed by Hospers (1965). It is sufficient to say here that, beyond the continental edge, the Oligocene and younger delta sediments are extremely thick. Geophysical evidence suggests a thickness possibly in the order of 26,000 feet (8,000 meters; Hospers, 1965, p. 417). New marine deposition was taking place mainly on the forward slope represented today by the continental slope off the Niger delta, which is a depositional feature and not a primary structural feature (Hospers, 1965).

Within this thick, rapidly deposited sedimentary lens, conditions were right to promote the movement of large masses of sediment under the influence of gravity. Much of the resultant fault movement was contemporaneous with deposition, producing an abrupt thickening of the sedimentary section across the fault line on the downthrown block. Faults causing such increases in thickness are named "growth faults" and are similar to those described in many publications from the Gulf Coast area.

Many growth faults are crescent-shaped, both in cross section and in plan view, with the concave side toward the downthrown block. Because of this curved form, the downthrown block tends to rotate along an axis roughly paralleling the fault. If sufficient movement takes place along such a curved fault plane, an elongate anticline forms in front of the fault, which in the Gulf Coast has been termed a "rollover" structure. In the Niger delta such anticlines are small, ranging up to 4–5 miles across the long axes and 2–3 miles along the short axes. They may be cut by minor growth faults sharing the movement of the main fault, or by antithetic compensatory faults. In a few places subsidiary growth faults may form on the flanks of the main rollover, possibly accompanied by the formation of satellite rollovers. A section through a simple growth fault and its associated structures is shown in Figure 8. In addition to the rollover anticline visible in the shallower levels on the downthrown side of the fault, the main growth fault commonly is associated with a deep-seated anticline behind and below the fault. Because of their depth, none of these underlying anticlines has yet been penetrated deeply with the drill. However, taking into account the stratigraphic succession described previously, one might think of such an underlying structure as taking the form of a shale "bulge" genetically connected with the process of growth faulting.

Subsurface structural maps demonstrate that,

[3] The term *sensitive* refers to the transitional environment in which this formation was deposited—an environment in which relatively abrupt changes from deltaic to marine conditions, and from sandstone to shale deposition, took place. Under these conditions, any unusual increase in the relative content of sandstone becomes easily recognizable.

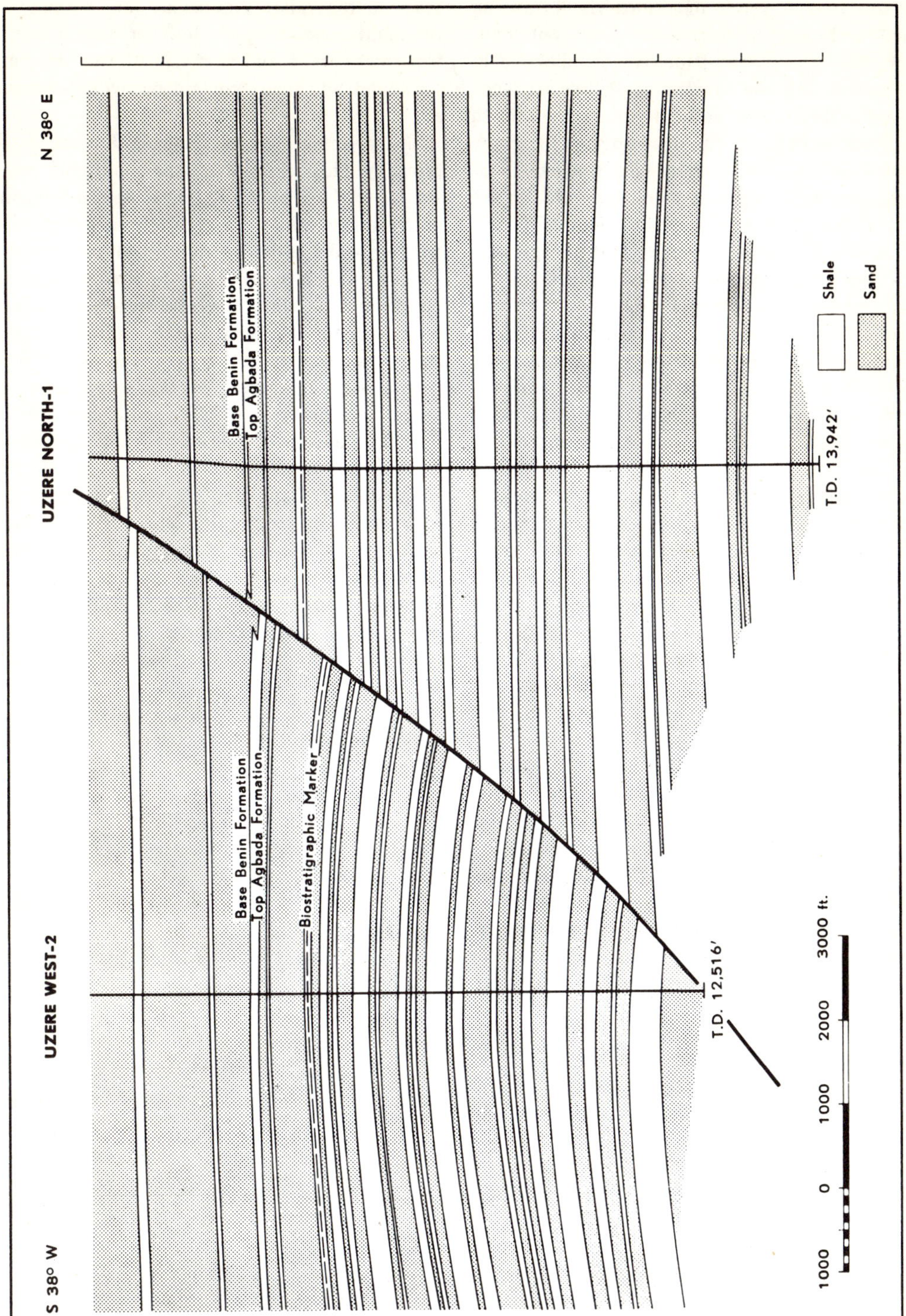


FIG. 8.—Diagrammatic section through simple growth fault and associated structures.

in many places, series of growth faults are on strike with one another forming a continuous trend, which may extend for tens of miles roughly parallel with the present coast line. In such cases the deep-seated anticline under the growth fault, where present, also may extend for many miles. The strike of such trends is markedly at variance with the tectonic trends observed in the older rocks north of the Niger delta (Fig. 7); this fact adds to the evidence that the growth-fault trends are related to growth stages of the delta and not to any orogenic process.

Subsurface evidence so far available suggests that the growth-fault trends die out laterally, and are replaced by long, comparatively unbroken flanks. Drilling along growth-fault trends also has demonstrated a tendency for the expanded section on the downthrown sides of the growth faults to thin toward the ends of the growth-fault trends. This in turn suggests that a relation exists between the structural growth-fault trend and a depositional subdeltaic unit. Similar areas of thicker deposition have been termed depocenters in the Gulf Coast area (Murray, 1949). In the Niger delta such thickening occurs mainly in the Agbada Formation. The section concerned tends to be more sandy in the center and less sandy toward the margins of the depocenter. Probably a depocenter is formed by the persistence in that area of a major river distributary during a long period of time. The resultant relatively heavy loading of sediment makes the formation of growth faults more likely, thus promoting the association described. However, evidence for this proposed hypothesis is still incomplete because of the lack of wells drilled on the intervening, comparatively unbroken flanks between the trends.

## Oil and Gas Occurrence

Almost all hydrocarbon accumulations in the Niger delta have been found in the sandstones of the Agbada Formation, mostly trapped in rollover anticlines fronting growth faults. The extent of the accumulation may or may not be restricted by subsidiary growth faults or antithetic faults cutting the anticline. This restriction becomes more evident on the larger anticlines, which, because of the size and extent of their crestal area, tend to form a less efficient focus for migration. In addition to these anticlinal traps, hydrocarbons have also been found in fault traps that are not closed on all sides by dip. Fields, particularly those in the rollover anticlines, are normally of the multi-reservoir type. However, few, if any, of the reservoirs found are full to the structural spill-point and many contain no hydrocarbons at all.

The reservoirs of the Agbada Formation are typically channel and barrier sandstone bodies similar to those of the Recent delta. Porosity and permeability values are generally high (up to 40 per cent and 1–2 darcys, respectively).

Oil found in the Agbada Formation is of the paraffinic type with very low sulfur and asphaltene contents. The wax content ranges from less than 1 per cent to 10 per cent, and up to 20 per cent in some reservoirs, whereas pour points may range from below −35°C. to more than +30°C. Specific gravities of oils range from 15° to 50° API. Great variations in one or more of these characteristics have been observed within a single multi-reservoir field. Little or no evidence of a well-developed pattern of distribution of oil types has emerged from discoveries made so far, beyond the conventional tendency toward lighter oil and more gas at greater depths.

This large variation in the hydrocarbon content of Niger delta fields, and the discontinuous nature of the sandstone reservoirs evidenced by the difficulties of inter-field correlation, argue strongly against any long-range migration of oil. The most obvious source rocks are the shales of the Agbada Formation itself and of the upper part of the Akata Formation, which lie close to the sandstone reservoirs. The oil would then migrate a relatively short distance updip into the nearest trap available at the time the oil migrated. Such a simple picture would explain most easily the distribution of hydrocarbons through the Niger delta sediments insofar as they are known today.

## References Cited


Allen, J. R. L., 1963, Sedimentation in the modern delta of the River Niger, West Africa: Proc. 6th Internatl. Conf. Sedimentologists, p. 26–34.

——— 1965, Late Quaternary Niger delta, and adjacent areas: sedimentary environments and lithofacies: Am. Assoc. Petroleum Geologists Bull., v. 49, no. 5, p. 547–600.

Bain, A. D. N., R. Bullen Newton, and A. C. Seward, 1924, The Nigerian coal fields, Section 1: Enugu area: Geol. Surv. Nigeria Bull., v. 6.

Barber, W., 1957, Lower Turonian ammonites from north-eastern Nigeria: Geol. Surv. Nigeria Bull., v. 26, 64 p.

Buser, H., 1965, Über den Einflusz von Paläostrukturen auf die Kreide-und Tertiär-Ablagerungen im westlichen Teile Afrikas: Neues Jahrbuch für Geologie und Paläontologie, no. 26, p. 400–409.

Clayton, W. D., 1958, Erosion surface in Kaba Province, Nigeria: Jour. W. African Sci. Assoc., v. 4, no. 2, p. 141.

Cratchley, C. R., and G. P. Jones, 1965, An interpretation of the geology and gravity anomalies of the Benue Valley, Nigeria: British Overseas Geol. Surveys, Geophys. Paper 1.

Hospers, J., 1965, Gravity field and structures of the Niger delta, Nigeria, West Africa: Geol. Soc. America Bull., v. 76, p. 407–422.

Murray, G. E., 1949, Geosynclinal sedimentation in central Gulf region of United States: Oil and Gas Jour., v. 48, no. 25, October 27, p. 87.

Nedeco (Netherlands Engineering Consultants), 1954, Western Niger delta, Report on investigation: The Hague, 143 p.

——— 1961, The waters of the Niger delta: The Hague, 317 p.

Parkinson, J., 1907, The post-Cretaceous stratigraphy of southern Nigeria: Quart. Jour. Geol. Soc. London, v. 63, p. 311–320.

Pugh, J. C., and L. C. King, 1952, Outline of the geomorphology of Nigeria: S. African Geogr. Jour., v. 34, p. 30–37.

Reyment, R. A., 1954, The stratigraphy of the southern Cameroons: Geol. Fören. Stockholm Förh., Bd. 76, h. 4, p. 661–683.

——— 1955, The Cretaceous Ammonoidea of southern Nigeria and the southern Cameroons: Geol. Survey Nigeria Bull., v. 25, 112 p.

——— 1965, Aspects of the geology of Nigeria: Ibadan Univ. Press.

Stolk, J., 1963, Contribution à l'étude des corrélations microfauniques du Tertiaire inférièur de la Nigeria méridionale: Colloque Internatl. de Micropaléontologie, p. 247–275.

**Reprinted from:**
BULLETIN OF THE AMERICAN ASSOCIATION OF PETROLEUM GEOLOGISTS
VOL. 53, NO. 1 (JANUARY, 1969), PP. 30-54, 12 FIGS.

# Depositional Systems in Wilcox Group (Eocene) of Texas and Their Relation to Occurrence of Oil and Gas[1]

W. L. FISHER[2] and J. H. McGOWEN[2]
Austin, Texas 78712

**Abstract** Regional investigation of the lower part of the Wilcox Group in Texas in outcrop and subsurface indicates seven principal depositional systems: (1) Mt. Pleasant fluvial system developed updip and in outcrop north of the Colorado River; (2) Rockdale delta system, present primarily in the subsurface, chiefly between the Guadalupe and Sabine Rivers; (3) Pendleton lagoon-bay system in outcrop and subsurface largely on the southern flank of the Sabine uplift; (4) San Marcos strandplain-bay system in outcrop and subsurface mainly on the San Marcos arch; (5) Cotulla barrier-bar system in the subsurface of South Texas; (6) Indio bay-lagoon system developed updip and in outcrop in South Texas; and (7) South Texas shelf system, an extensive system entirely within the subsurface. The Rockdale delta system consisting of large lobate wedges of mudstone, sandstone, and carbonaceous deposits, is the thickest and most extensive of the lower Wilcox depositional systems. It grades updip to the thinner terrigenous facies of the Mt. Pleasant fluvial system. Deposits of the Rockdale delta system were the source of sediments redistributed by marine processes and deposited in laterally adjacent marine systems. Delineation of depositional systems and, more specifically, delineation of component facies of the several systems permit establishment of several regional oil and gas trends which show the relation of producing fields and distribution of potentially productive trends.

## Introduction

The Wilcox Group (early Eocene) is a thick sequence of predominantly terrigenous clastic sedimentary rocks which is volumetrically a significant part of the large terrigenous fill of the western Gulf Coast province. It is an economically important group of rocks, providing oil, gas, and freshwater reservoirs, as well as deposits of lignite, ceramic clay, and industrial sand.

The study area is approximately 40,000 sq mi in the Texas Gulf coastal plain, extending from the Sabine River on the east to the Rio Grande on the south in a 100- to 300-mi-wide belt including outcrop and subsurface. Subsurface control consisted of approximately 2,500 electric logs, supplemented by lithologic control based on outcrop descriptions, and cuttings and cores of strategically located wells. Fifty stratigraphic dip sections and 15 strike sections were prepared at spacings of 8 to 10 mi to determine three-dimensional geometry and relations of principal lithologic units. Emphasis in outcrop study was on the mapping of principal rock units, interpretation of primary sedimentary structures, and determination of fossil composition.

This paper is part of a continuing, larger regional study of the Wilcox Group in Texas which has been under way at the Bureau of Economic Geology for the past three years. Herein, consideration is restricted to the lower part of the Wilcox Group, including the sequence underlain by marine mudstone and claystone of the Midway Group and normally overlain by marine or lagoonal mudstone of the middle part of the Wilcox Group. The Carrizo Formation and the upper part of the Wilcox Group are not considered.

## Terminology and Depositional Systems

The stratigraphic terminology used by the writers is informal. Reference is made to defined, formal stratigraphic units (chiefly in outcrop) where applicable. The terminology used derives from the concept of depositional systems or complexes, which has been employed recently, for example, in the study of modern

[1] Manuscript received and accepted, February 2, 1968. Parts of this paper were presented by W. L. Fisher to the Gulf Coast Association of Geological Societies, San Antonio, Texas, 1967. Present paper is republished with minor modifications from paper by Fisher and McGowen in Gulf Coast Assoc. Geol. Socs. Trans., v. 17, p. 105–125, 1967. This paper was nominated by the Gulf Coast Association of Geological Societies for republication in the AAPG *Bulletin;* publication authorized by the Director, Bureau of Economic Geology, The University of Texas at Austin, and by the Gulf Coast Association of Geological Societies.

[2] Bureau of Economic Geology, The University of Texas at Austin.

The writers thank M. J. Casey, L. F. Brown, Jr., and P. T. Flawn, Bureau of Economic Geology, and Alan J. Scott, Department of Geology, The University of Texas at Austin, for reading the original manuscript and offering many helpful suggestions. Gerald H. Baum, Texas Water Development Board, made available the extensive well-log library of that agency; numerous Gulf Coast operators provided records of specific wells. Peter U. Rodda and J. Stewart Nagle, Bureau of Economic Geology, The University of Texas at Austin, have been helpful in many ways during course of the project. Illustrations were prepared under direction of J. W. Macon, Bureau of Economic Geology, The University of Texas at Austin.

sediments and depositional environments of the South Texas coast (Hayes and Scott, 1964) and in reference to the Frio barrier bar system of South Texas (Boyd and Dyer, 1964). In studies of modern sediments emphasis generally is placed not only on description of specific environments and two-dimensional distribution of sediments and environments, but also on analysis and integration of facies; fabrication of facies tracts and three-dimensional depositional models of known depositional complexes or systems are used. These approaches are significant in the interpretation of ancient sediments, especially in regional analysis. The writers place more emphasis on lithologic and fossil composition, sedimentary structures, and distribution, geometry, and relations of component facies than on regional correlation and persistence of locally established formal stratigraphic units.

The principal unit here employed is the depositional system, recognized by specific criteria and designated by a genetic term, *e.g.*, delta system, fluvial system, barrier-bar system, *etc.* Components of a depositional system are referred to simply as facies, *e.g.*, the delta-front facies or distributary-channel facies of a delta system. Smaller scale components are designated by such terms as units, beds, or deposits. Basically, the concept of depositional systems in ancient sediments, fabricated principally from integration of facies, involves recognition of large-scale natural genetic elements, comparable to modern depositional systems readily apparent from physiographic characteristics.

## Principal Depositional Systems

Seven principal depositional systems are recognized in the lower part of the Wilcox Group of Texas (Fig. 1). The dominant depositional element is the large Rockdale delta system, which comprises large lobate wedges of mudstone, sandstone, and carbonaceous deposits, chiefly lignite (Fig. 1). Maximum lower Wilcox sedimentation, with sequences as much as 5,000 ft thick, took place during deposition of this system. It is present principally in the subsurface and grades up the depositional slope and up the structural dip to thinner terrigenous clastic facies of the Mt. Pleasant fluvial system, present in outcrop north of the Colorado River. A large and probably main part of the original fluvial system was deposited updip of the present outcrop, and only a part of the system is preserved in the East Texas basin.

Southwestward transport of sand from the delta system, principally by longshore currents during periods of delta destruction, and the presence of local embayments marginal to the delta system resulted in distinct depositional systems laterally associated with the delta system. These systems include, on the east, the Pendleton lagoon-bay system developed in embayments marginal to the delta system; on the southwest, the San Marcos strandplain-bay system developed proximal to the delta system; and the Cotulla barrier bar and complementary Indio lagoon-bay systems developed more distally from the main area of deltation. Contemporaneous sedimentation in systems marginal and lateral to the delta system was relatively slow, and the sequences generally are less than one-fourth the thickness of the delta system. Gulfward from the Cotulla barrier bar system and partly lateral to the Rockdale delta system is the South Texas shelf system, marked chiefly by mudstone and relatively minor sandstone.

### Mt. Pleasant Fluvial System

The Mt. Pleasant fluvial system includes most of the lower part of the Wilcox Group in outcrop north of the Colorado River and in the northern half of the Sabine uplift (Figs. 1, 2). It extends down structural dip to a line running approximately from central Bastrop County through central Lee, northern Burleson, southern Robertson, central Leon, northern Houston, southern Cherokee, and central Nacogdoches Counties. South and downdip from this line the system grades into the Rockdale delta system (Fig. 3); laterally it grades into embayed marine facies developed marginally to the fluvial and delta systems. Three fairly distinct facies are preserved, recognition of which is based on (1) sandstone-body geometry, (2) sandstone-isolith pattern, (3) internal sedimentary structures, (4) vertical sequence in structures and textures, (5) grain size, and (6) relative amounts of channel and overbank deposits (Figs. 1–3). These facies reflect different fluvial environments.

#### *Tributary-Channel Facies*

In northernmost Wilcox outcrops, from central Van Zandt County northeast to Texarkana, and in outcrop in the northern part of the Sabine uplift, sandstone deposits are present in definite, separated areas (*e.g.*, Mt. Pleasant, Reilly Springs, Simms deposits); claystone and mudstone beds crop out in intervening areas. Isoliths of sandstone units outline elongate bodies with an overall dendritic tributary pat-

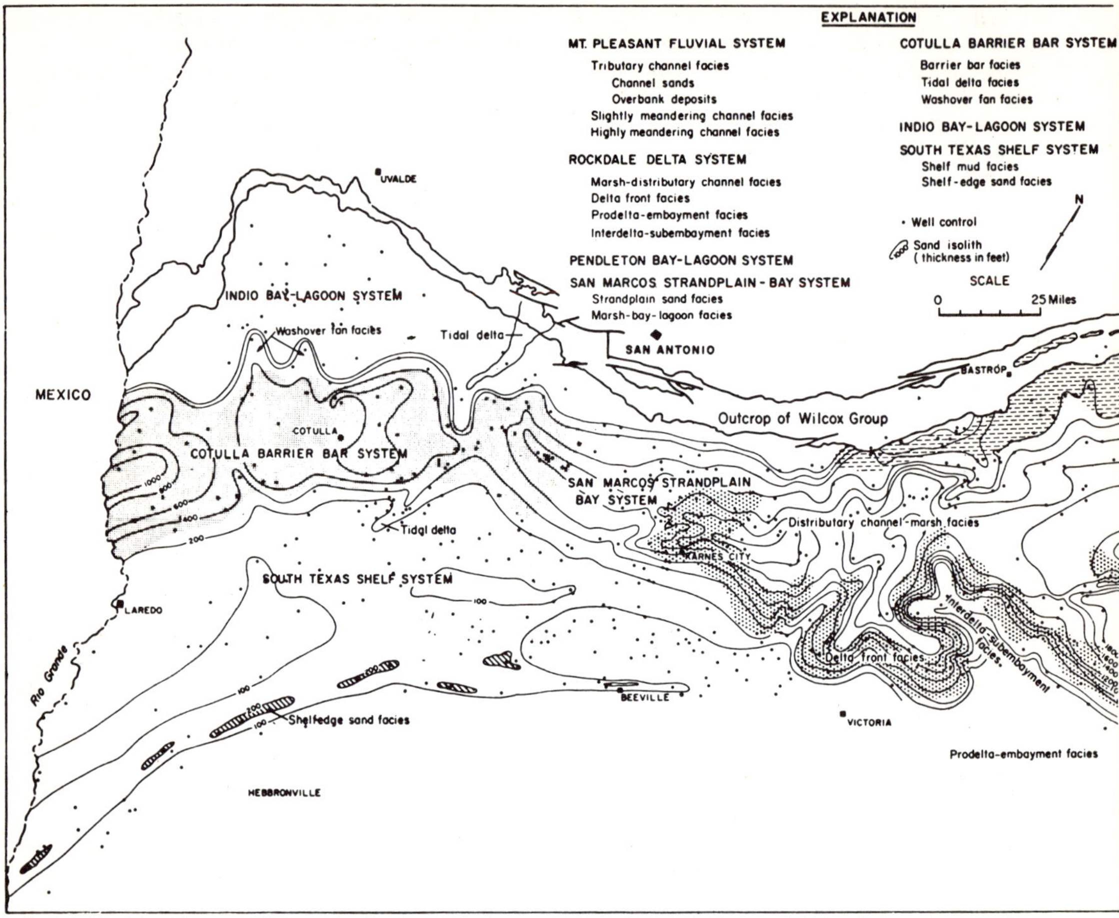


FIG. 1.—Principal depositional systems,

tern (Fig. 2). The fluvial facies defined chiefly by this pattern extends from the outcrop southward in the subsurface to a line from southwestern Smith County to west-central Rusk County. The thickness of the tributary-channel facies ranges from about 300 to 800 ft and commonly increases downdip and toward the axis of the East Texas basin; the facies consists of about 40 percent sandstone or channel deposits and about 60 percent mudstone or overbank deposits. The overbank (or nonchannel) mudstone beds are mostly laminated, light to moderate gray, and contain minor amounts of carbonaceous matter. Individual sandstone units range from 5 to about 80 ft in thickness and are channel deposits with flat tops and convex-downward bases. Bases of channels are characterized by channel lag of granules to fine pebbles, and a few angular clay pebbles. Sandstone in lower parts of the channel deposits contains trough or festoon cross-beds and in upper parts contains wedge sets of cross-beds. Upper parts of channel sandstone are horizontally bedded. Sandstone in this facies is mostly fine grained to locally medium grained; mean sand grain size averages about 0.14 mm, and range in mean is about 0.11 to 0.25 mm; channel sandstone typically becomes finer upward. Channels are cut into underlying and laterally adjacent overbank mudstone layers. Sinuous, elongate, narrow lignite deposits commonly are found laterally adjacent to channel sandstone deposits; they apparently accumulated as peat bogs in abandoned stream courses. Claystone beds within and adjacent to sandstone channel deposits are chiefly kaolinitic; feldspar is mostly weathered. Claystone of overbank deposits, by contrast, is largely nonkaolinitic—mostly illitic, chloritic, and montmorillonitic; feldspar of overbank mudstone deposits gener-

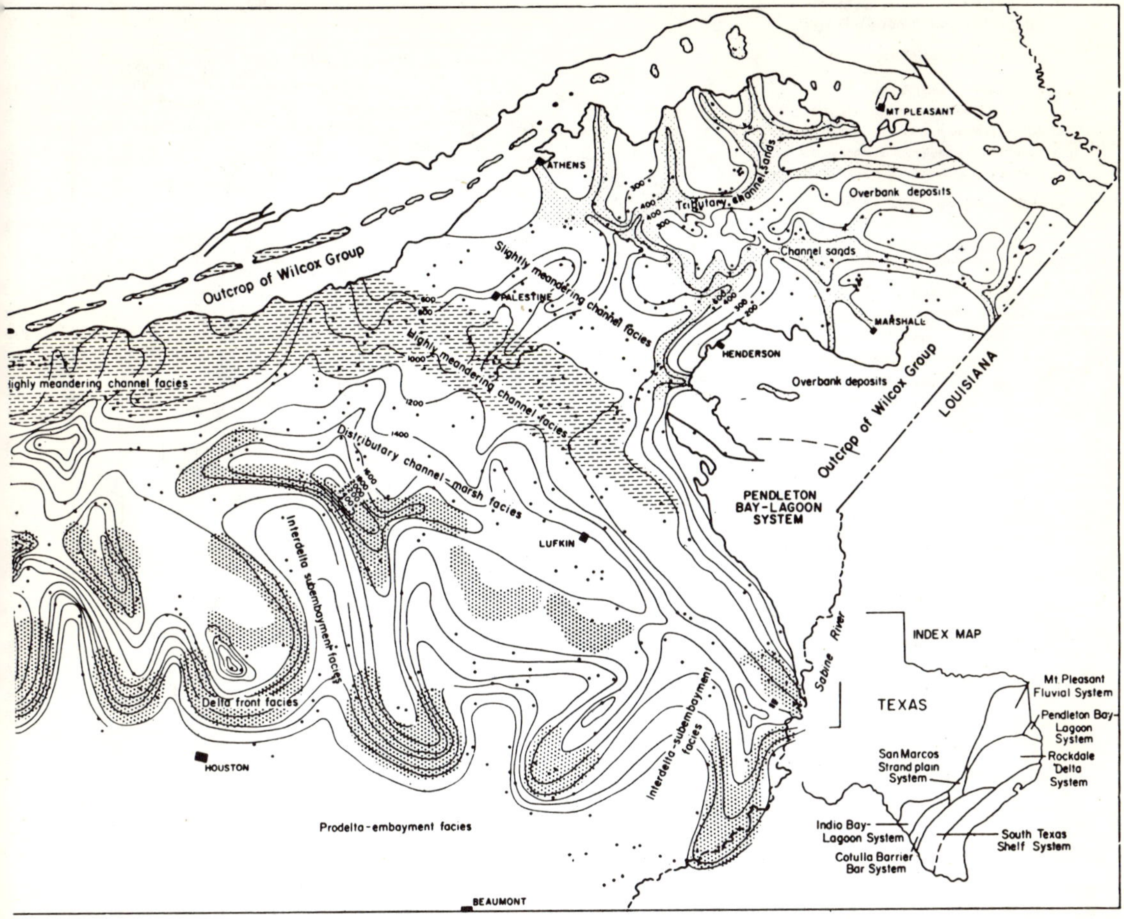


lower part of Wilcox Group (Eocene), Texas.

ally is unweathered. Ceramic clay plants and numerous abandoned lignite mines of East Texas are adjacent to the local Wilcox sandstone deposits because of the close association of kaolinitic clay and lignite to the channel deposits.

*Slightly Meandering Channel Facies*

Down the depositional slope in the fluvial system a transitional facies is characterized by uniform to slightly dendroid sandstone isolith patterns (Figs. 1, 2), which are less dendritic than those in the tributary-channel facies. This change in pattern probably represents a slightly greater meandering of the original stream system. Channel sandstone deposits are more numerous within this facies than in the tributary-channel facies; they make up about 60 percent of the total facies, and overbank deposits compose 40 percent. The facies ranges in thickness from about 500 to 800 ft; it crops out from central Freestone County, through western Henderson County, into southern Van Zandt County and extends downdip in the subsurface to a line from central Freestone County through central Anderson and central Cherokee Counties, and into northwestern Nacogdoches County. In addition to containing a higher percentage of channel deposits, individual sandstone bodies in this facies are slightly more extensive, and mean grain size is somewhat larger than that of the tributary-channel facies. Primary sedimentary structures of sandstone and relation of channels to clay-mineral distribution and lignite deposits are similar for the two facies.

*Highly Meandering Channel Facies*

A distinct facies is developed in the Mt. Pleasant fluvial system just downslope deposi-

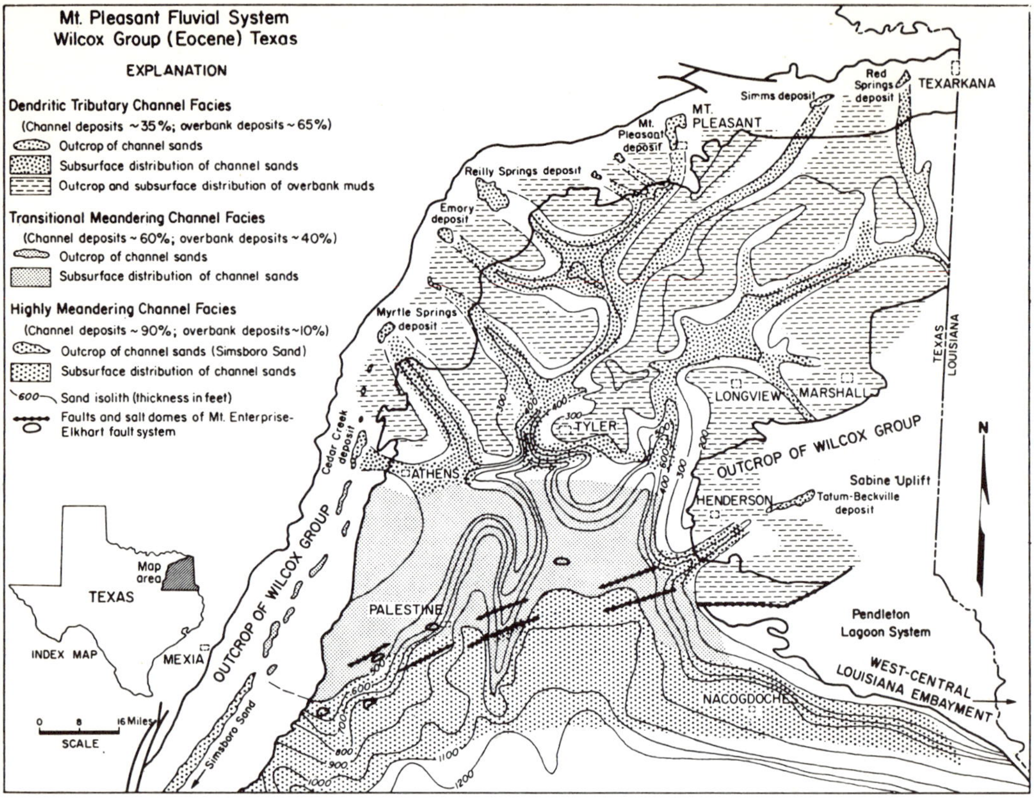


FIG. 2.—Outcrop and subsurface distribution, Mt. Pleasant fluvial system, Wilcox Group, Texas. For well control, see Figure 1.

tionally and downdip structurally from the Mt. Enterprise-Elkhart fault zone and associated salt domes on the southwest (Figs. 1, 2). Within this facies sandstone predominates and overbank mudstone composes less than 10 percent. Sandstone isolith lines generally conform and sandstone in the facies (*e.g.*, Simsboro sandstone outcrop) forms areally persistent units. For such units as these, Pettijohn *et al.* (1965) suggested the term multilateral. The bodies are made up of multiple channels which trend at various angles to the strike of the sandstone isolith lines. Aggregate thickness of sandstone in the facies ranges from 600 to 1,100 ft. Individual sandstone units, comprising numerous superposed and multilateral channel deposits, are as much as 30 mi wide. Areas along the strike between the more extensive sandstone units are made up of greater amounts of overbank mud deposits. Sand grains of the facies have a mean average size about 0.22 mm and are much coarser than sand grains in upslope fluvial facies. Sandstone channel deposits that compose the multilateral sandstone bodies extend down and laterally into underlying and adjacent channel sandstone. Channels are broadly convex downward and are characterized by a basal, locally conglomeratic zone with festoon or trough cross-bedding which represents channel-lag deposition. This zone is overlain by a thin unit with tabular cross-beds, in turn overlain by a zone of wedge-set cross-beds (epsilon bedding of Allen, 1963). The tabular and wedge cross-bedded sandstones probably represent lateral deposition as the toe and lower parts of point bars. The horizontally bedded and rippled zones, characteristic of the upper parts of many point-bar deposits, are rarely preserved in this facies, as the channel-lag and lower point-bar deposits are succeeded disconformably by channel deposits of a similar sequence. Within the preserved channel sequence there is only slight, if any, upward fining. The relatively small amount of clay in this facies is very kaolinitic. Lignite deposits, typically associated

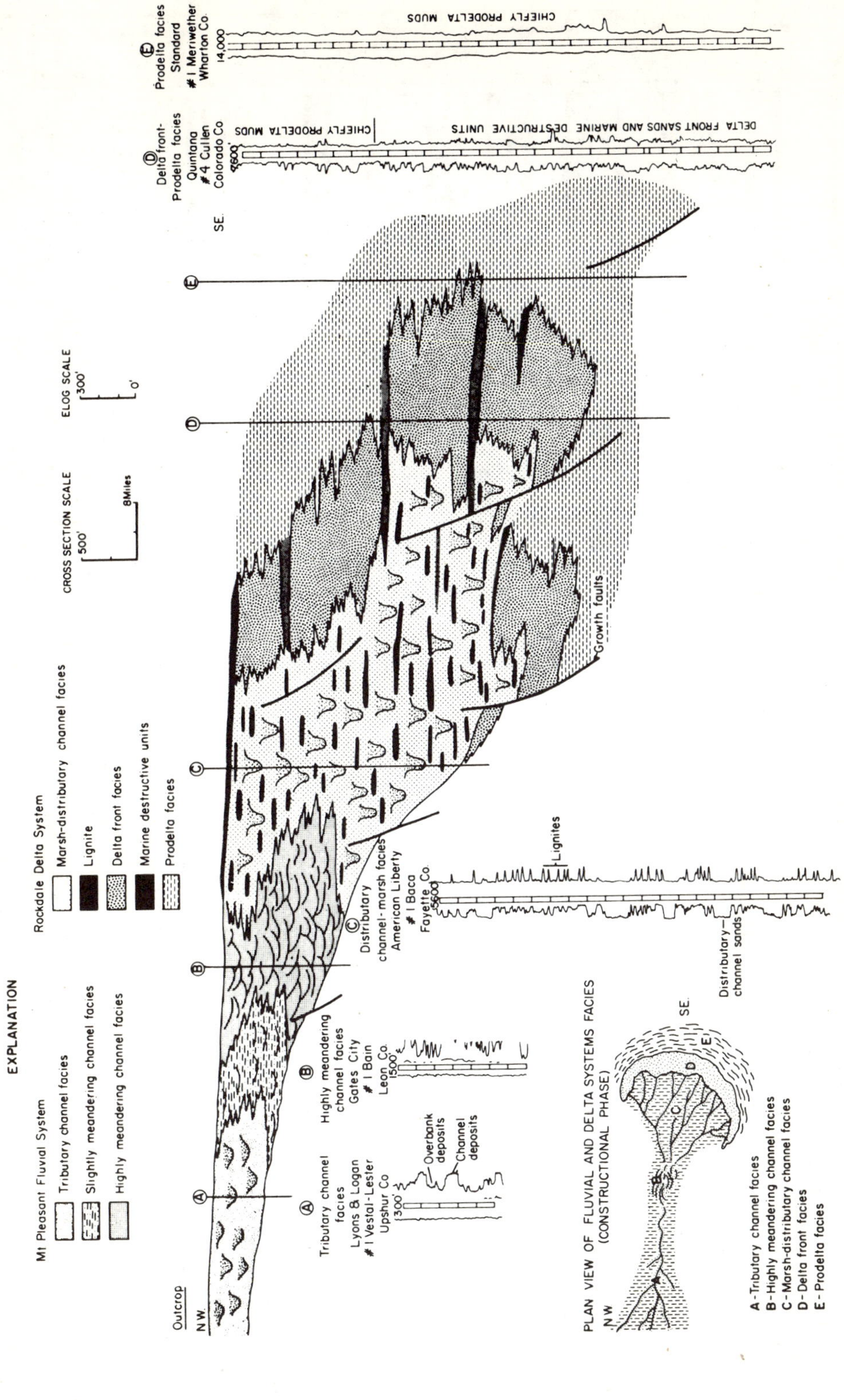


FIG. 3.—Diagrammatic stratigraphic dip section of Mt. Pleasant fluvial system and Rockdale delta system shows relation and character of principal component facies.

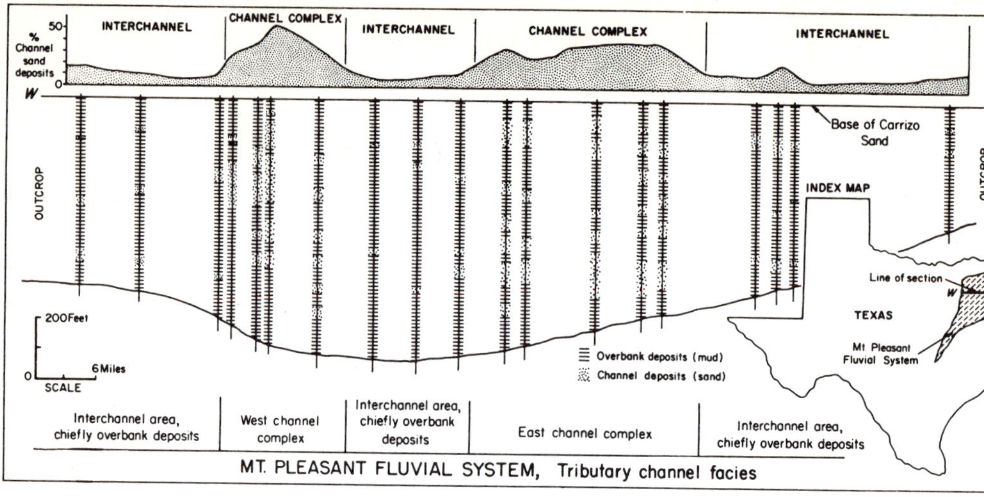


FIG. 4.—Diagrammatic stratigraphic strike section of Mt. Pleasant fluvial system, Wilcox Group, Texas.

with channel deposits in the other fluvial facies, are not preserved in the highly meandering channel facies. Lateral continuity and permeability of the sandstone make this facies a significant aquifer.

The highly meandering channel facies is the farthest downslope facies of the preserved part of the Mt. Pleasant fluvial system, grading downslope depositionally or southeastward to the distributary channel-marsh-swamp facies of the Rockdale delta system. It thus heads the delta system, extending in outcrop from central Bastrop County northward to southern Freestone County, and in subsurface from central Bastrop County northeastward to western Nacogdoches County (Fig. 1). Undoubtedly, it originally extended a considerable distance northwest of the present outcrop, probably coincident with the general courses of the main Wilcox stream systems. In central Bastrop and western Nacogdoches Counties the fluvial facies changes abruptly to the marginally embayed marine mudstone of the San Marcos strandplain-bay and Pendleton lagoon-bay systems, respectively. Within narrow belts marking these transitions, fluvial sandstones of the facies, which were modified by marine processes, have very well-sorted and fine grains and were locally redeposited as beach, barrier-bar, and strandplain deposits.

*Multistory and Multilateral Sandstone Bodies*

A characteristic feature of sandstone bodies in the fluvial facies containing relatively high proportions of overbank muds (tributary-channel and slightly meandering channel facies) is vertical persistence or stacking (Fig. 4). Successive sandstone bodies are superposed roughly, and complementary interchannel areas are void of significant channel sand deposits. This feature, termed multistorying by Feofilova (1954), has been noted in several facies containing channel sand deposits (*e.g.,* Mueller and Wanless, 1957; Friedman, 1960; Potter, 1963; Brown *et al.,* 1967). There is variation in degree of stacking, but vertical persistence is sufficient to show a fluvial drainage pattern of composite sequences similar to patterns of individual sandstone bodies. Where point-bar sequences are developed as parts of these facies, the sequences are characteristically complete. In contrast, sandstone bodies in fluvial facies containing low proportions of overbank muds (highly meandering facies) are markedly persistent laterally. Lateral and vertical repetition of channel sandstone units results in multilateral sandstone bodies as much as 30 mi wide. The channel units in these multilateral bodies commonly show an incomplete point-bar sequence; the coarse-grained channel-lag and lower point-bar deposits are bounded erosively vertically and laterally by similar channel sequences. Controlling factors in the formation of these distinct fluvial sandstone units apparently are differences in stream regime, differences in amounts of mud and sand in the system affording different degrees of compaction, and regional structure and tectonics.

## Rockdale Delta System

The Rockdale delta system (as in the nomenclatural designation of Echols and Malkin,

1948) is the dominant depositional system of the lower Wilcox Group in Texas, comprising areally about 60 percent and volumetrically about 80 percent of the known lower Wilcox (Fig. 1). It was the ultimate depositional site of most sediments transported through the fluvial system and the principal source of sediments redistributed and deposited in associated barrier-bar, strandplain, lagoon-bay, and shelf systems.

The Rockdale delta system is downdip structurally from the fluvial system. The southern limit is marked by a line extending from north-central Karnes County north to central Bastrop County, and the eastern limit is marked by a line from central Nacogdoches County through northern San Augustine and Sabine Counties to the Sabine River. The downdip limit is beyond present well control (Fig. 1). Maximum thicknesses of 3,500–5,000 ft coincide with the central axis of the system which is chiefly along drainage of the modern Brazos River and roughly coincident with structural axis of the Houston embayment. Laterally and along the present structural strike the delta system generally, but not uniformly, thins to about 1,500 ft in the southwestern part of the system and to about 1,000 ft in the eastern part.

Criteria for recognition of this part of the lower Wilcox Group as a delta system include (1) composition and boundary relations of component facies, (2) external geometry of skeletal sand facies, (3) position in relation to laterally associated depositional systems, and (4) contrast in rates of deposition, with thick, rapidly deposited terrigenous clastic sequences (constructional units) alternating with thin, slowly deposited, relatively persistent marine and marsh deposits (destructional units).

### *Principal Facies*

Four regional facies define the Rockdale delta system. Downslope depositionally and downdip structurally from the highly meandering facies of the Mt. Pleasant fluvial system, the lower Wilcox consists of an extensive facies of sandstone units alternating with mudstone, siltstone, and numerous beds of lignite (Fig. 3). Farther downdip this facies changes rather abruptly to thick sandstone, which in turn grades down dip and along structural strike to thick, uniform mudstone sequences. The alternating sandstone and mudstone-lignite sequences are interpreted to be distributary-channel and interdistributary deposits accumulated on a delta plain; the sandstone facies represents delta-front deposits laid down chiefly as distributary-mouth bars; the downdip thick mudstone sequence represents prodelta deposition. Mudstone sequences along strike of the delta-front sandstone facies are interpreted to be interdelta facies.

*Distributary channel-marsh-swamp facies.*—The facies consisting of alternating sandstone and mudstone-siltstone-lignite units is an areally extensive facies of the Rockdale delta system (Fig. 3). It contrasts markedly with the immediately upslope fluvial facies, chiefly in sandstone-body geometry and relative proportion of channel and nonchannel deposits. The fluvial facies contains cross-bedding structures and textural features indicative of channel-lag and lateral point-bar deposition in highly meandering streams. The result is relatively persistent, uniform, multilateral sandstone units with very few nonchannel deposits preserved. The sandstone changes downdip structurally and downslope depositionally to isolated, elongate, narrow sandstone bodies interspersed in nonchannel mudstone and associated lignite. In these respects it resembles the tributary-channel and slightly meandering facies of the Mt. Pleasant fluvial system. These facies differ significantly, however, in that the channels in the delta system are thicker, are symmetrical in cross section, and show a distributary rather than a tributary pattern in plan. Symmetry in cross section, thickness, and compactional features of underlying mudstone suggest accumulation of the sandstone as bedload deposits in relatively straight, stabilized streams, in which accumulation by downward accretion was facilitated by compactional subsidence of the channel sandstone in a framework of interchannel mudstone. Such features characterize certain modern delta-plain distributary channels and contrast with the lateral deposition in point bars of meandering, very sinuous streams. Mudstone and siltstone, composing about 60 percent of this facies, accumulated as subaerial levee, crevasse-splay, lacustrine, and flood-basin interdistributary deposits.

Two kinds of lignite are present within the distributary channel-marsh facies, both distinct from the elongate lignite deposits of the fluvial system. One lignite, which is tabular in shape but only local in extent, probably represents peat accumulation in small interdistributary depressions similar in origin to interdistributary peat in the Mississippi delta system, as discussed by Fisk (1960) and Frazier (1967). A second type is also tabular in shape but much more extensive; this type developed as regional, landward facies of marine delta destruction. Other distinctions between fluvial and delta-plain lignites are made on the basis of grade and composition. Fluvial lignites are more variable in composition than delta lignites and contain high percentages of woody materials consistent with a swamp origin. Delta lignites are slightly higher in grade and have higher specific gravity, and the relatively low percentages of woody material are more characteristic of marsh growth. Regional variations in Wilcox lignites (Fisher, 1963) reflect differences in depositional origin.

*Delta-front facies.*—Downdip the distributary channel-marsh-swamp facies changes to a thick sandstone facies in which lignite is absent and mudstone content is low (Fig. 3). The sandstones mostly have fine grains, generally much better sorted than those of fluvial and distributary channel sandstones. Commonly, the sandstone contains thin, multidirectional ripple cross-laminae and other current features. Plant and detrital lignite fragments are present locally. Individual sandstone units may be 200 ft or more thick; in plan, external geometry is lobate or lunate. The units are interpreted to be distributary-mouth bars representing delta-front deposition. The high percentage of sandstone in this facies resulted from deposition of bedload of delta-plain distributary channels debouching into relatively quiet marine water; the muds were carried in suspension farther seaward and accumulated as prodelta deposits. Relative cleanness, good sorting, and multidirectional current features indicate winnowing and reworking of sands in this environment by both channel and open-water currents under shoal conditions in the delta-front environment. Redistribution of these sands and lateral coalescence of the constructive delta-front deposits formed a fairly continuous delta-front sandstone facies. The relative amount of redistribution and lateral coalescence within the delta-front facies in the different Rockdale deltas resulted in delta lobes ranging from elongate to rounded.

*Prodelta facies.*—The farthest downdip sections of lower Wilcox penetrated by wells generally consist of relatively uniform, thick sequences of dark mudstone (Fig. 3). They contain moderately large amounts of very finely comminuted and disseminated organic matter, laminations marked commonly by slight textural and color variations, and a marine fauna exceedingly low in number and species diversity. This mudstone facies interfingers with, and grades updip to, the delta-front sandstone facies and represents deposition from suspension seaward from the delta front and river mouth as prodelta facies. Thickness and downdip extent of the facies are difficult to determine, primarily because few wells have penetrated significant thickness of the lower Wilcox in the downdip areas. However, the few deep wells indicate that the facies is probably the thickest in the delta system; approximately 4,000–5,000 ft has been recorded in some deep wells.

*Interdelta, subembayment facies.*—Sequences in areas between principal delta lobes are similar in some respects to the prodelta facies (Fig. 1). They consist principally of mudstone, much of which is laminated, but are thinner than the prodelta facies and in addition contain several thin, burrowed mudstone units or thin units of fossiliferous siltstone and very fine-grained sandstone with small amounts of glauconite. These thin units accumulated more slowly than the laminated mudstone, in areas not directly affected by deltation. The interdelta facies grades updip structurally to marsh facies of the delta plain and laterally along strike to delta-front sandstone facies.

### *Delta-Destruction Units*

Delineation of a thick terrigenous clastic sequence such as the Rockdale delta system into component and individual deltas is a time-consuming task. Basically, it involves recognition and correlation of thin marine units formed in the process of marine destruction of a constructive delta mass. During periods of great sediment influx, the tract of delta facies (distributary channel-marsh-swamp, delta front, prodelta) is established and, during progradation and basinal and compactional subsidence, a thick terrigenous sequence accumulates. With cessation of sediment influx due either to diminishment of the source area or, more commonly, to upstream avulsion resulting in abandonment of the previous drainage system and establishment of a new network, compaction of the water-saturated sediments of the constructed delta mass exceeds the rate of deposition. The delta lobe, or at least marginal parts of it, is subjected to marine encroachment involving reworking of the constructive delta surface and underlying delta sediments. Rate of sedimentation during destruction is very low and as a consequence units are relatively thin. In the Rockdale delta system, delta-destruction units consist of very well-sorted, fine-grained, slightly glauconitic, commonly bioturbated sandstone, shell detritus, and mudstone either burrowed or containing a marine fauna. These units were deposited chiefly on the lower parts of the abandoned delta plain. Landward destruction of the plain involved extensive marsh growth, coincident with compactional subsidence, which resulted in areally extensive lignite deposits. This lignite contrasts with the thinner, local lignite bodies that formed as interdistributary peat. The delta-destruction units apparently represent deposition in environments comparable with delta-destruction environments in the modern Mississippi system. The marine sandstone originated under conditions similar to those of the Chandeleur Islands, and the burrowed or fossiliferous mudstone originated as open-bay deposits similar to those of Breton Sound.

Recognition of delta-destruction units and their facies relations in delta systems should be followed by studies of their areal extent by careful correlation of closely spaced wells in a large area. By this means the delta system can be subdivided and smaller delta units mapped. Determination of sand content of such subdivisions permits isolith mapping of skeletal elements of the delta (distributary-channel and delta-front facies) and, with mapping of component facies, allows delineation of specific delta lobes. Most delta-destructive units are restricted areally to a specific, underlying, constructive delta lobe. Because delta construction is a relatively rapid process and delta destruction a slow process, destruction of one delta lobe may persist during the construction, subsidence, and subsequent initiation of destruction of an adjacent lobe; thus destructional units of two or more lobes may be partly contemporaneous. Recognition of widespread destructional units permits larger scale, though only approximate, subdivision of delta systems.

### *Individual Deltas in Rockdale Delta System*

The writers have mapped and delineated 16 principal deltas within the lower Wilcox Rockdale delta system (Fig. 5). These have the component facies and facies pattern heretofore described. Striking, and significant in another context, are the vertical stacking of individual deltas and the persistence of interdelta subembayments within the delta system. Further, the Rockdale deltas show a marked coincidence with modern coastal-plain streams (Fig. 5).

Maps of several delta-destruction units have made possible the delineation of three main phases of deltation in the Rockdale system, in addition to the outlining of principal individual deltas. The initial phase of deltation resulted in three relatively small deltas situated roughly along drainage of the modern Colorado, Brazos, and Trinity Rivers (Fig. 5). They range in area from 200 to 350 sq mi, and thicknesses are on the order of 300 to 500 ft. A second or intermediate period of delta building resulted in the largest, thickest, and most prograded deltas of the system. The larger deltas of this phase, as much as 2,000 sq mi in area, were constructed along the axes of the modern Guadalupe, Colorado, and Brazos Rivers; the smaller deltas, 500–700 sq mi in area, developed along the axes of the modern Sabine and Neches Rivers. The Brazos and Colorado deltas (Fig. 5) are the thickest, with maximum thickness of 1,500 ft, exclusive of the prodelta facies. A terminal phase involved delta construction in all areas of intermediate-phase deltation, as well as construction of a large delta situated primarily in Angelina, Houston, and Trinity Counties (Angelina delta). These terminal deltas are comparable in size and thickness with the underlying, intermediate-phase deltas, but do not extend as far downdip structurally.

Although individual deltas within the Rockdale system are varied in size and sediment volume, a more significant variation is in the shape of the delta lobes (Fig. 6). One type, shown by the intermediate-phase delta constructed in the area of the Trinity River, the four intermediate and terminal deltas constructed in the area of the Neches and Sabine Rivers, and, to some extent, the three initial-phase deltas, has the elongate, narrow shape of the modern birdfoot delta and certain initial deltas of older complexes delineated by Frazier (1967) in the Mississippi delta system. A second type, shown by the other deltas of the Rockdale system, has a generally lobate or rounded shape similar to that of premodern, abandoned deltas of the Mississippi system. If a delta line is drawn in the manner in which bayline maps commonly are made (Lowman, 1949), the elongate, narrow birdfoot deltas of the Trinity, Neches, and Sabine Rivers extend considerably beyond the line; thus, progradation into deeper waters, as in the modern deep-water birdfoot delta of the Mississippi system (Fig. 6), is suggested. These deltas are associated with a sequence relatively high in mudstones, a feature that strengthens the analogy with the sand-poor, deep-water birdfoot delta of the Mississippi. The initial-phase deltas similarly represent progradation over un-

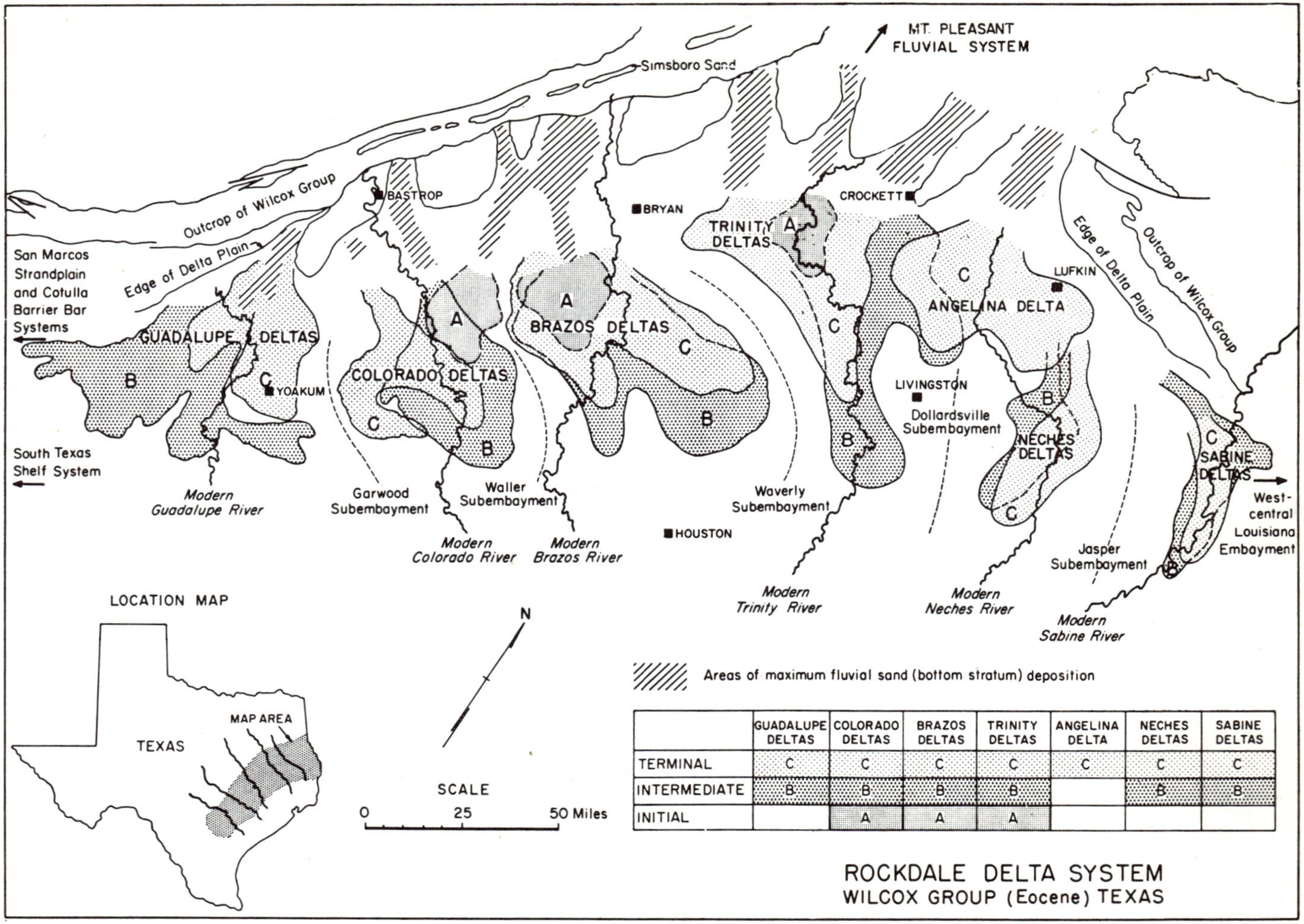


FIG. 5.—Distribution of principal deltas, Rockdale delta system, Wilcox Group, Texas; lobal outline drawn at distal margin of delta front facies. For well control, See Figure 1.

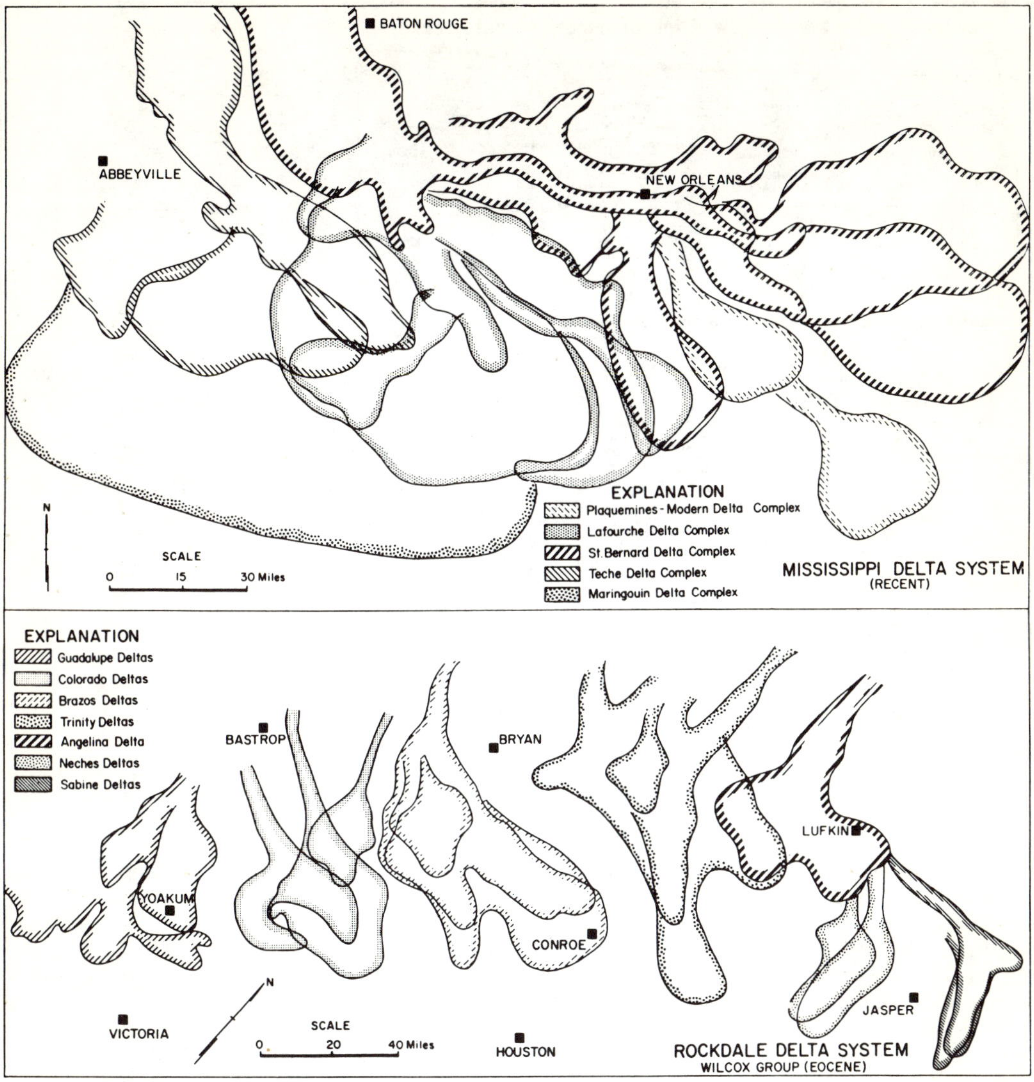


FIG. 6.—Comparison of size, distribution, and arrangement of principal delta lobes of the Holocene Mississippi delta system, southeastern Louisiana (modified from Frazier, 1967), and Rockdale delta system, Wilcox Group (Eocene, Texas).

derlying thick mudstone. By contrast, the lobate or rounded deltas of the Rockdale system consist of relatively large proportions of sandstone, a characteristic that suggests construction as shoal-water deltas analogous to the sand-rich, lobate, shoal-water deltas of the modern Rhone (Kruit, 1955) and premodern Mississippi (Fisk, 1955, 1961; Frazier, 1967).

Vertical stacking or approximate superposition of individual deltas within the system, persistence of interdelta subembayments, and coincidence of Rockdale deltas and modern drainage have been mentioned. The controlling factor or factors, like those of the multistoried channel sandstone bodies in upslope fluvial facies of the Mt. Pleasant system, are not fully understood. Such features apparently are common in fluvial systems containing relatively high percentages of mud and may be common in delta systems, although few ancient delta systems have been mapped and studied. Maps of delta systems in the Pennsylvanian of the Eastern Interior basin (Wanless *et al.*, 1963) and the modern and premodern deltas of the

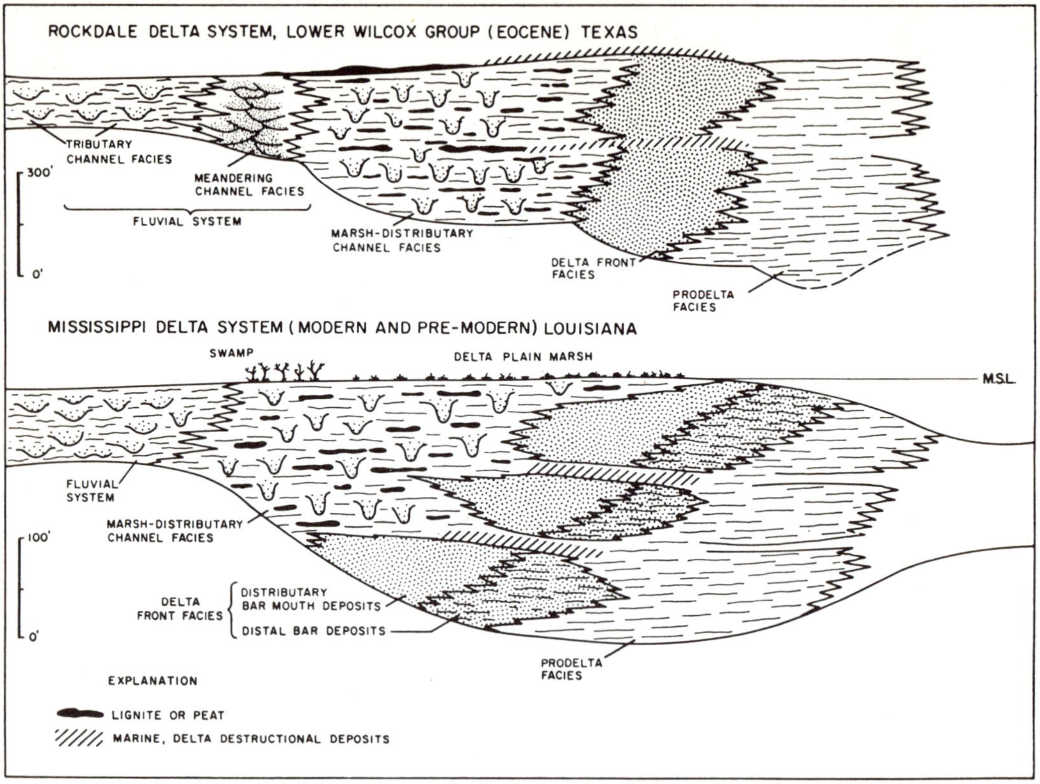


FIG. 7.—Comparison of component facies of Mississippi delta system (modified from partly hypothetical cross section of Coleman and Gagliano, 1964) and Rockdale delta system, Wilcox Group, Texas.

Mississippi system (Kolb and van Lopik, 1966; Frazier, 1967; Fig. 6) indicate some vertical stacking, though perhaps not as pronounced as that in the Rockdale delta system. Significant in this regard is not only the approximate coincidence of Rockdale deltas and modern coastal-plain drainage, but also the coincidence of other Eocene delta systems (*e.g.*, Yegua and Fayette) and Pleistocene delta systems along these axes. Certainly, compactional subsidence is a possible factor, although the more important control probably is tectonic. Structural control is indicated by the presence of growth-fault zones coincident with the updip limits of the thicker facies of the Rockdale delta system (Fig. 3).

### *Comparison with Mississippi Delta System*

The Eocene Rockdale and the Holocene Mississippi delta systems, both built into the Gulf Coast basin, are similar in shape, occurrence, position of shoal-water lobate deltas and deep-water birdfoot deltas, and vertical stacking of individual deltas. The two systems are comparable in area (Fig. 6). A most significant similarity of the two systems is the distribution of component facies (Fig. 7), in which the high-sandstone delta-front facies is bounded up-delta or landward by a relatively high-mudstone and organic facies (distributary-channel and marsh-swamp facies) and off-delta or seaward by a high-mudstone prodelta facies. These detrital lenses are bounded vertically by marine and landward-equivalent destructional units. The significance of bounding facies in delta sequences has been emphasized by Coleman and Gagliano (1964); delineation of component facies, together with determination of external geometry of the skeletal elements (distributary-channel and delta-front sandstone bodies) and delineation of delta-destruction units, are judged to be the best regional criteria for recognition of ancient deltas or delta systems.

## Pendleton Lagoon-Bay System

Marginal to the Rockdale delta system and along the southern part of the Sabine uplift (chiefly Shelby and northern parts of San Augustine and Sabine Counties), the Wilcox Group is predominantly a mudstone facies. This facies includes parts or all of several formally defined stratigraphic units (Pendleton, Marthaville, Hall Summit, Lime Hill, Converse, Cow Bayou, Dolet Hills, and Naborton Formations). It is present in outcrop and in subsurface. The facies, designated informally as the Pendleton lagoon-bay system (Fig. 1), is approximately 1,000 ft thick in Texas. It is comparable in thickness with fluvial facies along the strike, but thicker than updip fluvial facies and thinner than downdip delta facies.

The Pendleton system consists chiefly of laminated to locally bioturbated mudstone, and local lenses and beds of very fine- to fine-grained, sparingly glauconitic, massive to cross-bedded sandstone and fine-grained, broken, massive to cross-bedded, lignitic sandstone. Certain of the mudstones are gypsiferous; clay-ironstone concretions are common. Lignite is locally common, but relatively impure and in thin, discontinuous beds. The system contains a few fairly discrete units with marine fossils. Faunal composition is chiefly molluscan, with a few foraminifers and ostracods; fossils are present mostly in the sparingly glauconitic units. Species diversity is moderate to low, at least for fossiliferous units, but not so low as to suggest very abnormal salinity. Mollusks comprise largely a bank and mud fauna, characteized by several thick-shelled forms. The fauna of the Pendleton system has been described by Barry and LeBlanc (1942).

The system is underlain by marine clay of the Midway Group (Porters Creek Formation) and overlain by sandstone of the Carrizo Formation. A large part of the facies is contemporaneous with delta facies (Sabine deltas of the Rockdale system) on the west and south, although the upper part is contemporaneous with other Wilcox depositional systems not considered here. The Pendleton system grades west and north into the various fluvial facies of the Mt. Pleasant system; it extends east into Louisiana for an undetermined distance.

The lithologic and biologic composition of the Pendleton system and its situation marginal to the Rockdale delta system suggest that deposition took place in low-energy lagoons, lakes, bays, and mudflats flanking the area of maximum deposition in the delta system, much like the marginal environments of the Mississippi delta system. Deposition was chiefly marine and circulation ranged from restricted to open; the system opened seaward primarily on the southeast, and there was greater brackish-water and freshwater influence on the west and north, where the system grades into fluvial facies of the Mt. Pleasant system.

### San Marcos Strandplain-Bay System

In outcrop south of the Colorado River, the highly meandering channel sandstone facies of the Mt. Pleasant system changes abruptly to a facies primarily of mudstone which locally contains marine fossils (Fig. 1). This facies persists in outcrop through southern Bastrop, Caldwell, Guadalupe, and northern Wilson Counties and extends a short distance downdip into the subsurface. It includes the Hilbig facies of Claypool (1933). The faunal and lithologic composition of the facies is very similar to that of the Pendleton system. A similar origin is suggested, with deposition in embayments and on mudflats along the opposite margins of downslope fluvial facies of the Mt. Pleasant system and delta facies of the Rockdale system. In outcrop, mud deposits of this facies are continuous below the downslope facies of the Mt. Pleasant system (highly meandering channel facies or Simsboro sandstone) about as far north as midway between the Brazos and Trinity Rivers, where the mudstone grades into fluvial and delta facies of earliest Rockdale deltation (initial phase, Fig. 5). The northern extent of this mudstone facies includes at least parts of units formally designated the Hooper and Seguin Formations. Southwestward, displacement of the facies is contemporaneous with large-scale deltation in the lower Wilcox Rockdale system (intermediate and terminal phases, Fig. 5).

Downdip and marginal to the shoal-water deltas of the southwestern part of the Rockdale system (Guadalupe deltas), mudstone grades into a facies consisting of interbedded sandstone and mudstone. The sandstone units generally have fine, well-sorted grains, are calcareous and sparingly glauconitic, and commonly contain *Ophiomorpha* burrows, a few detrital shell fragments, and small armored mud balls. They are present as elongate bodies parallel with the depositional strike and as tabular or sheet sandstone characterized by low-angle cross-bedding. Individual sandstone and mudstone units range up to 50 ft in thickness. Mudstone units are similar to those of the updip mudstone facies, but are commonly pyritic and slightly gypsiferous. Carbonaceous materials are present locally as thin lignitic stringers. This interbedded sandstone and mudstone facies is approximately 1,200 ft thick, and is present chiefly in southwestern Gonzales, Wilson, southwestern Karnes, and eastern Atascosa Counties. Composition of the facies and its position marginal to shoal-water deltas of the Rockdale system suggest an origin at least mechanically analogous to that of the Holocene chenier system of southwestern Louisiana. The mud accumulated in low-energy environments ranging from mudflat to open bay, contemporaneous with periods of active deltation. During marine destruction of the adjacent sand-rich, shoal-water deltas, sands were redistributed laterally by prevailingly southwesterly longshore currents; elongate and sheet sands were deposited in a strandplain marginal to the delta system. Transgression and regression of relatively thin marine units in this facies thus were related to periods of active deltation and delta destruction. Subsurface sandstone units such as the Poth and Bartosch are typical of this facies. In reference to their origin and area of principal occurrence, the updip mudstone and downdip sandstone-mudstone facies along the southwestern margin of the Rockdale delta system are designated the San Marcos strandplain-bay system.

### Cotulla Barrier-Bar and Indio Lagoon-Bay Systems

Southwestward, sheet and elongate sandstone bodies of the San Marcos system grade into facies of elongate sandstone bodies arranged parallel with the depositional strike. The sandstone bodies generally have fine, very well sorted grains and are locally glauconitic; internal structures are not known because of lack of cores. Thickness of individual sandstone units ranges as much as 100 ft; aggregate sandstone thickness is 400–1,000 ft for the facies. Sandstone isolith lines show an extensive elongate sandstone unit oriented parallel with the depositional strike. This sandstone belt trends northeast-southwest in western Atascosa, southern Frio, northwestern LaSalle, southeastern Dimmit, and northwestern Webb Counties (Fig. 8). Updip from the sandstone belt, in southern Bexar, northern Atascosa, northern Frio, Zavala, western Dimmit, and northwestern Webb Counties and extending to the outcrop, is a predominantly mudstone facies, including most of the Indio Formation (Fig. 8). The principal lithologic component of this facies is mud, although locally sand is predominant. Mudstone is mostly gray to buff, calcareous to locally gypsiferous, and pyritic, and some is glauconitic or dolomitic. Structurally, it ranges from distinctly laminated to extensively burrowed. Marine fossils are chiefly thick-shelled mollusks; a few units contain large amounts of a very few species of foraminifers, mostly textulariids and globigerinids. Concentric mounds with internal structures, probably of algal origin, commonly are associated with the mudstone. Lignite beds of this facies are thin, discontinuous, and commonly impure, in contrast with the lignite of the fluvial and delta systems. Sandstone in this facies is of varied texture, but most has fine, moderately well-sorted grains; it is developed locally in relation to distinct depositional features, including tidal deltas, washover fans, and small-scale delta and fluvial units.

The lower Wilcox sandstone and mudstone facies of South Texas are judged to have originated as extensive barrier-bar and complementary lagoon-bay systems and are designated here, respectively, the Cotulla barrier-bar system and the Indio lagoon-bay system. The sand was transported laterally from shoal-water deltas of the Rockdale system on the northeast by prevailing southwesterly longshore currents. In the San Marcos strandplain system, situated intermediate to the delta and barrier-bar systems, the adjacent delta system served not only as a source of sand but also, through periods of active and inactive deltation, as a control for the distribution of the sheet sand. Accumulation of sand in the barrier-bar system was too distant from the delta system for the latter to have had direct control on sandstone-body geometry. The principal distinction in sand accumulation between these two systems is that sheet or tabular units formed in the San Marcos and elongate bodies formed in the Cotulla. Similar patterns of sand dispersal are present in modern and premodern marginal-marine systems southwest of the Mississippi delta system. Accumulation of sand in relatively narrow elongate belts in the barrier-bar system, as opposed to sheet-sand accumulation in the strandplain, facilitated development of extensive, landward mud facies—the Indio lagoon-bay system. Development of a distinctive lagoon-bay system is thus a complement of the barrier-bar system.

Other features substantiate interpretation of the South Texas, lower Wilcox facies as barrier-bar and lagoon-bay systems. First, sandstone-isolith patterns depicting the lagoonal side of the barrier-bar system are generally more irregular than those of the seaward side; further, the transition from sandstone to mud-

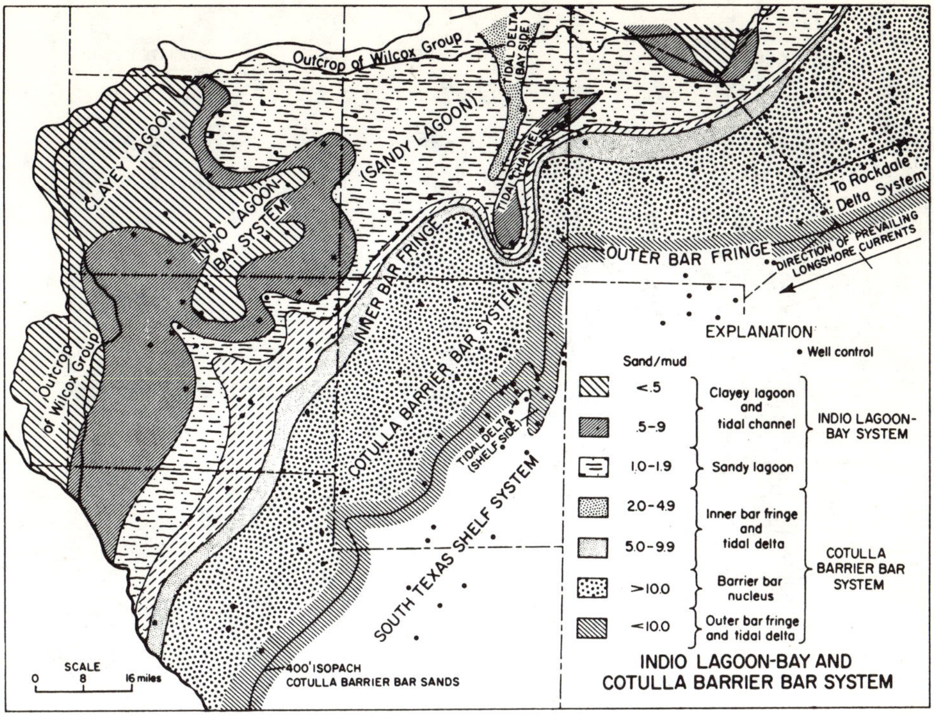


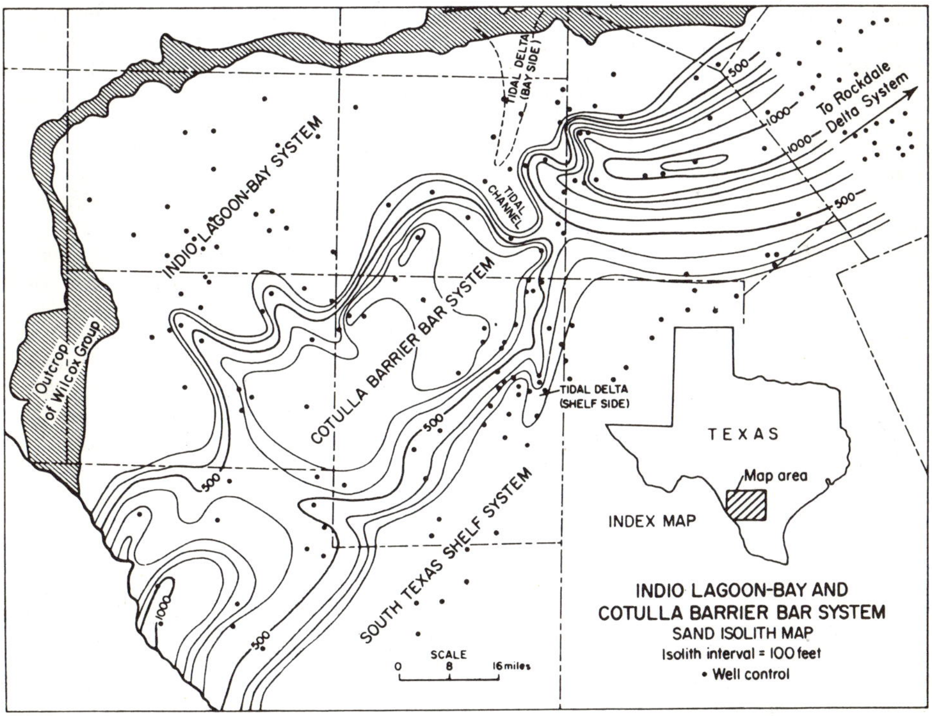


FIG. 8.—Sand-mud ratio map and sand isolith map of Cotulla barrier-bar system and Indio lagoon-bay system, Wilcox Group, Texas.

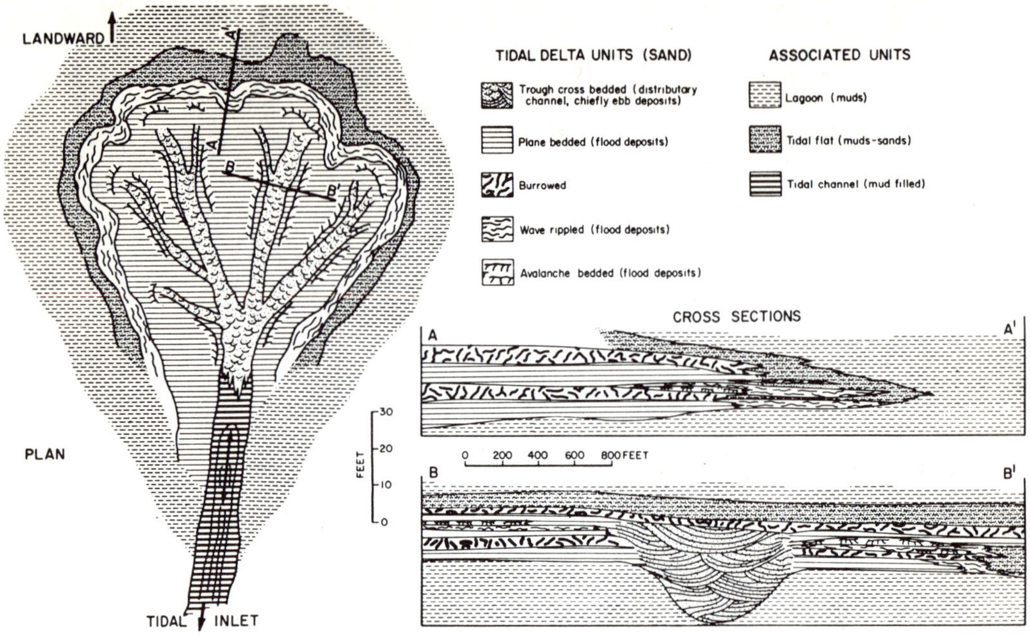


FIG. 9.—Principal rock units, lagoon-side tidal delta, Indio lagoon-bay system, Wilcox Group, southeastern Medina (outcrop) and northeastern Frio (subsurface) Counties, Texas.

stone is more abrupt on the lagoonal than on the seaward side (Fig. 8). Such features are typical of modern barrier-bar systems. Second, in southeastern Frio and northeastern LaSalle Counties, sandstone isolith lines of the Cotulla barrier-bar system are indented sharply on the lagoon side; a less pronounced indentation is present on the seaward side. Position and configuration of these indentations suggest the former presence of a tidal channel, a feature common to modern barrier bars. Flood and ebb tidal surges move through a channel or inlet in the bar, accreting sand in the form of tidal deltas both within the lagoon and onto the foreshore or shelf side of the bar. Presence of an anomalous thickness of sandstone in the Indio lagoon-bay system updip from the terminus of the inferred tidal channel (in northeastern Frio and southeastern Medina Counties and extending into outcrop), as well as sedimentary structures and units within the sandstone bodies, indicate accumulation as a lagoonal tidal delta (Fig. 8). Principal units of the lagoonal tidal deltas developed in outcrop of southern Medina County include flood and ebb deposits (Fig. 9). Deposition occurred both within tidal distributary channels and on interchannel flats. Channel deposits are trough or festoon cross-bedded sandstone, with axes of cross-bed sets trending generally south; deposition was principally during ebb flow. Toward the barrier bar, and in the subsurface, the tidal channel becomes relatively straight and nondistributive; this part of the channel is filled with mudstone and apparently was filled after the channel became inactive. Deposits of the interchannel flats are chiefly plane-bedded sandstone grading marginally through megarippled sandstone to tidal-flat and lagoonal mudstone; locally, thin, avalanche-bedded sandstone is associated with the plane-bedded deposits. Bedding, structures, and directional features of the interchannel deposits indicate that deposition took place mainly during flood surge. Plane-bedded sandstone shows graded bedding with a vertical increase in the amount of clay. The upper units are extensively burrowed, apparently as the result of destruction while parts of the tidal delta became emergent. Lobate extension of sandstone on the foreshore side of the bar and just south of the inferred tidal channel (east-central LaSalle County) probably represents a delta constructed by ebb-tide transport of sand through the tidal channel. On the assumption that these features originated as a tidal channel and complementary tidal deltas, the position of the deltas in respect to the channel, as well as the shape and configuration of the tidal channel, substantiate the inference of a prevailing southwestward transport of sediment by longshore currents. Lobate lagoonward extensions of sandstone in northeastern and north-central Dimmit County (Fig. 8), without associated isolith indentations indicative of tidal channels, possibly represent deposition as washover fans, with sand carried into the lagoon by storm surges. This feature is common in modern barrier bars (Andrews, 1966).

### South Texas Shelf System

Downdip and seaward of the Cotulla barrier-bar system, a predominantly mudstone system characterizes the lower Wilcox of South Texas. This system extends across about 3,000 sq mi in southwestern Karnes, northern Goliad, northern Bee, north Live Oak, McMullen, northwestern Duval, Webb, Zapata, and western Starr Counties (Fig. 1). It is relatively uniform in thickness through much of its extent, ranging from about 1,000 to 1,500 ft, but thickens northeastward toward the Rockdale delta system and southward along the Rio Grande. Principal lithologic component is mudstone; aggregate thickness of sandstone in the system is generally less than 200 ft (Fig. 1). Individual sandstone units are varied in number, but commonly 8–10 are present. Thickness ranges generally from 10 to a maximum of 100 ft; average thickness of sandstone units is about 30 ft. Mudstones are of varied composition, but generally are light or medium gray to olive or buff, and laminated to extensively burrowed; marine fossils and glauconite pellets are common. Mudstones are commonly calcareous, though some are pyritic or carbonaceous. The sandstone is very fine to fine grained, well sorted to muddy, slightly glauconitic, and commonly contains detrital shell fragments. Faunal diversity is moderate; composition is chiefly molluscan, but thick-shelled forms are not predominant as in other lower Wilcox marine facies. Foraminifers are numerous and a few echinoderms are present.

The comparative uniformity of the system, the predominantly mudstone composition, marine fauna, and position relative to other depositional systems suggest that deposition took place on an extensive, shallow-water continental shelf; hence the system is designated the South Texas shelf system. Part of the Falls City Formation, as proposed and described by Hargis (1962), is included in this system. Vertical variations in abundance and diversity of marine fossils, amount of glauconite, extent of burrowing, and presence of sandstone units probably are related to variations in rate of deposition coincident with filling and foundering of the shelf; these cycles presumably relate to phases of delta construction and destruction in the Rockdale delta system along strike on the northeast.

Origin of the relatively thin sandstone units in the South Texas shelf system is not clear. Their external geometry is not well known, but present control suggests that there are both laterally persistent sheet sandstones, more numerous updip toward the Cotulla barrier-bar system, and elongate, NE-SW-trending sandstone bodies. The source of sand in the South Texas shelf system, like that of updip sandstone systems, probably was the Rockdale delta system on the northeast, but mode of transport and dispersal is unknown. Relatively sharp indentations in sandstone isolith lines of the southwestern lobes of the Guadalupe deltas of the Rockdale delta system (Fig. 1) suggest that submarine channeling of the deltas may have occurred during periods of delta destruction, with possible dispersal of sand across the shelf by southward-flowing turbidity currents. (Submarine channeling in the Wilcox as reported by Hoyt in 1959 is associated with an upper Wilcox system.) A more probable explanation of the origin of these sandstone units, is accumulation as barrier bars during periods of initial marine destruction of gulfward-projecting delta lobes and subsequent to shelf filling coincident with principal delta construction. The origin thus is similar to that of the Cotulla barrier-bar system, but there is only minor accumulation downdip and principal accumulation updip in the Cotulla system. Sheet sandstone of the shelf system may be related to transgression associated with periods of continued delta destruction; shelf-edge marine reworking of foundered barrier-bar sands during this period was likely. Extensive mud deposition in this system presumably was contemporaneous with periods of active deltation in the Rockdale delta system, although mudstones of the two systems are compositionally distinct. The amount of mud deposition on the shelf was small in comparison with the amount of contemporaneous prodelta mud deposition. Differences in composition and amount of mud in the two systems probably are related to marked differences in rates of accumulation.

## Sand-Dispersal Patterns

The dispersal pattern of coarser fraction sediments (generally sand) in a terrigenous clastic system provides significant clues to the origin of specific facies and also provides a basis for the integration of various facies and systems.

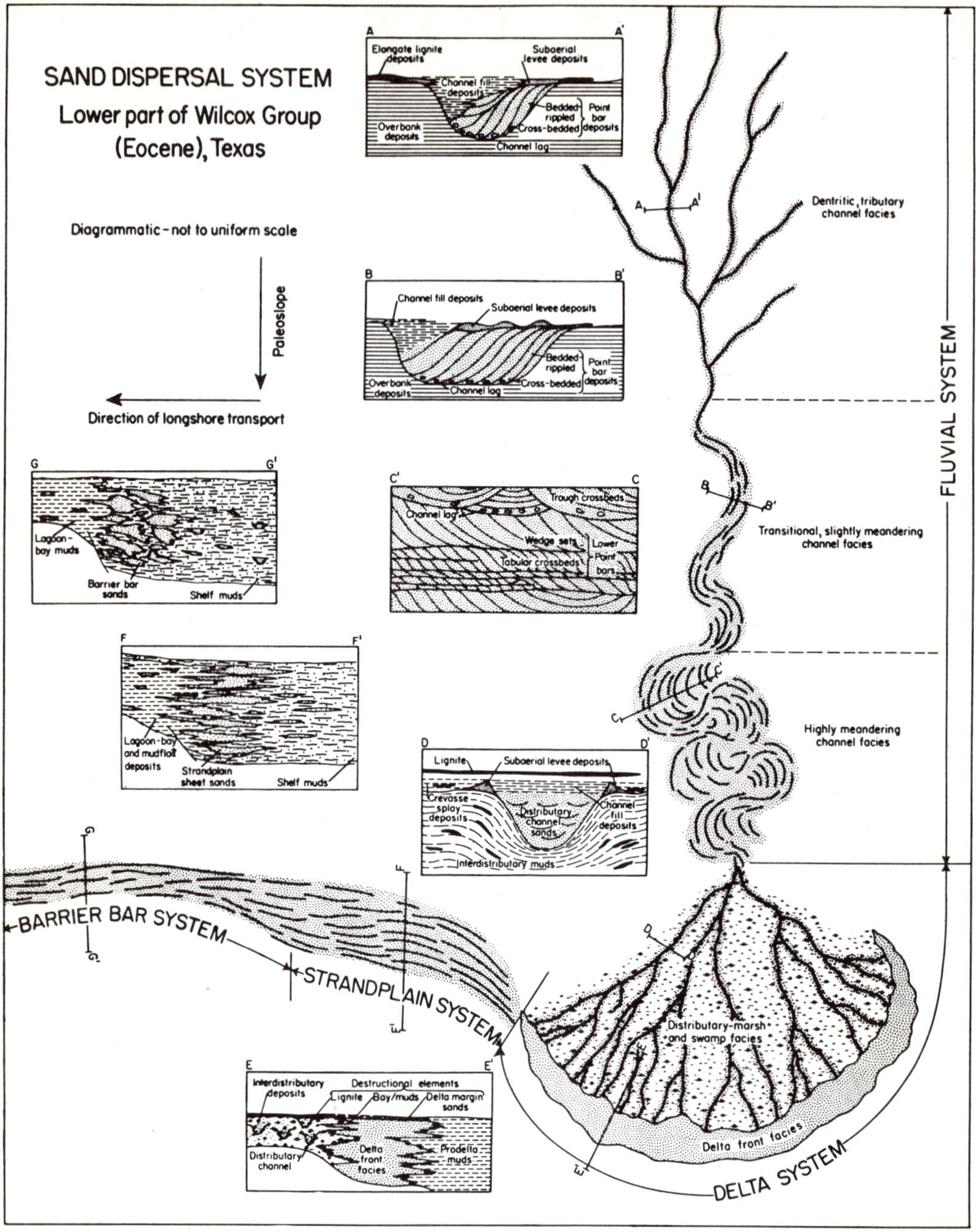


FIG. 10.—Idealized sand-dispersal system in various depositional systems, Wilcox Group, Texas.

Transport, dispersal, and deposition of the coarser fraction sediments normally involve higher energy elements, and thus the geometry of sandstone bodies effectively outlines skeletal units of the system. Within the various depositional systems of the lower Wilcox, several modes of sand transport and deposition occurred; each resulted in sandstone units of distinct geometry, structure, and texture (Fig. 10). The pattern involves dispersal through several fluvial facies, in which only a relatively small part of the sand was permanently stored or preserved, and through separate delta facies in which most deposition involved permanent storage, especially in the deep-water, elongate deltas. Marine redistribution of a relatively small part of the sand deposited in the delta-front facies, chiefly in shoal-water deltas, resulted in accumulation of sand in marginal strandplain, barrier-bar, and shelf systems (Fig. 10). Sandstone bodies of the fluvial system and of distributary channels of the delta system have regional trends normal to the depositional strike; sandstone bodies of the delta-front facies and those in the shelf system roughly parallel the depositional strike; those of the barrier-bar and strandplain systems closely parallel the depositional strike.

The highest upslope element preserved in the lower Wilcox sand-dispersal pattern, apparently only a minor part of the original fluvial system, consists of elongate channel sandstone with dendritic tributary distribution; ratio of nonchannel or overbank mudstone to channel sandstone is high. Downslope depositionally, the ratio of nonchannel mudstone to sandstone decreases, and individual sandstone bodies increase in width as a result of deposition by increasingly meandering streams. A zone of highly meandering stream deposition, as judged from the dominance of a point-bar sequence, resulted in a fluvial facies composed chiefly of extensive, multilateral sandstone bodies with very little associated nonchannel mudstone. This fluvial facies is present downslope and as the preserved part of the main fluvial system. Upslope this facies is bounded by a fault system (present Mt. Enterprise-Elkhart system) or threshold (Tyler basin threshold) which influenced deposition in other Cenozoic and Mesozoic units. It suggests the possible presence of a fall line, or at least a local gradient change that resulted in increased sand deposition. Once such a zone was established, weak bank conditions of streams traversing it would contribute to a high degree of meandering or lateral shifting. A similar high-sandstone fluvial facies in sand-dispersal patterns of Upper Pennsylvanian units in the Eastern Interior basin is reported by Potter (1963).

The next downslope element in the sand-dispersal pattern consists of elongate sandstone bodies in a distributive network; these are associated with relatively large amounts of nonchannel mudstone (Fig. 10). Channel sand accumulation was chiefly by downward accretion through compaction, as indicated by the nonerosive bases of the distributary channels and compactional features of underlying mudstone.

Principal deposition of sand in the delta system occurred in the delta-front facies, built up of distributary-mouth bars. Where delta-front sands form elongate, narrow sand facies (bar-finger sands of Fisk, 1961) in relatively high-mud frameworks and under deeper water conditions (birdfoot deltas), sand storage is generally permanent. This is apparently true of the elongate delta fronts characteristic in the eastern part of the Rockdale system where they are not associated with marginal barrier-bar and strandplain sands; interdelta subembayment facies contain very few sandstone bodies. Where the delta front is constructed as a lobate to rounded facies of sand-rich, shoal-water deltas (typical of deltas in the southwestern part of the Rockdale system), some of the sand may be redistributed laterally by marine destruction of the deltas and transported by longshore currents. In such situations two main types of sandstone bodies accumulated, though both have a common source. Adjacent to the delta system, accumulation generally resulted in relatively thin, areally persistent sheet sands which made up a strandplain system. Down the longshore current and more distal from the delta system, sand accumulated in more narrowly restricted, elongate barrier bars with the complementary development of a distinct landward lagoon-bay system.

A final pattern of sand dispersal, not shown on Figure 10 but possibly a part of lower Wilcox depositional systems, is transport and deposition by turbidity currents. Sharp indentations in sandstone isolith lines of certain of the shoal-water deltas in the Rockdale system (Guadalupe and Colorado deltas) may reflect mud-filled, submarine canyons that were cut and subsequently filled during specific stages of delta construction and destruction. Such a feature has been documented in a higher Wilcox system (Hoyt, 1959), and submarine channeling is well known from several modern delta systems. Comparable features in lower Wilcox systems are probable. Under such conditions delta sands eroded during channeling would be carried down depositional slope by turbidity currents and deposited as fans at the first major flattening of the slope. Mode of dispersal would be similar to that outlined by Walker (1966, Fig. 18) for near-source submarine fans associated with the Pennine delta in the upper Carboniferous of northern England. Verification of such fans in the lower Wilcox awaits the drilling of more deep downdip wells.

### Depositional Systems and Basin Tectonics

The writers have used the concept of depositional systems not only as an approach to interpreting certain rock units and facies of the Wilcox Group, but also as a basis for establishing criteria for recognizing large-scale, regionally developed depositional units. Many sedimentary basins, and notably the Gulf Coast basin, contain very thick and large masses of terrigenous clastic rocks. Application of diagnostic criteria for recognition of specific depositional environments and the utilization of detailed depositional models are facilitated first by outlining major depositional systems involved in the basin fill.

The nature of a sedimentary basin, especially its structural or tectonic attitude, is critical to the concept of depositional systems because of its effect on (1) volume of sediment or magnitude of sequences involved, and (2) variation in distribution of component facies. First, in a relatively unstable basin where subsidence is generally a response to sediment load (as is judged to be the case of the Gulf Coast basin during Wilcox deposition), sites of deposition of particular facies tend to be maintained. Thus a barrier-bar system, for example, can be constructed of several individual barrier bars, the main features of which are similar to those of the larger system. Essential difference between the two is size. In magnitude, the Cotulla barrier bar system here recognized in the lower Wilcox of South Texas is closely comparable to other barrier-bar systems recognized elsewhere in the Gulf Coast basin—Frio barrier-bar system (Neogene) of South Texas (Boyd and Dyer, 1964) and Terryville barrier bar (Jurassic) of southern Arkansas and northern Louisiana (Thomas and Mann, 1966).

Second, the tectonic nature of the basin markedly influences the distribution of component facies within a depositional system, although composition and character of component facies of a given depositional system may be very similar in different structural basins. For example, delta systems developed in relatively stable basins, such as the Pennsylvanian of North Texas (Eastern Interior basin) and the western interior United States, commonly show fairly extensive progradation. Sections within such systems contain a vertical succession of component facies with prodelta mudstone overlain by delta-front sandstone, in turn succeeded by delta-plain and fluvial facies. By contrast, delta systems formed in rapidly subsiding, unstable basins, such as the Gulf Coast basin, commonly do not contain a pronounced vertical succession of facies, but rather have pronounced lateral facies that generally persist vertically. Within any particular part of the delta system, therefore, a vertical sequence generally shows only a single component facies of the system.

### Oil and Gas Occurrence

A well-established generalization in Gulf Coast petroleum geology is that occurrence of oil and gas is controlled regionally by depositional facies, and local entrapment is by structures coincident with depositional facies that contain optimum reservoirs. The environmental control of oil and gas occurrence in terrigenous clastic rocks has been emphasized particularly by Rainwater (1963). Distribution of depositional facies in the context of the regional system of which they are a part explains control of known accumulations and provides a basis for outlining regional, potentially productive trends. In the lower Wilcox of the Texas Gulf Coast, five or six main oil- and gas-producing trends (Fig. 11) are defined in relation to depositional systems established herein: (1) delta trend (chiefly delta-front constructional and marine destructional sandstone), (2) strandplain trend, (3) barrier-bar trend (including tidal deltas), (4) shelf trend, (5) shelf-edge trend, and (6) a possible submarine-fan trend not yet defined.

Oil and gas accumulation in the lower Wilcox from north-central Karnes County northeastward to the Sabine River is controlled closely by distribution of specific deltas of the Rockdale delta system. Principal reservoirs are delta-front sandstone and closely associated marine-reworked sandstone from delta destruction; known fields are largely coincident with a relatively narrow belt determined by the more distal fringes of delta-front sandstone of specific delta lobes (Fig. 11). Interfingering of delta-front sandstone with adjacent marine prodelta mudstone provides the multipay reservoirs typical of many of the large delta-front fields (*e.g.*, Falls City, Sheridan, Columbus, Lake Creek, New Ulm, and Quicksand Creek). Marine clay deposited over delta-front sand during stages of marine destruction of delta lobes locally provides an effective seal for sandstone reservoirs. Numerous large-scale structures along the structural trend of lower Wilcox pro-

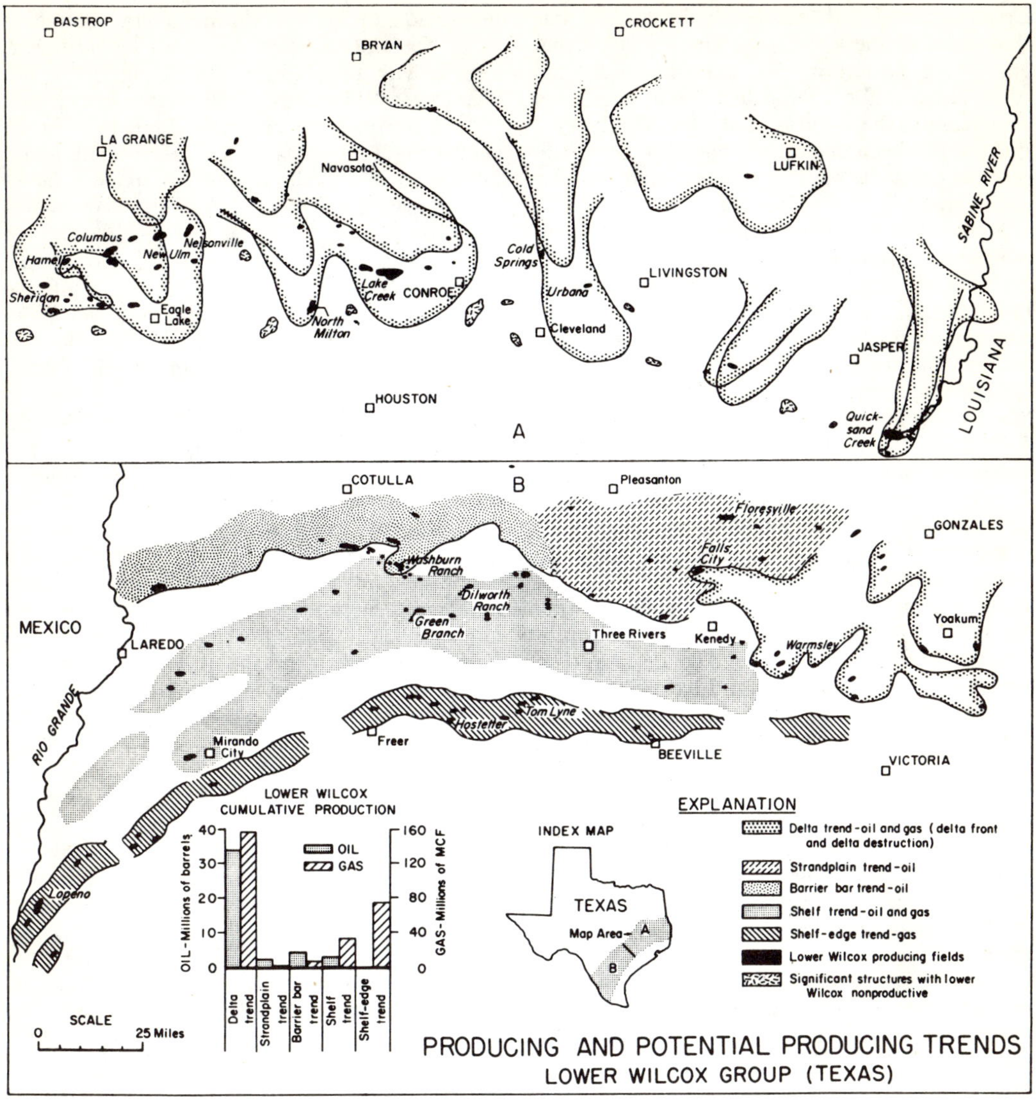


FIG. 11.—Productive and potentially productive oil and gas trends, lower part of Wilcox Group, Texas. For well control, see Figure 1.

duction within the delta system are notably nonproductive because of their location in sandstone-poor interdelta subembayment facies (Fig. 11). Growth faults and diapiric structures are associated with zones of thick deposition, such as the delta front; structures associated with these features are locally important traps.

Marginal to the delta trend is the standplain trend, controlling oil accumulation chiefly in the lower Wilcox of northwestern Gonzales, Wilson, northwestern Karnes, and eastern Atascosa Counties. Areally extensive sheet sandstone interbedded with dominantly marine mudstone forms a wide productive trend, which includes the farthest updip lower Wilcox production.

Sandstone units on the seaward side of the Cotulla barrier-bar system constitute the barrier-bar trend which extends in a narrow zone from northern McMullen County southwestward across LaSalle County through western Webb County. Most of the established production from this trend is in east-central LaSalle

County, associated mainly with a shelf-side tidal delta of the Cotulla system. Provided that entrapping structures are present, additional discoveries in the barrier-bar trend are likely; significantly, the trend is at shallow depth. Production has been chiefly oil; however, some significant gas wells have been completed recently in this trend in Webb County.

Several lower Wilcox fields have been developed in the sandstone reservoirs of the South Texas shelf system. With the exceptions of Green Branch and Dilworth Ranch, however, fields are small. Individual sandstone units within this extensive system are sporadic in distribution and not easily definable; many sandstone units are muddy with low permeability. Production has been primarily gas with small amounts of oil.

A final, main lower Wilcox trend is referred to here as the shelf-edge trend (Fig. 11). Sandstone bodies of this trend are part of the South Texas shelf system, though their origin is not fully understood. They form a fairly well-defined, narrow zone extending from the southwestern margin of the Rockdale delta system through central Bee, central Live Oak, northwestern Duval, southeastern Webb, eastern Zapata, and western Starr Counties. Productive sandstone units presumably originated as barrier bars similar to elongate sandstone bodies updip on the remaining part of the shelf system and in the Cotulla system. Their characteristics and downdip occurrence suggest that shelf-edge reworking and cleaning may have been a result of the higher energies associated with such a zone.

If the presence of submarine-fan deposits is indicated by deep drilling, an additional lower Wilcox trend may be established. Most likely area of such sandstone development would be downdip of the major shoal-water deltas of the Rockdale system, chiefly in Goliad, Victoria, Jackson, and Wharton Counties. Identification of such a trend will require extensive deep drilling.

### Comparison of Lower Wilcox and Northwest Gulf of Mexico Depositional Systems

Comparison of the lower Wilcox with the Holocene of northwest Gulf of Mexico shows many similarities both in the kinds and distribution of depositional systems and in the component facies of these systems (Fig. 12). This is expected, because both are developed in the same structural basin. Prominent in the northwest Gulf is the extensive Mississippi delta system made up of five distinct delta complexes (Fig. 6) and at least 16 subdeltas (Frazier, 1967), with total area of about 20,000 sq mi. The lower Wilcox Rockdale delta system consists of seven delta complexes and at least 16 subdeltas (Fig. 5) with a total area of about 24,000 sq mi. Similarities in component facies and shapes of specific deltas in the two systems have been noted.

The Mississippi delta system is fed by the large, interior-draining fluvial system of the Mississippi River. Much of the analogous fluvial system of the lower Wilcox (Mt. Pleasant fluvial system) extended inland from present outcrop and is not preserved. On the basis of comparable size of the Rockdale and Mississippi delta systems, a large, interior-draining fluvial system is presumed to have been present during lower Wilcox deposition. Maximum thickness of the preserved part of the Mt. Pleasant fluvial system is between the modern Colorado and Trinity Rivers, and presumably the larger streams entered the Gulf in that area. The principally preserved part of the Mt. Pleasant system in northeast Texas was probably a minor or secondary part of the system, with streams comparable to modern streams east of the Mississippi River such as the Pearl and Tangipahoa Rivers.

Marginal to the area of maximum development of the gulfward-extending Mississippi River and delta systems are relatively low-energy basins or depressions in which freshwater to marine deposition occurs (Lake Pontchartrain, Lake Borgne, and part of Mississippi Sound on the east, Vermillion Bay, West Cote Blanche Bay, White Lake, and Grand Lake on the west). Lower Wilcox analogues include the predominantly marine mudstone systems of the Pendleton lagoon-bay system developed along the east flank of the Rockdale and Mt. Pleasant systems and the updip part of the San Marcos system on the southwest or opposite side.

A net southwestward longshore transport of sediments from both the modern Mississippi and the Wilcox Rockdale delta systems formed comparable depositional systems along strike of these systems. The southwestern Louisiana chenier system, adjacent to the Mississippi delta system, is at least mechanically analogous to the San Marcos strandplain system, which developed just down the longshore drift of the Rockdale delta system. More distally along the strike of the Holocene Mississippi delta system is a barrier-bar system, of which the modern Bolivar-Galveston barrier bar is a part. This system is considered analogous to the bar-

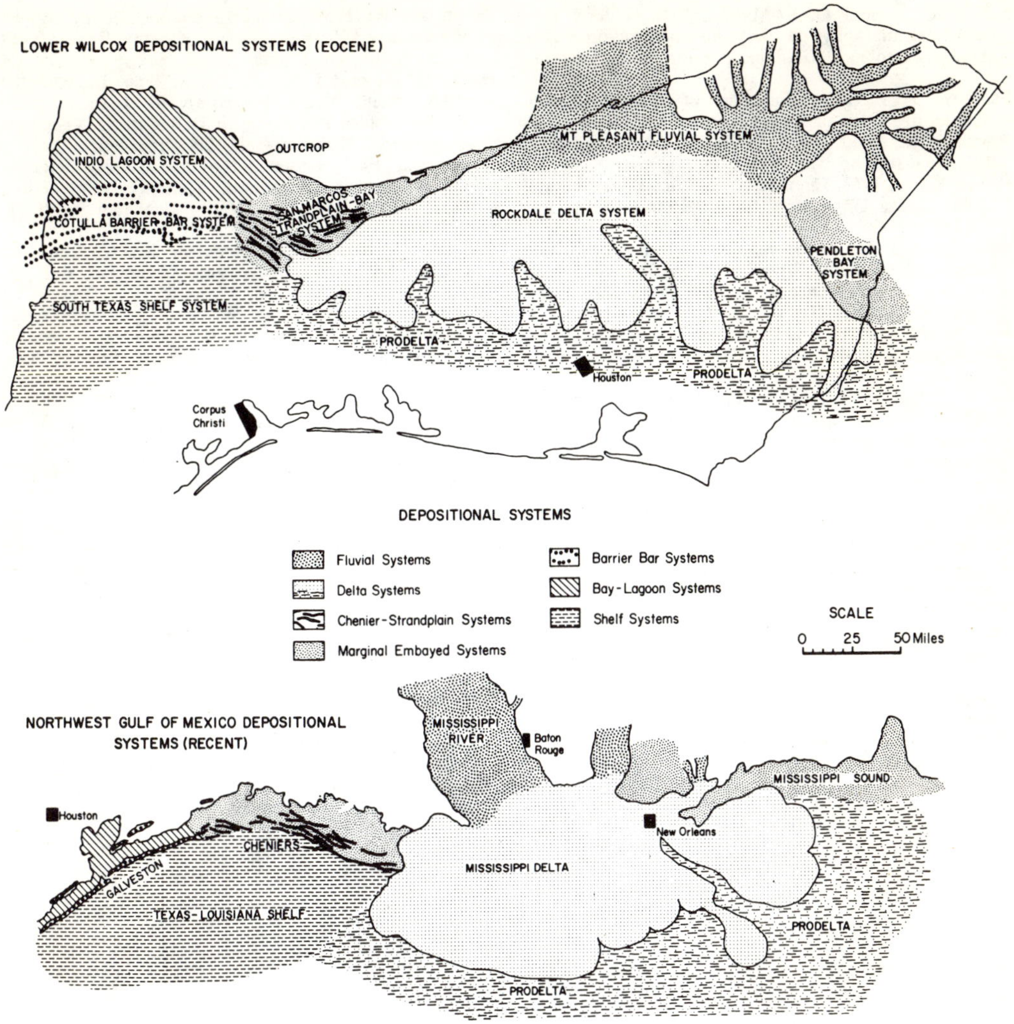


FIG. 12.—Comparison of lower Wilcox (Eocene) and northwestern Gulf of Mexico (Holocene) depositional systems.

riers making up the Cotulla barrier-bar system of the South Texas lower Wilcox. Landward complement of the modern barrier bars is a bay-lagoon system, represented at present by Galveston Bay. Deposits of this system are the modern analogue of the Indio lagoon-bay system. The extensive Louisiana-Texas continental shelf, situated gulfward of the barrier bar and chenier system and partly marginal to the Mississippi delta system, is the modern counterpart of the lower Wilcox South Texas shelf system.

## REFERENCES CITED

Allen, J. R. L., 1963, The classification of cross-stratified units, with notes on their origin: Sedimentology, v. 2, p. 93–114.

Andrews, P. B., 1966, Facies and genesis of a hurricane washover fan, St. Joseph Island, central Texas coast: Unpub. Ph. D. dissert., Texas Univ. at Austin, 238 p.

Barry, J. O., and R. J. LeBlanc, 1942, Lower Eocene faunal units of Louisiana: Louisiana Geol. Survey Bull. 23, 208 p.

Boyd, D. R., and B. F. Dyer, 1964, Frio barrier bar system of South Texas: Gulf Coast Assoc. Geol. Socs. Trans., v. 14, p. 309–322.

Brown, L. F., *et al.*, 1967, Role of compaction in development of geometry of superposed elongate sandstone bodies (abs.): Am. Assoc. Petroleum Geologists Bull., v. 51, p. 455–456.

Claypool, C. B., 1933, The Wilcox of central Texas: Abs. of Ph. D. thesis, Illinois Univ., 12 p.

Coleman, J. M., and S. M. Gagliano, 1964, Cyclic sedimentation in the Mississippi River deltaic plain: Gulf Coast Assoc. Geol. Socs. Trans., v. 14, p. 67–80.

Echols, D. J., and D. S. Malkin, 1948, Wilcox (Eocene) stratigraphy, a key to production: Am. Assoc. Petroleum Geologists Bull., v. 32, p. 11–33.

Feofilova, A. P., 1954, The place of alluvium in sedimentation cycles of various orders and time of its formation: Akad. Nauk SSSR Geol. Inst. Trudy, v. 151, Coal ser., no. 5, p. 241–272.

Fisher, W. L., 1963, Lignites of the Texas Gulf coastal plain: Texas Univ. Bur. Econ. Geology Rept. Inv. 50, 164 p.

——— and J. H. McGowen, 1967, Depositional systems in the Wilcox Group of Texas and their relationship to occurrence of oil and gas: Gulf Coast Assoc. Geol. Socs. Trans., v. 17, p. 105–125.

Fisk, H. N., 1955, Sand facies of recent Mississippi delta deposits: 4th World Petroleum Cong. Proc., Sec. 1–C, p. 377–398.

——— 1960, Recent Mississippi River sedimentation and peat accumulation: 4th Internat. Cong. Carboniferous Stratigraphy and Geology, Heerlen 1958, Compte Rendu, p. 187–199.

——— 1961, Bar-finger sands of the Mississippi delta, *in* Geometry of sandstone bodies: Am. Assoc. Petroleum Geologists, p. 29–52.

Frazier, D. E., 1967, Recent deltaic deposits of the Mississippi River: their development and chronology: Gulf Coast Assoc. Geol. Socs. Trans., v. 17, p. 287–311.

Friedman, S. A., 1960, Channel-fill sandstones in the middle Pennsylvanian rocks of Indiana: Indiana Geol. Survey Rept. Prog. 23, 59 p.

Hargis, R. N., 1962, Stratigraphy of the Carrizo-Wilcox of a portion of South Texas and its relationship to production: Gulf Coast Assoc. Geol. Socs. Trans., v. 12, p. 9–25.

Hayes, M. O., and A. J. Scott, 1964, Environmental complexes, South Texas coast: Gulf Coast Assoc. Geol. Socs. Trans., v. 14, p. 237–240.

Hoyt, W. V., 1959, Erosional channel in the middle Wilcox near Yoakum, Lavaca County, Texas: Gulf Coast Assoc. Geol. Socs. Trans., v. 9, p. 41–50.

Kolb, C. R., and J. R. van Lopik, 1966, Depositional environments of the Mississippi River deltaic plain—southeastern Louisiana, *in* Deltas in their geologic framework: Houston Geol. Soc., p. 17–61.

Kruit, C., 1955, Sediments of the Rhone delta; I, Grain size and microfauna: Koninkl. Nederlandsch Geol. Mijnbouwk Gen. Verh., Geol. Ser. v. 15, p. 397–499.

Lowman, S. W., 1949, Sedimentary facies in the Gulf Coast: Am. Assoc. Petroleum Geologists Bull., v. 33, p. 1939–1997.

Mueller, J. C., and H. R. Wanless, 1957, Differential compaction of Pennsylvanian sediments in relation to sand-shale ratios, Jefferson County, Illinois: Jour. Sed. Petrology, v. 27, p. 80–88.

Pettijohn, F. J., *et al.*, 1965, Geology of sand and sandstone: Indiana Geological Survey and Dept. Geology, Indiana Univ. Conf., 205 p.

Potter, P. E., 1963, Late Paleozoic sandstones of the Illinois basin: Illinois Geol. Survey Rept. Inv. 217, 92 p.

Rainwater, E. H., 1963, The environmental control of oil and gas occurrence in terrigenous clastic rocks: Gulf Coast Assoc. Geol. Socs. Trans., v. 13, p. 79–94.

Thomas, W. A., and C. J. Mann, 1966, Late Jurassic depositional environments, Louisiana and Arkansas: Am. Assoc. Petroleum Geologists Bull., v. 50, p. 170–182.

Walker, R. G., 1966, Shale grit and Grindslow shales: transition from turbidite to shallow water sediments in the upper Carboniferous of northern England: Jour. Sed. Petrology, v. 36, p. 90–114.

Wanless, H. R., *et al.*, 1963, Mapping sedimentary environments of Pennsylvanian cycles: Geol. Soc. America Bull., v. 74, p. 437–486.

**Reprinted from:**
BULLETIN OF THE AMERICAN ASSOCIATION OF PETROLEUM GEOLOGISTS
VOL. 55, NO. 8 (AUGUST, 1971), PP. 1155-1169, 8 FIGS., 1 TABLE

# Deltaic Sedimentation in Athabasca Tar Sands[1]

**MAURICE A. CARRIGY**[2]
**Edmonton, Alberta**

**Abstract** The Athabasca tar sands of northern Alberta underlie an area greater than 26,000 sq km and contain more than 700 billion bbl of heavy oil. Excellent outcrops of the oil-impregnated Lower Cretaceous strata (McMurray and Clearwater Formations) are present in the valley of the Athabasca River and its tributaries near Fort McMurray. The middle and upper members of the reservoir consist of many small deltas of limited thickness (less than 30 m) built by rivers flowing northward into a lake or lagoon. The water occupied an elongated depression formed by extensive salt removal from evaporite beds beneath the Devonian limestone which underlies the tar sands. Lithofacies associated with the changing environments of deposition during the infilling of this depression include, successively, fluvial sand; delta-front sand and silt; deltaic-platform sand, silt, and clay; and nearshore marine sand.

The distribution of oil within the tar-sands reservoir is strictly controlled by the original porosity and permeability of the sediments. Thus, the maximum amount of oil (18–20 percent by weight) is present in well-sorted, clean sands that are most commonly of fluvial origin. No oil is present in the poorly sorted, argillaceous parts of the nearshore marine sands, or in the silt beds within the deltaic complex.

## Introduction

Probably the most difficult environment of deposition to recognize in the subsurface is the ancient delta. Because of its affinities with both alluvial and marine deposits and the delicate balance between forces of construction and destruction, it is only under unusual circumstances that all the constructional elements of the deltaic cycle will be preserved in the stratigraphic record. Perhaps it is even more unusual for an ancient deltaic deposit to contain immobile oil, be visible in outcrop, and have abundant core and electric log data available from the adjacent subsurface. Because the Athabasca tar sands have most of these attributes, they provide a unique opportunity to study the relation between deltaic sedimentation and petroleum occurrences.

[1] Manuscript received, July 14, 1970; accepted, November 20, 1970. Presented before the Association at Calgary, Alberta, June 24, 1970. Research Council of Alberta Contribution No. 503.

[2] Research Council of Alberta.



## Location

The Athabasca tar sands are in northeastern Alberta, adjacent to the valley of the Athabasca River, about 300 mi (483 km) northeast of Edmonton (Fig. 1). They underlie an area in excess of 26,000 sq km between 55° and 58°N lat. and 110° and 114°W lang., and crop out extensively in the valleys of the Athabasca River and its tributaries north of the town of Fort McMurray. It is estimated that the Athabasca tar sands contain more than 700 billion bbl of heavy oil averaging about 10.5° API (Pow *et al.*, 1963).

## Geologic Setting

### Stratigraphy

The Athabasca tar sands were deposited during Early Cretaceous (Aptian-Albian) time in an elongated depression, 200 mi (322 km) long and 80–100 mi (130–160 km) wide, developed on Devonian limestones adjacent to the Precambrian shield in northeastern Alberta (Fig. 1). The succession of strata unconformably overlying the Devonian rocks in this area is divisible into three major formational units, which are, in ascending order:

*McMurray Formation,* 0–300 ft (0–100 m) thick, consisting of fresh- to brackish-water, lenticular sandstone and laminated gray siltstone with thin interbeds of dark gray foram-bearing shale in the upper part; disconformable with underlying Devonian limestones, conformable with overlying Clearwater Formation.

*Clearwater Formation,* 250 ft (80 m) thick, consisting of marine dark gray, fossiliferous, silty shale and laminated siltstone with fine-grained, glauconitic, cherty sandstone at base and in upper part; gradational with overlying Grand Rapids Formation.

*Grand Rapids Formation,* 350 ft (115 m) thick, consisting of thick, fine- to medium-grained, feldspathic sandstone beds separated by intervals of laminated glauconitic siltstone and silty shale with thin coaly beds in upper part; shoreline to lagoonal in origin.

The tar (oil) sands of the lower Athabasca River comprise the McMurray Formation and, in places, the overlying basal glauconitic sandy beds of the Clearwater Formation (Wabiskaw Member), which together represent the continental and shallow marine phases of sedimentation associated with the transgression of the boreal Cretaceous sea over northeastern Alberta.

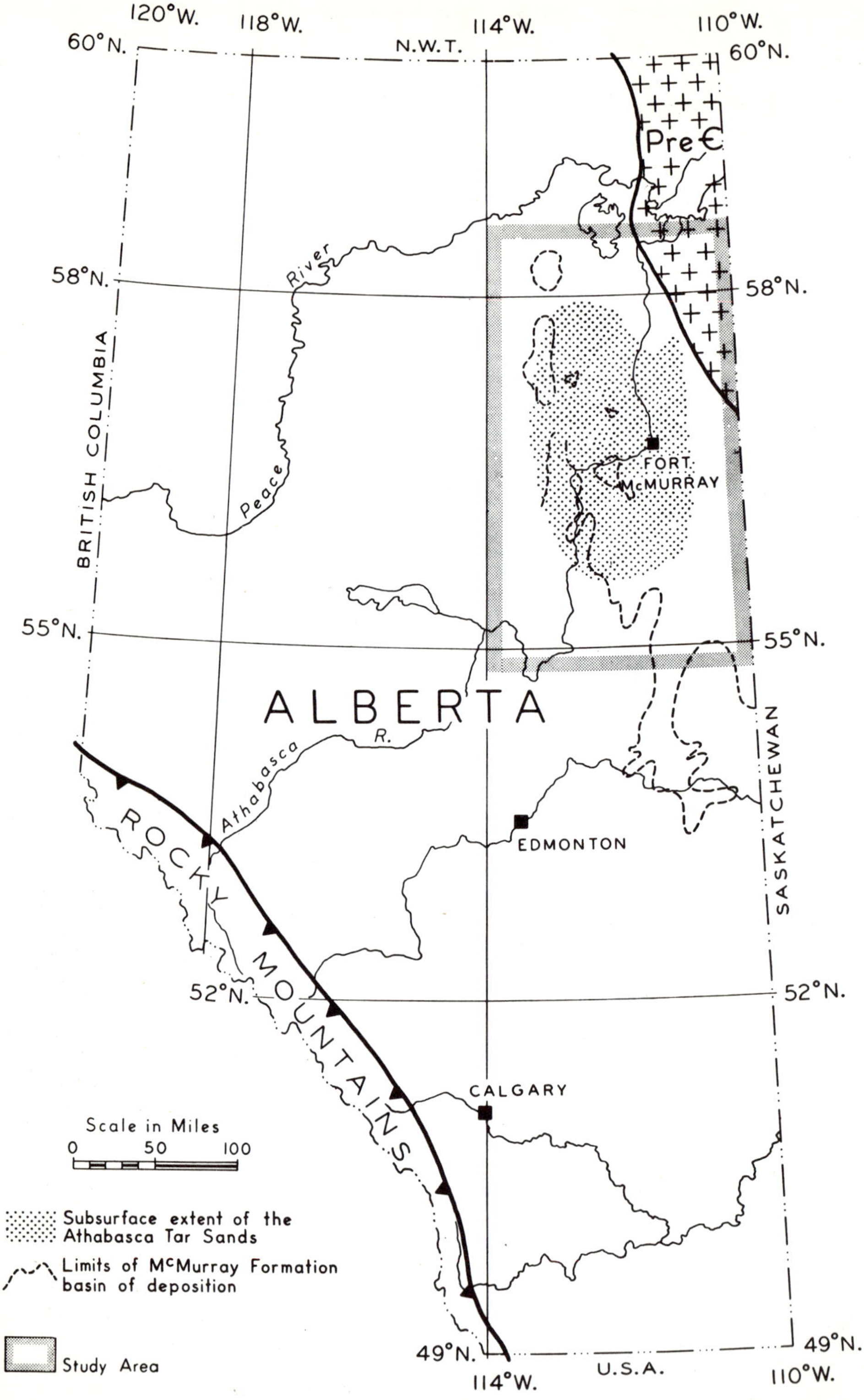


FIG. 1.—Index map showing subsurface extent of Athabasca reservoir.

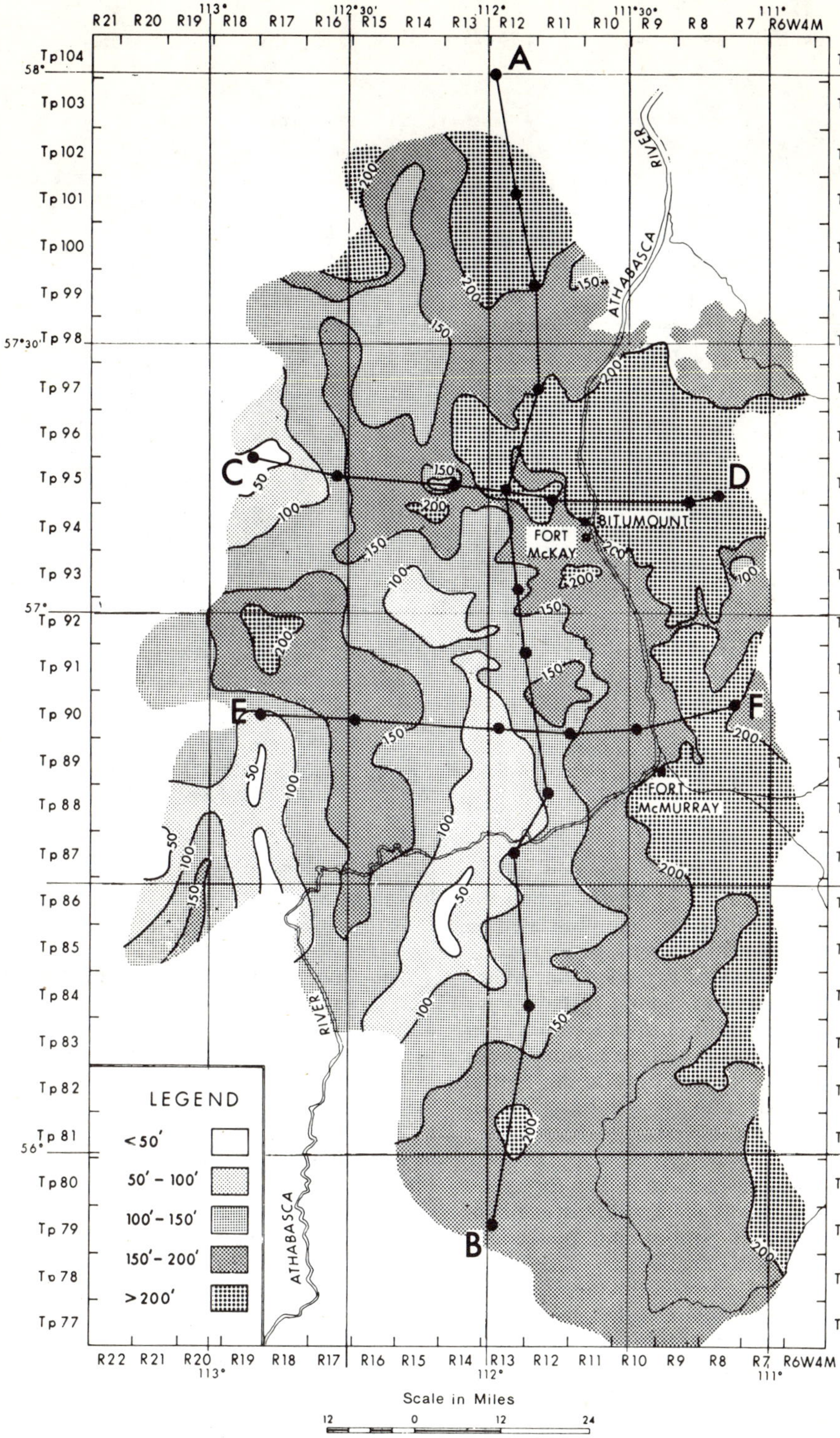


FIG. 2.—Generalized isopach map of McMurray-Wabiskaw stratigraphic units showing approximate structure of depositional basin (after Alberta Oil and Gas Conservation Board, 1963).

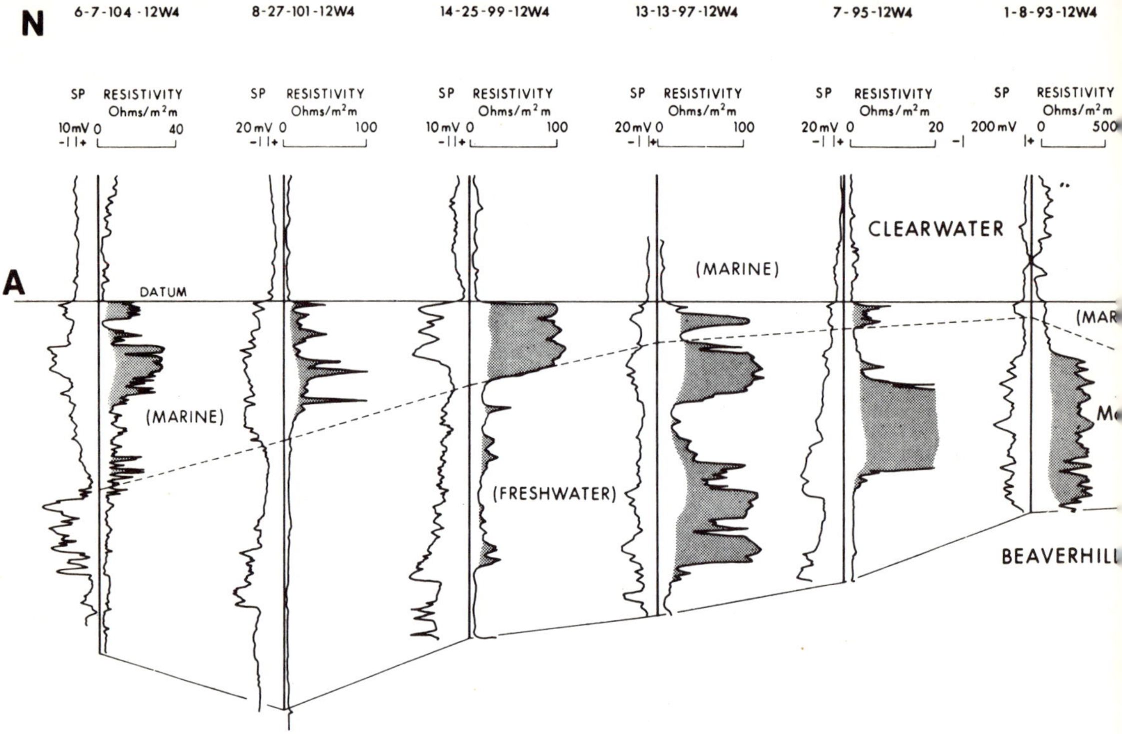


FIG. 3.—North-south electric log correlation with upper surface of Wabiskaw Member as datum. Oil-conten ern end of section. For location of section A-B

An isopach of the total thickness of oil-impregnated strata is shown in Figure 2, and the subsurface aspects of the beds are illustrated on the electric log correlation shown in Figures 3–5. The upper boundary of the Wabiskaw Member, a thin but persistent, poorly sorted clayey sandstone averaging 20 ft in thickness, has been used as a datum. The east-west correlations (Figs. 4, 5) can be carried only to the central part of the basin because the eastern half is believed to be barren (no wells have been drilled there).

The cross sections show that the formation consists of a series of collateral sandstone bodies, generally basal in position and interbedded with thick siltstones. One of the outstanding features of the McMurray Formation is the absence of correlatable units within the formation. The nature of the reservoir boundary is difficult to determine because of the lensing nature of the sandstone bodies. Cross section A-B (Fig. 3) shows oil-water contacts in the northern part of the section (well 7–95–12 W.4) and water-saturated sandstone under a gas cap in well 14–25–89–14 W.4 situated above a limestone ridge.

## Structure

The Cretaceous strata dip uniformly southwest toward the Rocky Mountains at a rate of 5 ft/mi with local exceptions. One of these is in a relatively small area of about 50 sq mi (130 sq km), 40 mi (65 km) north of Fort McMurray, where Cretaceous strata are found up to 200 ft (60 m) below their expected elevations as a result of post-Cretaceous collapse of Devonian limestones caused by salt removal from underlying strata (Carrigy, 1959a). The relation of salt tectonics to the accumulation of heavy oil in the Lower Cretaceous strata of western Canada is discussed by Vigrass (1966).

## Paleogeographic Reconstruction

The McMurray sedimentary basin, which owes its origin to leaching of soluble salts from underlying Middle Devonian evaporites, had a well-defined drainage system developed on its

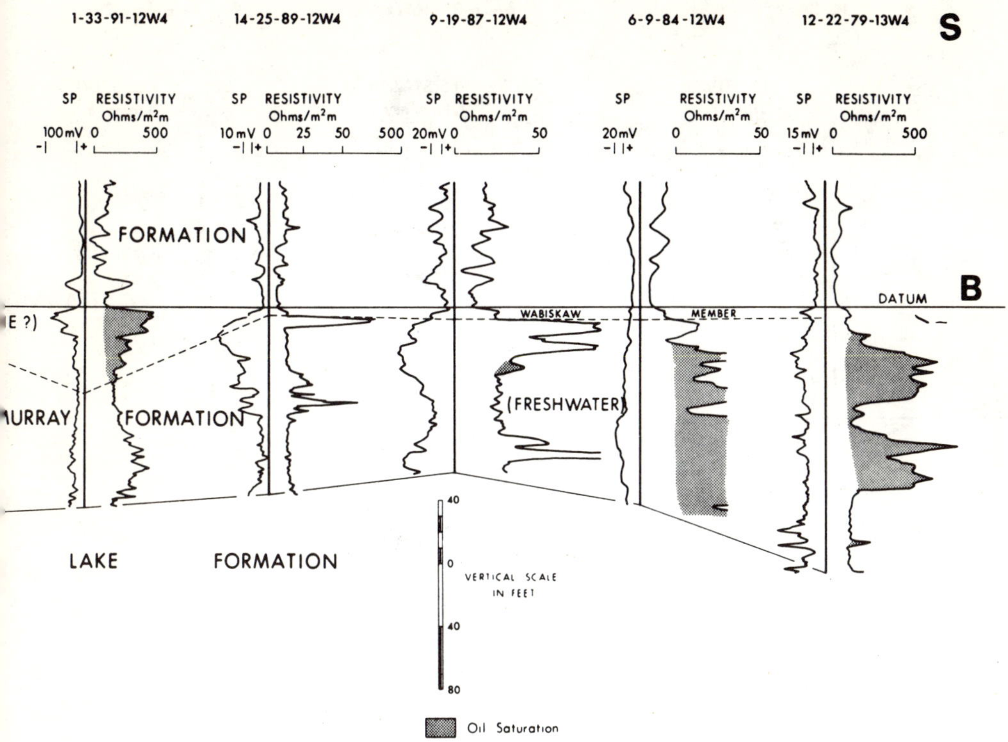


f sandstone is approximately proportional to resistivity. Note increase in thickness of marine sediments at north-
ee Figure 2. Distance A to B is 200 mi (320 km).

surface prior to the beginning of sedimentation in Early Cretaceous time. The topography was characteristic of a mature landscape, with wide valleys underlain by soft calcareous shales separated by ridges of resistant limestone (Martin and Jamin, 1963). The inferred drainage pattern (Fig. 6) strongly suggests that there were two areas of internal drainage just prior to the burial of this surface. These drainage channels commonly contain coarse-grained sandstone and conglomerate, some of which may be pre-McMurray in age (Carrigy, 1966). Probably as a result of continued salt removal, the drainage system was finally disrupted and the depression rapidly filled with water to form a large lake, at least 100 ft (30 m) deep, which stood at, or close to, sea level. Into this lake or lagoon a north-flowing river system built a series of small deltas (Fig. 6). When the depression was nearly filled with sediment, fingers of the boreal Clearwater sea, which had been slowly advancing from the north, began to encroach along the margins of the delta, as evidenced by thin shale beds containing brackish-water forams in the upper part of the McMurray succession. When the depression was filled with sediment, the boreal sea continued to advance south across the now swampy area. As the sea level rose, the limestone ridges became shoals over which the glauconite sandstone of the Wabiskaw Member was deposited as barrier-bar sand (Fig. 6). The final episode in this cycle was the inundation of the entire landscape by the Clearwater sea. The sedimentary structures and features associated with this sequence of environments (illustrated and described by Carrigy, 1967) are summarized in Table 1).

It can be deduced from the paleogeographic reconstruction presented in Figure 6 that deltaic sedimentation in the Athabasca deposit was controlled by the elongate shape of the basin of deposition and the relatively shallow depths of water in the basin. The deposit should possess the gross lithologic attributes associated with a valley-fill cycle of sedimentation, the relatively shallow depths of water resulting in homopyc-

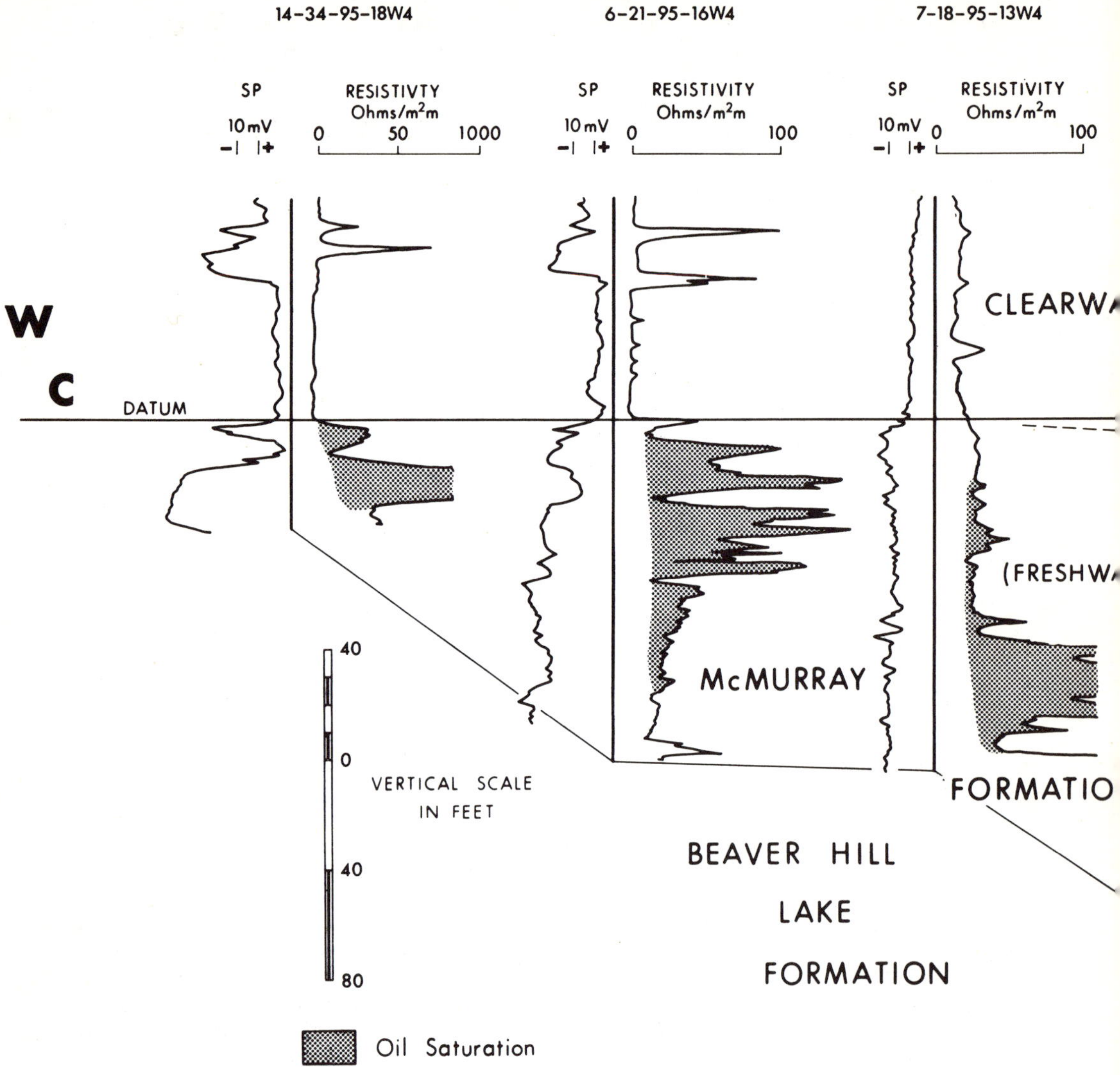


FIG. 4.—West-east electric log correlation with upper surface of Wabiskaw Member as datum. Only west Figure 2. Distance C to D

nal inflow (Bates, 1953), which theoretically deposits an arcuate symmetrical delta similar to the classic model of Gilbert (1885). Burial by the rapidly transgressing Clearwater sea subsequently preserved the lake deposits intact, and Holocene erosion has exposed a fairly complete cross section for study (Fig. 7).

## VERTICAL PROFILE

One of the objectives of this paper is to present the environmental evidence for shallow-water deltaic sedimentation in the Athabasca tar sands in a form that will be useful to the petroleum geologist. The rediscovery of Walther's Law, one expression of which states that the

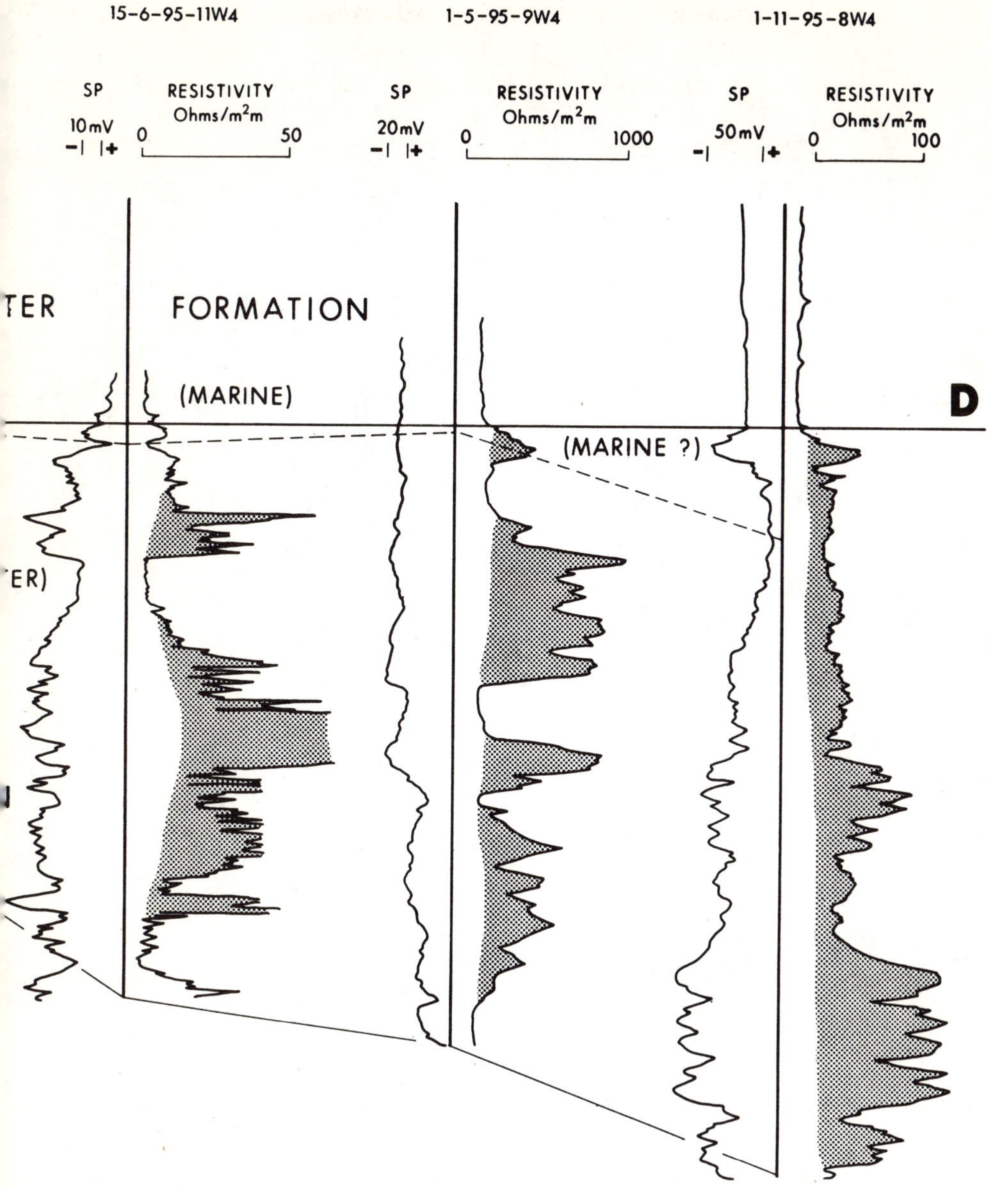


half of basin and axial location of lacustrine deltaic units rich in oil are shown. For location of section C-D, see is ± 125 mi (200 km).

vertical sequence of environmental products mirrors the lateral distribution of environments at the time of deposition (Visher, 1965; Masters, 1967), has stimulated petroleum geologists to construct vertical profiles based on environmental models (Visher, 1969). The acceptance of this concept gives rise to the expectation that, when sufficient well-documented vertical profiles of ancient deltas are available, comparison with profiles from unexplored sedimentary successions will give valuable clues to the environment of deposition much earlier in the exploration stage than is now possible. If we bear in mind the type of information that is

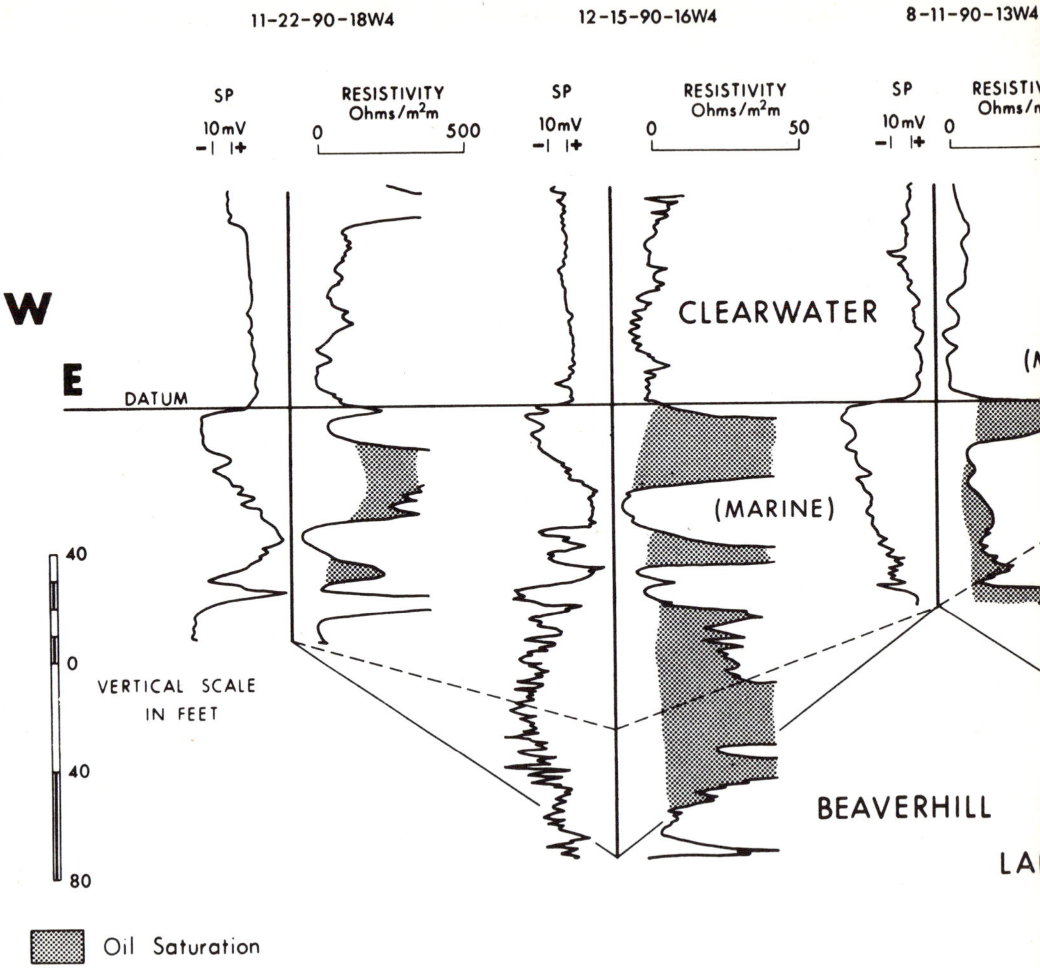


FIG. 5.—West-east electric log correlation with upper surface of Wabiskaw Member as datum. Note inte units rich in oil. For location of section E-F se

available and the type of information that is desirable, it would be ideal if the environmental information could be interpreted from various types of subsurface logs, singly or in combination.

Several workers have noticed that the self-potential curves of fluvial and marine sands seem to have certain characteristics which might be useful for this purpose (Visher, 1965, 1969; Cannon, 1966; Shelton, 1967; Hansen, 1969). Nevertheless, the environmental interpretation from a self-potential response can be regarded as tentative until confirmed by detailed study of a continuous core or observation of the same strata in outcrop. Several vertical profiles have been made on cores and outcrops from the Athabasca deposit (Carrigy, 1966). In this paper two wells have been chosen to illustrate the significant features of the assumed fresh and brackish shallow-water deltaic units of the McMurray Formation. Profiles of the median grain size, clay content, oil content, self potential, and resistivity have been plotted for each well (Fig. 8). The most obvious feature of these profiles is the uniform gradation from top to bottom in all five parameters. The grain-size data indicate a fining-upward profile coupled with an upward increase in clay-size material. Both these properties of the sediment are reflected in the self-potential curve, which

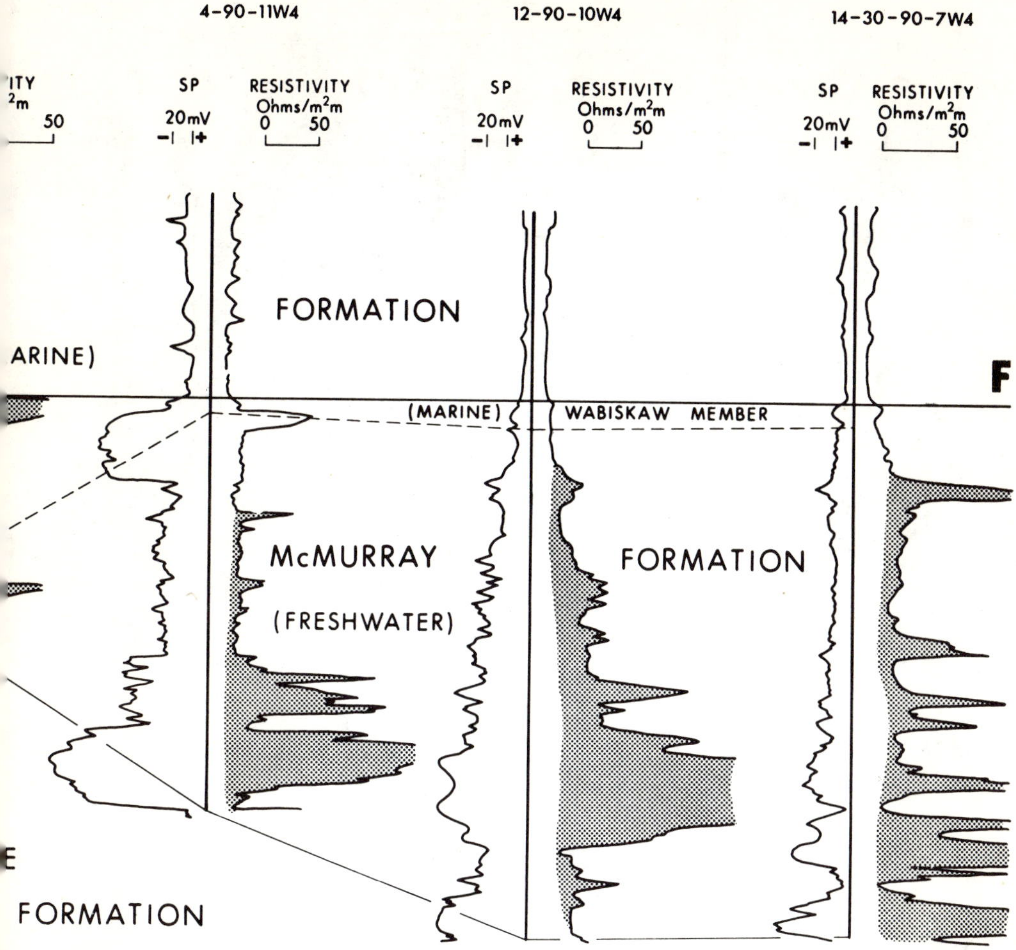


ngering of marine and freshwater sediments in western half of section and axial location of lacustrine-deltaic igure 2. Distance E to F is ± 125 mi (200 km).

shows its maximum negative deflection at the base and gradual return to the shale line at the top of the unit. The oil content of the unit also is greatest at the base, gradually decreasing toward the top. The oil saturation is picked up by the resistivity log, which traces a line showing maximum deflection at the base and a gradual return to the baseline at the top.

The sum of all these profiles gives a picture of sedimentation very similar to both the fluvial channel and valley-fill succession described by Visher (1965, 1969) and the fluvial-deltaic model of Nanz (1954) and Hansen (1969). Differentiation between the two environments on the basis of self-potential profiles does not seem possible. This picture bears out the theoretical expectation enunciated in an earlier section that the elongate shape of the basin of deposition would give these deltaic deposits strong fluvial affinities. Because of this effect, other lines of evidence would be required to identify unequivocally a similar unit as deltaic. Nevertheless, if it is assumed that the electric log responses characteristic of deltaic sandstone-siltstone complexes in this deposit are similar to those shown in Figure 8, a plot of the distribution of these self-potential curves from wells shows that these units are most abundant in the southern part of the deposit and in the axial part of the basin of deposition in the vi-

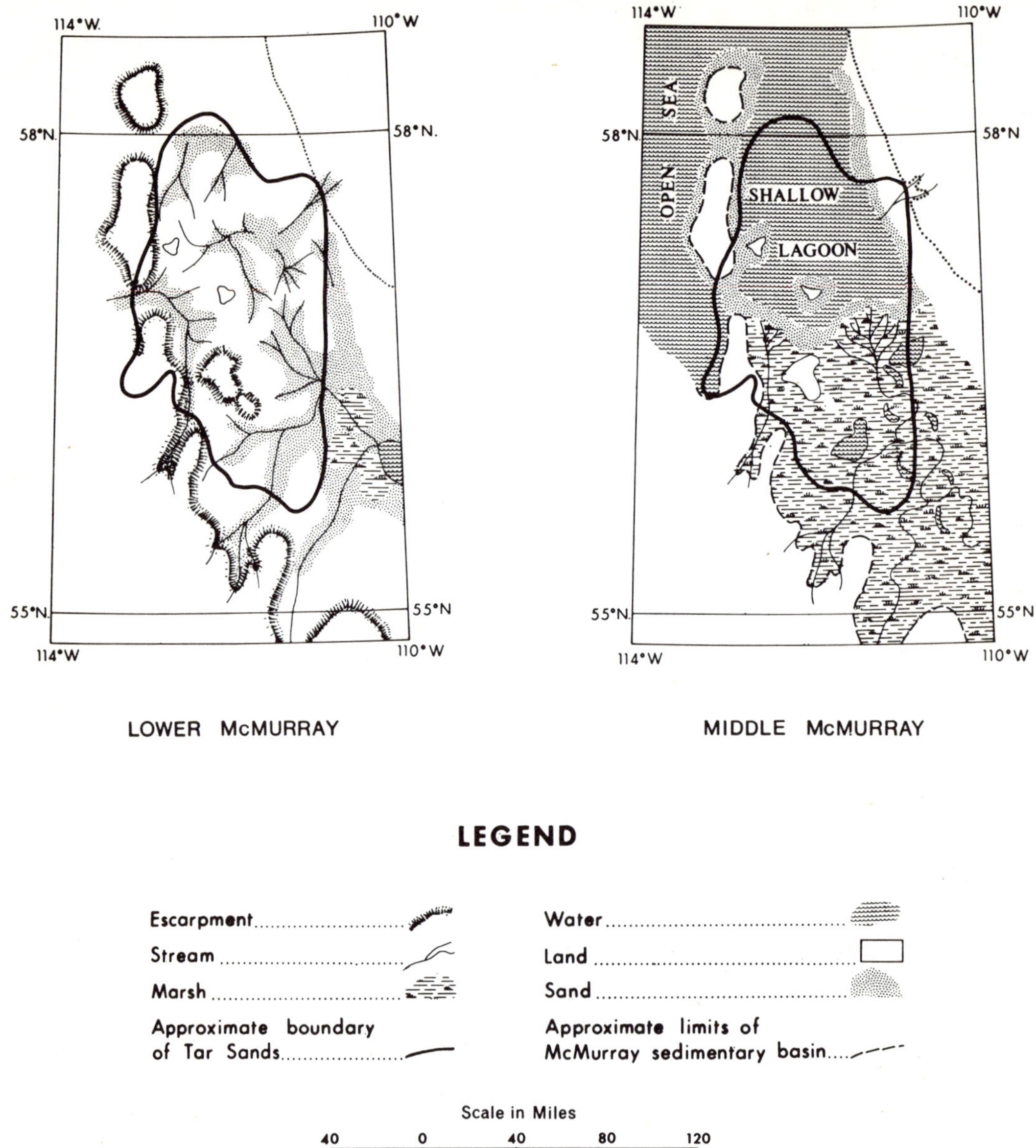


cinity of Fort McMurray. On the north and east they give way to the marine type self-potential curve (Table 1), which is the inverse of the fluvial model (Cannon, 1966; Visher, 1969) indicated in Figure 5.

### Criteria for Recognizing Freshwater Deltas

A review of the literature dealing with modern and ancient deltas has revealed few criteria which can be used universally for the recognition of ancient deltas. Shepard (1964) found five features characteristic of all large modern marine deltas. They are: (1) lamination consisting of very fine sand and silt alternating with clays; (2) a coarse fraction in which forams and mollusks are scarce, but wood fibers are abundant and ostracods possibly common; (3) unusually large amounts of mica in the sediments; (4) very minor discordance between

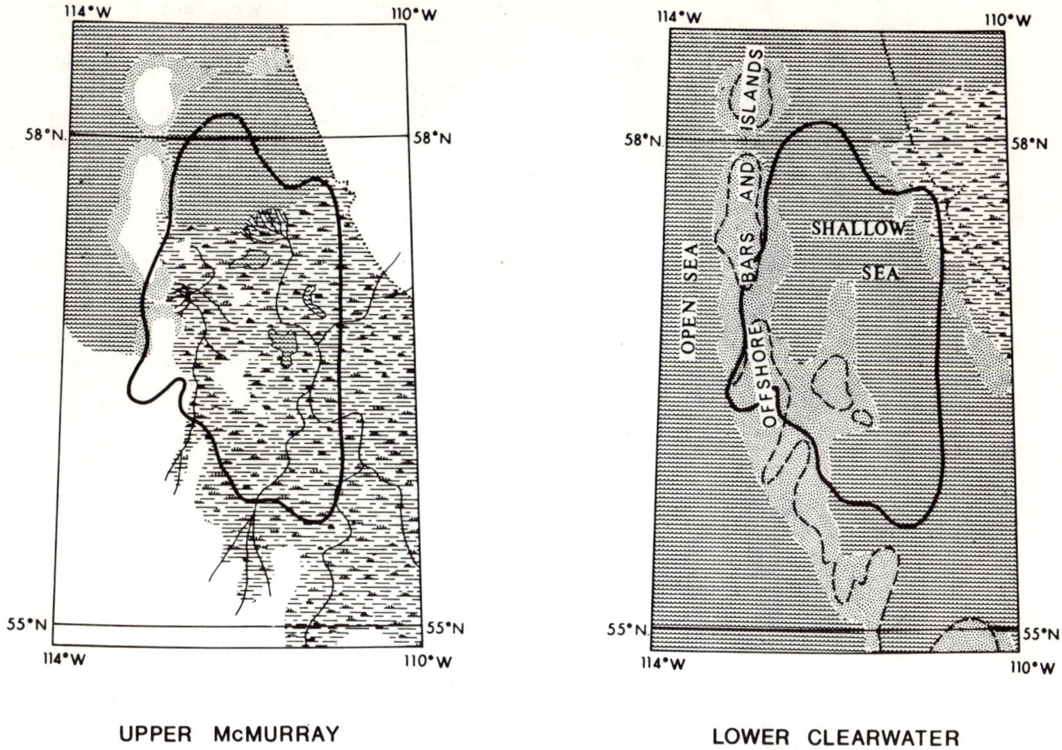


FIG. 6.—Paleogeographic reconstruction of depositional history of oil-impregnated Lower Cretaceous formations (lower McMurray, middle McMurray, upper McMurray, lower Clearwater) in northeastern Alberta.

topset and foreset beds; and (5) very minor discordance between foreset and bottomset beds.

Despite the small size of the McMurray deltas they meet the criteria of Shepard in all respects except for the absence of marine fossils in the foreset beds. It is pertinent at this point to speculate how a lacustrine-lagoonal delta might differ from a marine delta, if both were supplied by rivers of identical hydraulic characteristics carrying identical sediment loads.

Any classification of deltaic deposits must include criteria for distinguishing between deltas built by rivers carrying predominantly a bed load of medium- to coarse-grained sand, such as debris deposited by a young mountain stream, and those formed by mature rivers carrying predominantly a suspended load of fine sand and silt. The bed-load delta will be mainly subaqueous. The foreset beds will be deposited as avalanche deposits standing at or near the angle of repose, and their thickness will be controlled by the depth of water. This is the classical delta model of Gilbert (1885) and the one most often studied in the laboratory. Under the ideal conditions postulated above, this type of delta should be the same whether deposited in fresh or salt water.

A suspended-load delta is formed when the fine-grained sediment settles upon reduction of the velocity of a stream entering a standing body of water.

Bates (1953) demonstrated that the density of water has a marked effect on the current pattern at the mouth of a stream and hence the area of deposition of the suspended load. On entering a denser saltwater body, the flow from a stream does not mix in the vertical plane but tends to spread out horizontally. This results in a wider dispersal of sediment and hence a lower angle of slope for the foreset beds than in a freshwater body. Another effect of salt water on the suspended load is to flocculate the clays and organic matter, which results in poorer sorting of the fine-grained sediment deposited in a marine environment.

In addition to the effects of salt water on the discharge pattern of a stream and its flocculating effect on the fine-grained particulate matter,

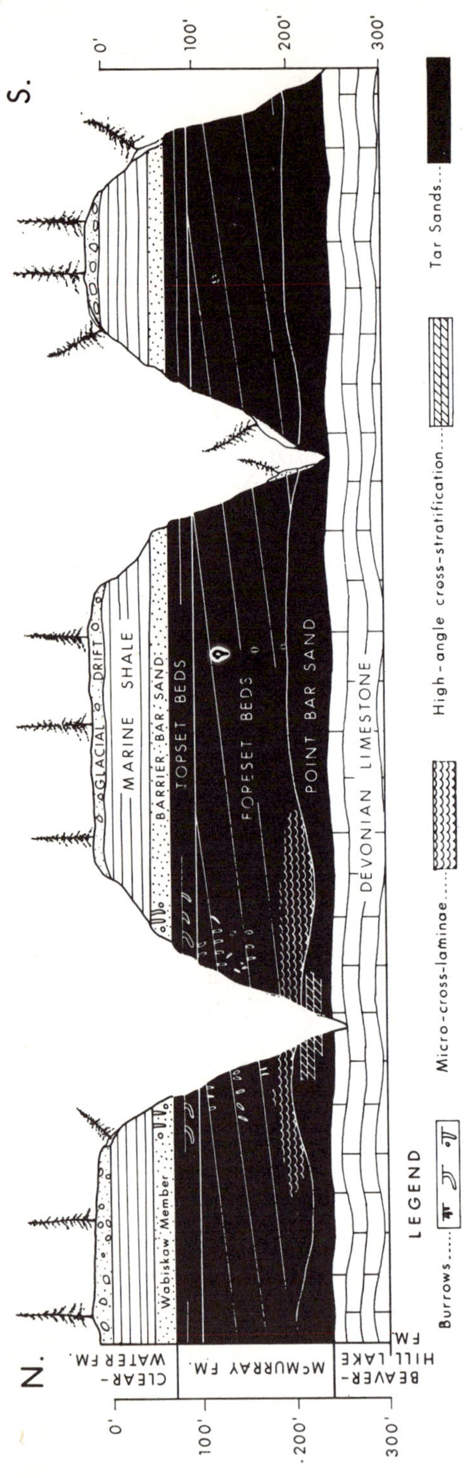


FIG. 7.—Idealized illustration of sedimentary structures and nomenclature of stratigraphic units within outcrops of oil-saturated cliffs on east bank of Athabasca River 3–4 mi (5–7 km) north of Fort McMurray.

the energy level of waves and tides in an open-sea environment also aids the wide dispersal of sediment by providing a greater variety of sub-environments and less distinction between topset and foreset beds than would be expected in a lake, where the areal extent and thickness of the deltaic complex are fixed. Owing to the lower energy of the destructive forces in lacustrine-lagoonal environment, the seasonal variations in the sediment load of streams are likely to be recorded as rhythmic successions of sand and silt in the foreset beds, as well as varves in the mud of the lake-floor bottomset beds. These features are more commonly preserved in the lake or lagoon environment, for sediment-burrowing animals are less abundant in lakes and lagoons than in a marine environment; however, they are not entirely absent in lakes, as Müller (1966) reports large populations of tubificidae in the delta-front area of the Rhine River in Lake Constance.

In summary, the features that distinguish an ancient lake or lagoonal suspended-load delta from an ancient marine suspended-load delta are the marked discordance between topset and foreset beds; the rhythmic intercalation of well-sorted fine sand and silt in the foreset beds as a result of seasonal variations in sediment load; relatively less burrowing activity and hence better preservation of major and minor sedimentary structures; the general absence of marine microfossils; the restricted stratigraphic extent of the foreset units, all of which are found within a single lenticular body with a flat upper surface and a concave lower surface; the lack of correlation of stratigraphic (lithologic) units within the delta which may extend over an area of several hundred to several thousand square miles; the absence of erosional features within the individual deltaic units; and the absence of marine intercalations.

### Relations Between Oil Impregnation and Environment of Deposition

The area of heavy oil impregnation in the lower Athabasca River area is superimposed on the paleogeographic maps in Figure 6. It is confined to the western half of the sedimentary basin, where it probably is trapped in place by hydrodynamic forces. Within the oil-impregnated area the thickest oil-rich sandstone bodies appear to be localized in the deepest depressions formed by pre-McMurray drainage channels (Stewart, 1963). Locally, the distribution of oil within the Athabasca reservoir is con-

**Table 1. Features of Environments of Sandstone Deposition in Athabasca Tar Sands**

| | PRE-DELTAIC | DELTAIC | | POST-DELTAIC |
|---|---|---|---|---|
| | Alluvial | Lacustrine — Lagoonal | | Marine |
| | Point bar and channel sands | Foresets | Subaqueous topsets | Barrier bar |
| Texture | Medium- to coarse-grained well-sorted sand | Very fine- to fine-grained well-sorted sand | Silt and very fine-grained sand | Sand, silt and clay poorly sorted<br>Locally clean and well-sorted |
| Sand composition | Quartz ≈95%<br>K-feldspar < 5%<br>Muscovite < 1% | Quartz ≈90%<br>K-feldspar < 5%<br>Muscovite < 5% | Quartz ≈90%<br>K-feldspar < 5%<br>Muscovite < 5% | Quartz ≈ 50%<br>Chert and volcanic rock fragments ≈ 25%<br>Glauconite ≈ 20%<br>Feldspar ≈ 5% |
| Clay composition | Kaolinite, Illite | Kaolinite, Illite | Kaolinite, Illite, Montmorillonite | Montmorillonite, Illite, Kaolinite, Chlorite |
| Organic matter | Mummified logs, abundant comminuted carbon<br>Some thin coal and lignite beds present | Abundant comminuted carbon | Macerated vegetable matter in clay matrix | |
| Mineral cements | Siderite (rare)<br>Pyrite nodules (rare)<br>(up to 6 inches diam.) | Siderite (rare) | Siderite (common) and calcite | Siderite (common) |
| Fossil content | Spores and pollen | Spores and pollen | Some lenticular beds of brackish water gastropods and pelecypods<br>Agglutinated foraminifera<br>Pollen and spores<br>Hystrichospherids | Calcareous foraminifera<br>Radiolaria (pyritized)<br>Sponge spicules (pyritized)<br>Hystrichospherids<br>Diatoms (pyritized) |
| Sedimentary structures | Medium scale (25 cms), straight, high-angle, cross beds (common) | Large scale (30 meters), low-angle (5 - 7°), foreset beds<br>Small scale, micro-cross-lamination<br>Burrows and castings (common) | Lenticular sand layers in silt<br>Crab? burrows (rare) | Horizontal and vertical tubes filled with clean white sand |
| Bedding | Lenticular cosets of cross stratified sets | Silt beds (1 cm thick) interbedded with very fine-grained sand beds (10 cm thick) | Laminated micaceous silt with lenticles of sand | Bedding destroyed by animal activity |
| Boundaries | Upper and lower boundaries sharp | Lower boundary sharp, upper boundary transitional | Upper and lower boundaries transitional | Upper boundary sharp, lower boundary transitional |
| Thickness | 2 - 7 meters | 20 - 30 meters | 5 - 10 meters | 2 - 10 meters |
| Ideal S.P. response | 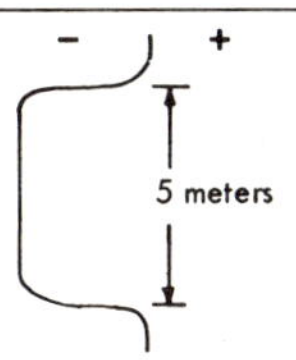  | 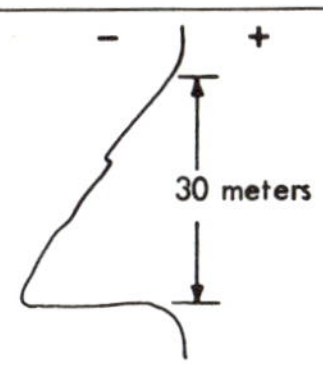  | 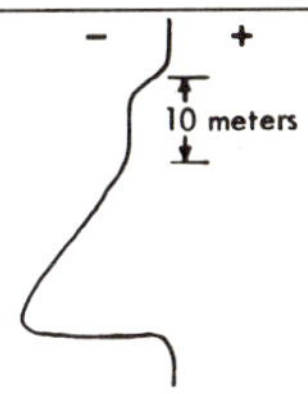  | 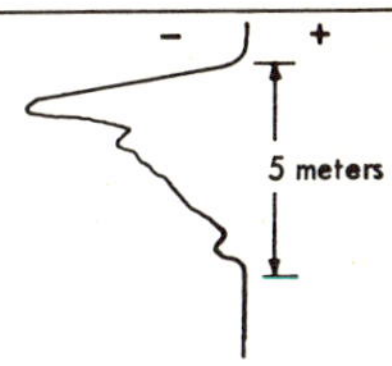  |
| Location | Lower one third of the McMurray Formation | Lower two thirds of the McMurray Formation | Upper one third of the McMurray Formation | Basal Clearwater Formation |
| Oil content | Very good, uniform (up to 35% by volume) | Fair to good, uniform (20 - 30% by volume) | Poor (5 - 10% by volume) | Very poor, variable (local bodies of rich tar sand) |

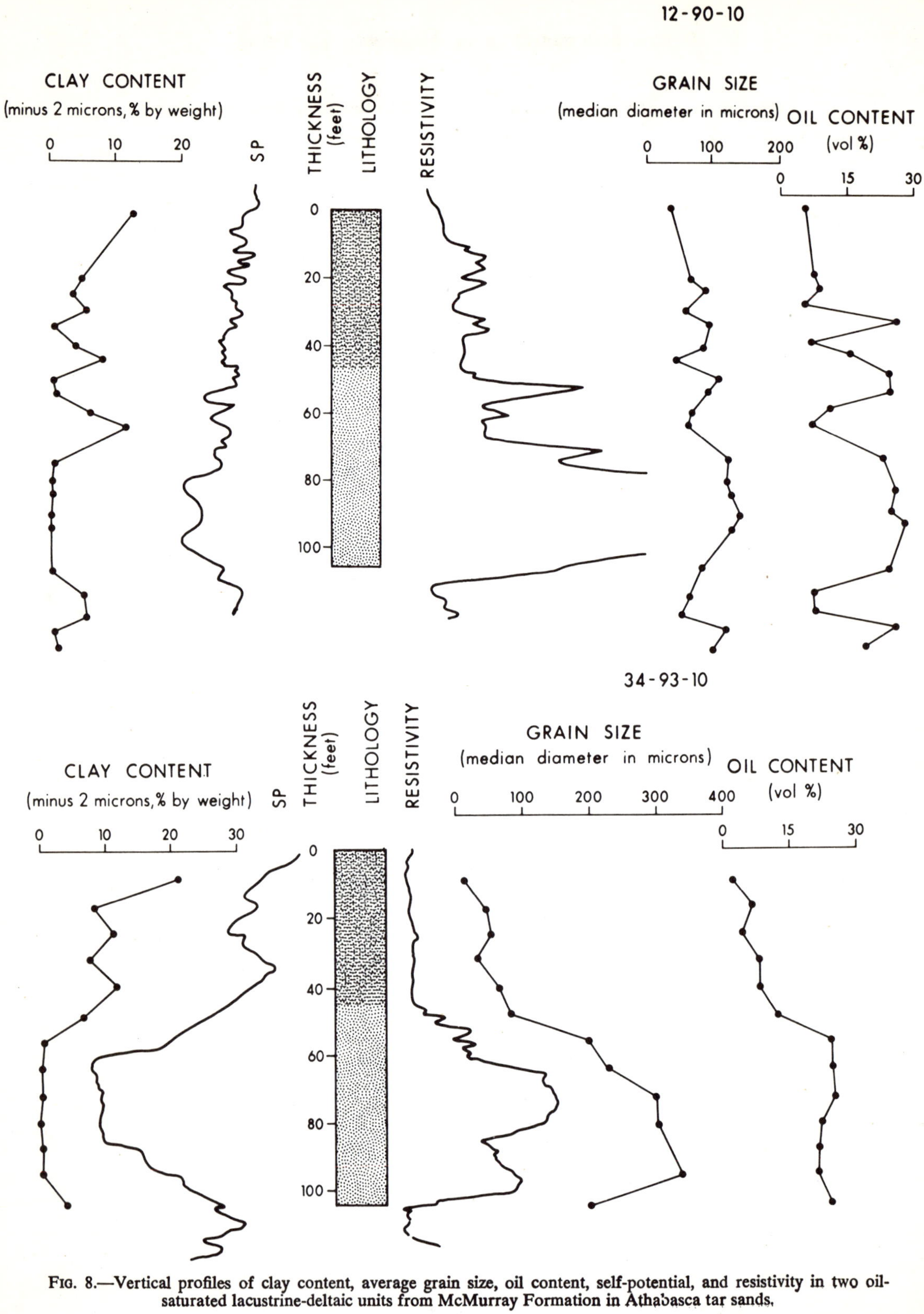


FIG. 8.—Vertical profiles of clay content, average grain size, oil content, self-potential, and resistivity in two oil-saturated lacustrine-deltaic units from McMurray Formation in Athabasca tar sands.

trolled strictly by the original porosity and permeability of the sediments (Carrigy, 1959b, 1962). Thus the maximum amount of oil possible (18–20 percent by weight) is present in well-sorted, clean sandstones which are most commonly of fluvial or lacustrine-deltaic origin. No oil is present in the poorly sorted argillaceous phases of the nearshore marine sandstones of the Wabiskaw Member of the Clearwater Formation, except in areas above buried hills on the limestone surface. In the deltaic units the oil content is controlled by the amount of clay present in the fine sandstone and siltstone, and is concentrated in the lower part of the foreset unit. The topset beds are very poor in oil (Table 1).

Sandstone bodies that might have been reservoirs if the gravity of the oil had been greater are very numerous. Stewart (1963) has outlined about 50 sandstone bodies greater than 100 ft (30 m) thick and averaging more than 8 percent oil by volume (4 percent by weight) within the Athabasca deposit; these bodies are within the channels of the ancient river system beneath the McMurray Formation. It appears that the sequence of environments outlined has produced a sedimentation pattern consisting of many multilateral sandstone bodies that are thickest in the axial part of the basin. In dealing with ancient deposits of this nature, it is obvious that a comprehensive regional study using all available evidence is necessary to make valid environmental reconstructions of ancient subsurface shallow-water delta deposits.

### References Cited

Alberta Oil and Gas Conservation Board, 1963, A description and reserve estimate of the oil sands of Alberta: Edmonton, Govt. Alberta, 60 p.

Bates, C. C., 1953, Rational theory of delta formation: Am. Assoc. Petroleum Geologists Bull., v. 37, p. 2119–2162.

Cannon, J. L., 1966, Outcrop examination and interpretation of paleocurrent patterns of the Blackleaf Formation near Great Falls, Montana: Billings Geol. Soc. 17th Ann. Field Conf. Guidebook, p. 71–111.

Carrigy, M. A., 1959a, Geology of the McMurray Formation, pt. 3, General geology of the McMurray area: Research Council Alberta Mem. 1, 130 p.

——— 1959b, The significance of a grain size classification of the sands of the McMurray Formation, Alberta: 5th World Petroleum Cong. Proc., v. 1, p. 575–590.

——— 1962, Effect of texture on the distribution of oil in the Athabasca oil sands, Alberta: Jour. Sed. Petrology, v. 32, p. 312–325.

——— 1966, Lithology of the Athabasca oil sands: Research Council Alberta Bull. 18, 48 p.

——— 1967, Some sedimentary features of the Athabasca oil sands: Sedimentary Geology, v. 1, p. 327–352.

Gilbert, G. K., 1885, The topographic features of the lake shores: U.S. Geol. Survey Ann. Rept. 5, p. 104–108.

Hansen, H. J., 1969, Depositional environments of subsurface Potomac Group in southern Maryland: Am. Assoc. Petroleum Geologists Bull., v. 53, p. 1923–1937.

Martin, R., and F. G. Jamin, 1963. Paleogeomorphology of the buried Devonian landscape in northeastern Alberta: Research Council Alberta, Inf. Ser. 45, p. 31–42.

Masters, C. D., 1967, Use of sedimentary structures in determination of depositional environments, Mesaverde Formation, Williams Fork Mountains, Colorado: Am. Assoc. Petroleum Geologists Bull., v. 51, p. 2033–2043.

Müller, G., 1966, The new Rhine delta in Lake Constance, *in* M. L. Shirley, ed., Deltas in their geologic framework: Houston Geol. Soc., p. 107–124.

Nanz, R. H., 1954, Genesis of Oligocene sandstone reservoir, Seeligson field, Jim Wells and Kleberg Counties, Texas: Am. Assoc. Petroleum Geologists Bull., v. 38, p. 96–117.

Pow, J. R., G. H. Fairbanks, and W. J. Zamora, 1963, Descriptions and reserve estimates of the oil sands of Alberta: Research Council Alberta Inf. Ser. 45, p. 1–14.

Shelton, J. W., 1967, Stratigraphic models and general criteria for recognition of alluvial, barrier bar, and turbidity-current sand deposits: Am. Assoc. Petroleum Geologists Bull., v. 51, p. 2441–2461.

Shepard, F. P., 1964, Criteria in modern sediments useful in recognizing ancient sedimentary environments, *in* L. M. J. V. van Straaten, ed., Deltaic and shallow marine deposits: Amsterdam, Elsevier, p. 1–25.

Stewart, G. A., 1963, Geological controls on the distribution of Athabasca oil sand reserves: Research Council Alberta Inf. Ser. 45, p. 15–26.

Vigrass, L. W., 1966, General geology of Lower Cretaceous heavy oil accumulations in western Canada: Canadian Mining Metal. Bull. 59, p. 87–94.

Visher, G. S., 1965, Use of vertical profile in environmental reconstruction: Am. Assoc. Petroleum Geologists Bull., v. 49, p. 41–61.

——— 1969, How to distinguish barrier bar and channel sands: World Oil, v. 168, p. 106–113.

**Reprinted from:**
BULLETIN OF THE AMERICAN ASSOCIATION OF PETROLEUM GEOLOGISTS
VOL. 55, NO. 8 (AUGUST, 1971), PP. 1179-1193, 9 FIGS.

# Petroleum Related to Middle and Upper Devonian Deltaic Facies in Central Appalachians[1]

**JOHN M. DENNISON**[2]
**Chapel Hill, North Carolina 27514**

**Abstract** Strata of the Devonian subaerial and prodelta complex are preserved in 110,000 sq mi (260,000 sq km) of the Appalachian basin. Petroleum is obtained only from the western half, principally from higher permeability marine sandstones, although some is from prodelta dark shales. Absence of production in the east presumably is due to the escape of hydrocarbon from folds which bring potential reservoir strata to outcrop level.

Outcrops along the Allegheny Front from Maryland to Virginia were studied to interpret physical stratigraphy and environments of deposition and to understand better the factors controlling petroleum occurrence.

The deltaic clastic sediment came from an eastern quartzose source located in latitude from New York to Virginia. Depositional patterns of petroliferous sands resulted from gradual infilling of a subsiding basin, interrupted by rather abrupt eastward transgressions, which produced nearly synchronous, winnowed sands traceable across many counties. The earliest prodelta (Needmore Shale of Onesquethaw Stage) was derived from an upland near Baltimore. Cazenovia Stage deltaic siltstones and sandstones accumulated in Pennsylvania and New York. In late Tioughnioga time the winnowing zone extended to West Virginia (Clearville Siltstone) while terrestrial redbeds were accumulating in the Catskill Mountains. Early in the Taghanic, hundreds of feet of eustatic sea level rise shifted the shoreline scores of miles eastward. Throughout the rest of the Late Devonian the shoreline prograded generally westward, with fairly sudden eastward shifts of tens of miles during times of eustatic sea level rise. Younger petroliferous sandstones of the Cassadaga and Bradford stages accumulated in the western part of the basin during winnowing that accompanied shoreline shifts. Maximum westward encroachment of the subaerial Catskill delta was during Bradford times, followed by Early Mississippian transgression which produced Pocono Group sandstones.

Eustatic sea level changes leave sedimentary imprints which should be traceable for long distances in shallow marine sediments. Specific correlations from Virginia to New York are developed by matching times of sea level fluctuation.

## Introduction

This paper summarizes general patterns of petroleum occurrence in the Devonian Catskill delta complex in the central Appalachians and relates the writer's outcrop studies in Maryland, West Virginia, and Virginia to known petroleum and future prospecting. A fresh look at a classic area provides insight into sedimentologic patterns of economic significance.

The 1859 Drake well at Titusville, Pennsylvania, was completed in the top of the Bradford Stage of Catskill delta strata (Dickey, 1941, p. 24), and tens of thousands of wells subsequently have penetrated the Upper Devonian deltaic rocks. Most of the hundreds of Devonian delta oil and gas fields in the region were discovered before modern geologic methods were available, and detailed information is scanty on older wells.

Only in the last decade have detailed stratigraphic studies been linked with a knowledge of sedimentology and paleoecology in outcrops on all sides of the Appalachian basin. My work has been concentrated on outcrops on the east side of the Appalachian basin, where exposures along the Allegheny Front from Maryland to Virginia provide a line of stratigraphic sections of the complete deltaic sequence as close as possible to the oil and gas production in the southern part of the basin. The distribution and character of the deltaic sediments in these outcrops yield interpretations that provide better understanding of the petroleum in the subsurface part of the basin.

## Delta Stratigraphy

The Middle and Upper Devonian Catskill delta and prodelta complex is preserved over about 110,000 sq mi (260,000 sq km) of the Appalachian basin, extending from New York southward to easternmost Tennessee. The general Devonian stratigraphy, lithofacies, and thickness patterns in the Appalachian basin have been summarized by Oliver *et al.* (1967, 1969). A generalized correlation chart (Fig. 1) provides sufficient terminology to review the

[1] Manuscript received, October 14, 1970; accepted, December 28, 1970. Read before the Association at Calgary, Alberta, June 24, 1970.

[2] Geology Department, University of North Carolina at Chapel Hill.

Parts of this research were supported by grants from the West Virginia Geological and Economic Survey and from the Research Council of University of North Carolina. I acknowledge the fruitful discussions with J. D. Glaeser, Gene Haney, L. S. Matteson, W. R. Wagner, and W. H. Wheeler.

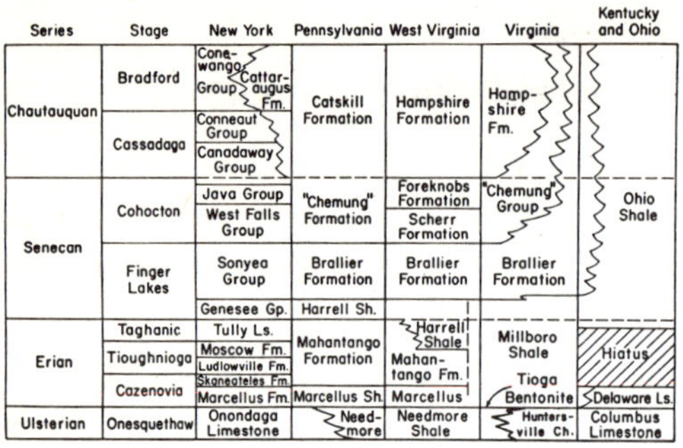


FIG. 1.—Generalized correlation chart of Catskill delta and prodelta strata in Appalachian basin.

stratigraphic framework associated with petroleum. The region with true delta or prodelta influence is probably confined to the area in which the total Devonian thickness exceeds 500 ft (Oliver *et al.*, 1967, Fig. 10).

### Ulsterian Series

At the approximate time of Deerpark deposition, seas withdrew from the continent to form the Wallbridge discontinuity (Wheeler, 1963). In the deeper parts of the Appalachian basin marine sedimentation was probably continuous between the Deerpark and Onesquethaw, but around nearly all of the basin margin a post-Oriskany unconformity occurs beneath the Onesquethaw Stage (Dennison, 1961). The shoreline oscillated during the lowered sea levels of latest Helderberg and Deerpark time, so that multiple unconformities mark the approximate position of the Wallbridge discontinuity in the Appalachian basin, in contrast to a distinct break of much longer duration in cratonic areas farther west (Dennison, 1970a).

The marine sediments of post-Oriskany age show evidence of a newly uplifted clastic source. The marine Needmore Shale is a prodelta mud marking the first appearance of Catskill delta clastics arcing around and coarsening toward an eastern source area in the latitude of Baltimore (from the site of the present continental shelf or Atlantic coastal plain, or in northwest Africa if one prefers continental drift interpretations). The dark mudstone and interbedded limestone of the Needmore Shale extend about 120 mi (190 km) westward toward the center of the Appalachian basin before changing facies into the Onondaga Limestone on the north or Huntersville Chert on the south and west (Dennison, 1961).

### Erian Series

The Erian Series in the classic Devonian reference region of western New York is a sequence of marine shale and siltstone punctuated stratigraphically by limestone marker beds. The most widespread of these is the Tully Limestone, which extends over much of the Appalachian basin (Heckel, 1966; Austin, 1969). Heckel (1969) described it in outcrop in Pennsylvania, and Hasson and Dennison (1968) extended the known distribution of the Tully bedded or concretionary limestone outcrops southward along the west side of the Ridge and Valley Province from southernmost Pennsylvania to Bath County, Virginia, in the type section of the Millboro Shale.

Generally thinner limestone marker beds punctuate stratigraphic division of the clastic Hamilton Group of New York into the Marcellus, Skaneateles, Ludlowville, and Moscow Formations. Eastward in New York all the Erian Series is reported to pass into the conspicuously reddish deltaic subaerial strata of the Catskill Mountains, which provide the type rocks of the Catskill delta. This transition eastward into nonmarine deltaic beds was summarized by Friedman and Johnson (1966). The New York stratigraphic names in Figure 1 apply to the western part of the state, where petroleum occurs in the Catskill delta complex; the nonmarine strata of the Catskill Mountain region are not indicated in the chart.

Southward in Pennsylvania, the Erian Series limestones are less conspicuous, and the Mahantango Formation encompasses time equivalents of several New York formations. The Mahantango is a southeastern-source siltstone-sandstone wedge which separates the black, deep-water prodelta shales of the Harrell and Marcellus Formations. Farther southwest the Mahantango grades to black shale, so that Millboro Shale deposition spans the entire time range of the Marcellus through Genesee dark shales of New York (Hasson and Dennison, 1968).

On the western edge of the Appalachian basin dark shales change facies into the Delaware Limestone of Ohio. An erosional hiatus precedes the preserved prodelta deposits of the Ohio Shale in Ohio (Hoover, 1960) and Kentucky. Paleoecologic interpretation over a broad area led Johnson (1970, p. 2083) to interpret an overlap over more than 800,000 sq mi (2,072,000 sq km) of North America. A eustatic sea level rise of up to hundreds of feet is consistent with the transgressive relations I have observed above the Mahantango shallow-water strata in central Appalachian outcrops.

### Senecan Series

Sea level rise at the beginning of the Senecan Epoch was accompanied by widespread black prodelta muds from western New York (Genesee Group) to Maryland (Harrell Shale) to southwest Virginia (Millboro Shale). Immediately thereafter, turbidite deposits of the Brallier Formation and Sonyea Group began to fill in the subsiding basin. In Pennsylvania and southeastern New York, delta front marine sands accumulated, and these merged eastward into delta plain redbeds. During Finger Lakes and especially Cohocton Ages subaerially deposited sediments displaced the shoreline westward, so that by the end of the Senecan Epoch the strandline was at the general position of west-central New York, westernmost Maryland, and Staunton, Virginia. Senecan prodelta turbidite siltstones reached the west border of West Virginia, and black Ohio Shale accumulated over the extreme west of the basin.

### Chautauquan Series

The conspicuous redbeds of the Catskill Formation of west-central Pennsylvania, the Cattaraugus Formation of western New York, and Hampshire Formation of Virginia and West Virginia represent the maximum western extent of subaerial deposits of the Catskill delta complex. These interfinger with marine sandy beds of the Canadaway, Conneaut, and Conewango Groups in western New York, a relation that extends in the subsurface across the western half of the Appalachian basin through westernmost Pennsylvania and west-central West Virginia, and can be examined on outcrop in Virginia near Roanoke, where the Hampshire Formation intertongues with the marine "Chemung" formation.

Paleocurrent studies suggest a Late Devonian continuous eastern landmass (Africa?) from the latitude of New York (Burtner, 1963; Woodrow and Fletcher, 1967; Sutton *et al.*, 1970) southward (McIver, 1970, p. 78–79) to the Virginia-Tennessee border (Jones and Dennison, 1970, p. 648), but there is no evidence of a significant eastern landmass shedding clastic material south of there.

Areas of persistent clastic influx causing westward extension of subaerial accumulations have been named the Snyder and Fulton lobes after counties in Pennsylvania (Willard, 1939, p. 273). A similar accumulation lobe in Virginia is the Augusta lobe, named for Augusta County (Dennison, 1970b). The lobes probably persisted near the mouths of major river systems draining the eastern source area. Between these accumulation lobes were rather persistent embayments—Perry bay between the Snyder and Fulton lobes in Pennsylvania (Willard, 1939, p. 273) and Grant bay (here named for Grant County, West Virginia) between the Fulton and Augusta lobes. Evidence for these embayments and lobes appears in the geographic distribution of clastic sediments through much of the Senecan and Chautauquan Epochs; however, the shoreline shifted westward intermittently as the sediment influx displaced the sea.

Lower Mississippian Pocono strata overlie the redbeds of the Devonian subaerial delta and its sandy marine equivalents. The Pocono abounds in better sorted, yellowish-gray sandstone and siltstone, and in West Virginia it has scattered reddish beds in the lower part. Pocono deposition resulted in part from marine transgression over the Catskill delta complex and in part from subaerial delta deposits centered in northeastern Ohio (Berea delta of Pepper *et al.*, 1954), Maryland (Gay-Fink delta), and southwest Virginia (Carolina delta).

## Petroleum Distribution

Petroleum production associated with the Catskill delta strata is limited to the western half of the Appalachian basin as a result of the structural setting. Hydrocarbons have escaped from the eastern part of the depositional basin, apparently because the folded Devonian delta strata are erosionally truncated in the Ridge and Valley province and in the higher anticlines in eastern parts of the Appalachian Plateau. The principal producing zones are summarized in Figure 2 according to age and geographic locale. The names in parentheses are potential producing zones in coarse clastics. Approximately 2 billion bbl of oil has been produced from the Upper Devonian strata of the basin, as well as important amounts of gas.

The chief production is from various shallow marine sandstone and siltstone reservoirs. These sands and silts accumulated just offshore (westward) from the Catskill redbed facies at several times in the Chautauquan Epoch. Specific oil and gas fields within the permeable extent of the sandstone and siltstone bodies are controlled structurally by gentle folds and by varying permeability in the Appalachian Pla-

| Series | Stage | New York | Pennsylvania | West Virginia | Ohio, Ky. and Va. |
|---|---|---|---|---|---|
| Chautauquan | Bradford | | Venango sands<br>Fourth<br>Fifth | Gantz<br>50 foot<br>30 foot<br>Gordon Stray<br>Gordon | |
| | Cassadaga | Bradford 1st<br>Sugar Run<br>Chipmunk<br>Bradford 2nd<br>Harrisburg Run<br>Bradford 3rd<br>Kane | Bayard<br>Elizabeth<br>Warren<br>Speechley<br>Tiona<br>Balltown<br>Bradford | Fourth<br>Fifth<br>Bayard<br>Elizabeth<br>Speechley<br>Balltown<br>Bradford<br>Riley<br>Benson | Ohio Shale |
| Senecan | Cohocton | | (Pound Ss.)<br>(Briery Gap Ss.) | (Valley Head Ss.)<br>(Elkins Ss.)<br>(Pound Ss.)<br>(Briery Gap Ss.)<br>Ohio Shale | |
| | Finger Lakes | | | (Parkhead Ss.) | |
| Erian | Taghanic | | | | |
| | Tioughnioga | | (Clearville Siltst.) | (Clearville Siltst.) | |
| | Cazenovia | | | | |

FIG. 2.—Age of Devonian delta petroleum sandstones and petroliferous Ohio Shale, plus (in parentheses) other permeable clastic units which should be considered potential reservoir strata.

teau. Published state maps show the location of individual fields (Graham, 1952; Johnson *et al.*, 1962; Pharr and Calver, 1963; Calvert, 1964; Lytle and Heeren, 1964; Cardwell *et al.*, 1970). Figure 3 shows the general areas of the southern Appalachian basin that contain the numerous pools in the Gordon, Fourth, Benson, and Riley sandstones. Within an area several counties in size, each of these sandstones contains scattered oil or gas pools. Many fields produce from more than one sandstone. Because of high carbon ratio, only gas is present in the eastern part of the region indicated for the Benson and Riley sandstones. In Pennsylvania major Venango sandstone petroleum production comes from postulated barrier bars trending north to northeast (Dickey *et al.*, 1943; Ingham *et al.*, 1949). The westernmost encroachment of Catskill delta terrestrial redbeds (Fig. 3) came somewhat later in Bradford time.

Some gas and even less oil are produced from prodelta black shales or silty shales at localities indicated by the Ohio Shale pattern in Figure 3. These shales generally require fracturing of hundreds of feet of shaly strata to yield commercial production.

The principal current Catskill delta petroleum exploration in the southern part of the Appalachian basin is in the southeastward extension of Benson sandstone gas production (Haught, 1965, 1966). The area of Ohio Shale gas and oil production continues to be enlarged slowly. There is also some drilling in the vicinity of old fields in other sandstones.

## ALLEGHENY FRONT STRATIGRAPHIC SECTIONS

For nearly a decade I have been studying facies changes in the Devonian delta strata exposed along the Allegheny Front from Maryland south to Tennessee (Dennison and Naegele, 1963; Dennison, 1963; Dennison *et al.*, 1967). Studies of the Upper Devonian subaer-

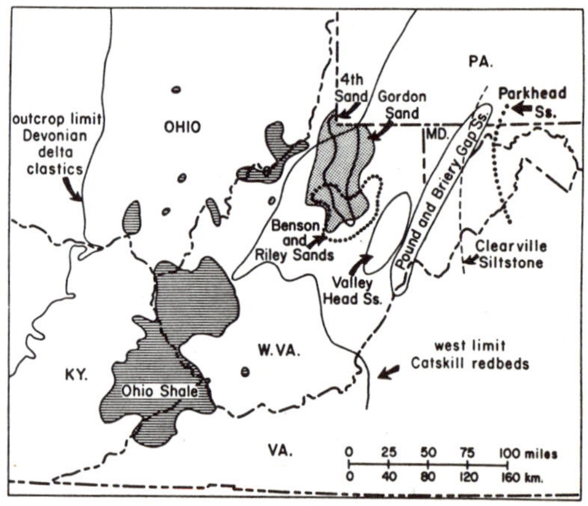


FIG. 3.—Oil and gas production occurs from individual fields within areas indicated for Ohio Shale and Gordon, Fourth, Benson, and Riley sandstones. Valley Head, Pound, Briery Gap, Clearville, and Parkhead strata are locally well sorted in indicated area of outcrop, and should be considered potential gas reservoirs in subsurface.

ial delta and associated delta front and prodelta marine beds concentrated on 11 sections (Fig. 4) along a 96-mi segment extending from Corriganville, Maryland, to U. S. Route 250 in Highland County, Virginia. Over 50,000 ft of sections were measured in bed-by-bed detail. Figure 5 is a stratigraphic cross section of the Devonian System in this region.

A revised stratigraphic nomenclature for the marine part of these Allegheny Front strata has been published (Dennison, 1970b). Strata formerly called the Chemung Formation along the Allegheny Front are named the Greenland Gap Group because Chemung is no longer used as a mapping formation in New York (Rickard, 1964; Oliver *et al.*, 1967, 1969). Along the Allegheny Front the Greenland Gap Group is over ½-mi thick and is divided into two distinct and mappable units (Scherr and Foreknobs Formations). In other outcrop belts of the folded Appalachians, stratigraphic division of the original Chemung has not yet been accomplished formally, so that away from the Allegheny Front in Virginia, Maryland, and West Virginia, I advocate use of the phrase "Chemung" formation (with quotes as indicated) until detailed stratigraphic and mapping studies progress far enough that the term "Chemung" formation can be supplanted by formal stratigraphic units.

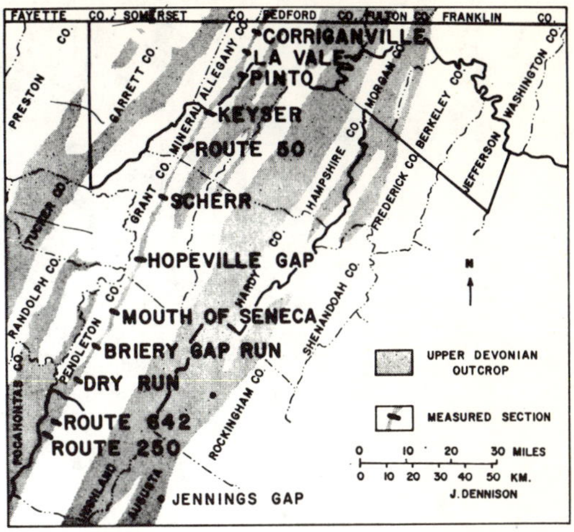


FIG. 4.—Location of stratigraphic sections measured along Allegheny Front in Maryland, West Virginia, and Virginia.

## Millboro Shale

The Millboro Shale in Figure 5 is prodelta black shale, split into lower (Marcellus) and upper (Harrell) black shale units in the northeast by the Mahantango Formation prodelta pulse of mostly silty shale to some very fine sandstone (Clearville Member at top).

## Brallier Formation

Brallier strata (Fig. 6) are characterized by sharply bounded, graded siltstones of turbidite origin scattered throughout dark-gray shale. The Brallier is 575–2,170 ft (175–661 m)

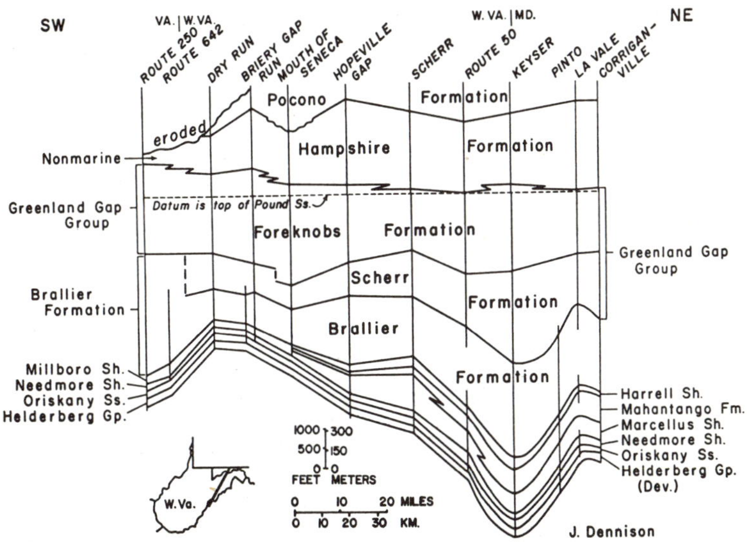


FIG. 5.—Stratigraphic cross section of Devonian System along Allegheny Front in region indicated in Figure 4.

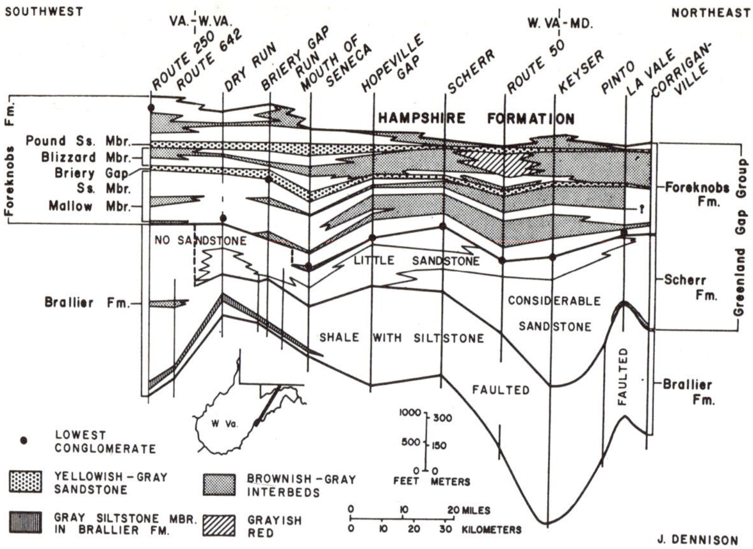


FIG. 6.—Details of Brallier Formation and Greenland Gap Group along Allegheny Front in region indicated in Figure 4.

thick, and locally is partly cut out by faulting of the sections. Near the base of the Brallier in the southwest is a prominent zone of coarser turbidite siltstones which were erroneously mapped by Tilton *et al.* (1927) as Chemung Formation.

### Scherr Formation

The base of the Scherr Formation is marked by the abrupt introduction of a coarser sequence of turbidite clastic units, including some sandstone. The base of the Scherr along the Allegheny Front coincides with the base of the Cohocton Stage which contains *Cyrtospirifer*. The Scherr Formation is characterized by the presence of some sandstone. At the approximate position of the West Virginia-Virginia boundary in Figure 6, sandstones disappear from the Scherr Formation, although a distinct pulse of coarser siltstone is still conspicuous on the southwest at the stratigraphic position of the base of the Scherr. The name Scherr cannot be extended southwestward into Virginia where Brallier is time-equivalent to the combined Brallier and Scherr Formations of West Virginia and Maryland.

A brownish-gray, marine "redbed" marker in Virginia is present at the approximate stratigraphic position of the base of the Cohocton Stage, corresponding almost identically with the stratigraphic position of other marine "redbeds" at the paleontologic base of the Cohocton Stage in Maryland. The Maryland redbeds reflect influence of the Fulton lobe, and in Virginia the conspicuous lower Brallier turbidite siltstone tongue and redbeds show influence of the Augusta lobe.

The Scherr Formation is 426–2,025 ft (130–617 m) thick along the Allegheny Front in West Virginia and Maryland. The formation accumulated as prodelta foreset beds in the Catskill delta complex, as indicated by prominent display of turbidity current criteria, including flute and groove casts, graded bedding, and sharply bounded sandstones and siltstones.

### Foreknobs Formation

The Foreknobs contains numerous massive sandstones; considerable siltstone; substantial "redbeds" of brownish-gray sandstone, siltstone and shale containing scattered marine fossils; and a few quartz-pebble conglomerate or conglomeratic sandstone beds. The Foreknobs "redbeds" are definitely more brownish in color than those of the Hampshire Formation. The marine fossils in the Foreknobs "redbeds" are the brachiopod *Camarotoechia* and small pelecypods, which constitute a nearshore faunal association (Bretsky, 1969, p. 52–54). The intervals designated on Figure 6 as brownish-gray interbeds actually contain less than half reddish-colored sandstone, siltstone, and shale; most of the interval consists of interbeds which weather light olive gray. Driftwood fragments are common in brackish water and open-marine fossiliferous strata of the Foreknobs Formation. True grayish-red "redbed" strata of nonmarine origin form a tongue in the Blizzard Member in the Route 50 section (Fig. 6).

The base of the Foreknobs is marked by an abrupt upward increase in percentage, thickness, and massiveness of sandstone beds, and northeast of Mouth of Seneca the base generally is marked by conglomeratic strata.

The top of the Foreknobs, at the highest occurrence of marine fossils, generally is marked by abrupt upward passage into dominantly grayish-red sediments of the type that characterize the Hampshire Formation. The upward change to nonmarine deposition is fairly abrupt at any one locality, but the formation boundary is probably time-transgressive, becoming younger toward the southwest, along the trend of the stratigraphic cross section in Figure 6.

Sedimentology and fauna of the Foreknobs Formation indicate a gradual, somewhat pulsating change upward from deeper water turbidite deposits to shallow marine to brackish estuarine deposits. The Foreknobs strata have the general characteristics of the somewhat younger oil-bearing sandstones of the upper "Chemung" of the western subsurface part of the Appalachian basin.

The Foreknobs has been divided along the Allegheny Front (Dennison, 1970b) into four named and one unnamed member as follows:

Foreknobs Formation (1,321–2,264 ft; 403–690 m)
- Unnamed member (0–700 ft; 0–213 m)
- Pound Sandstone Member (23–193 ft; 7–59 m)
- Blizzard Member (278–580 ft; 85–177 m)
- Briery Gap Sandstone Member (28–133 ft; 9–41 m)
- Mallow Member (741–1,100 ft; 226–335 m)

*Mallow Member*.—The Mallow Member is totally marine and consists of much olive-gray-weathering siltstone, some shale, and much fairly massive sandstone in units several feet thick; some brownish-gray, probably brackish-water siltstone and sandstone; and a few quartz-pebble-conglomerate and conglomeratic sandstone beds, usually less than a foot thick. Within the Mallow two nearly continuous units containing brownish-gray, marine "redbeds" persist from Route 250 to Corriganville, except for a general absence of "redbeds" in the Dry Run section.

The base of the Mallow Member probably is somewhat diachronous across the total distance represented in Figure 6, becoming slightly younger southwest of Mouth of Seneca. The top contact of the Mallow with the Briery Gap is isochronous so far as now can be resolved.

During Mallow deposition the sea became progressively shallower as the basin was filled to above wave-base. Turbidity current features abound in the lower Mallow, but shallow water current structures dominate near the top, particularly crossbedding and shale-chip conglomerates at the base of channeled sandstones.

*Briery Gap Sandstone Member*.—The Briery Gap Sandstone Member is massive, yellowish-gray-weathering sandstone with very little interbedded siltstone and shale. *Cyrtospirifer* is the most common marine invertebrate, and plant debris fossils are abundant. The top and bottom contacts are placed where a dominance of sandstone gives way to olive or reddish siltstone and shale interbedded with less than half sandstone.

The Briery Gap Sandstone is thought to represent barrier deposits rather than cheniers, because of its large geographic extent, thickness, and the steep inclination of cross-stratification (Hoyt, 1969). The barrier environment may merge with exceptionally winnowed delta-front sheet sands (Scott and Fisher, 1969, p. 17) in slightly deeper water. The Briery Gap is coarser, better sorted, and thicker than any of the massive sandstones in the underlying Mallow Member. The Briery Gap probably formed by winnowing removal of fine clastic materials, leaving rather clean sand. The apparent blanketlike nature of the sandstone along this outcrop belt probably results from a eustatic change in sea level which permitted winnowing of poorly sorted sediments into barrier deposits trending nearly parallel with the Allegheny Front.

*Blizzard Member*.—The Blizzard Member resembles the Mallow Member in its heterogeneity. The Blizzard consists of interbedded sandstone, siltstone, and some shale, which are greenish gray or light olive gray in color and some brownish gray, plus a few scattered beds of quartz-pebble conglomerate and conglomeratic sandstone. Sedimentary structures such as crossbedding and absence of flute casts suggest that the Blizzard formed in shallower water than did the lower part of the Mallow, which exhibits conspicuous turbidite features. At Route 50 the middle and upper parts of the Blizzard contain abundant reddish-gray, nonmarine redbeds which accumulated in a small Hampshire-type-lithology, subaerial distributary lobe (Fig. 6).

*Pound Sandstone Member*.—Like the Briery Gap Sandstone, the Pound Sandstone is a fine-to medium-grained, crossbedded sandstone with some conglomeratic layers. The Pound, which is better sorted than most Foreknobs sandstones, weathers light olive gray and has few siltstone partings. Fossils include plant stem fragments, brachiopods (*Cyrtospirifer*), and pelmatozoan plates. The upper and lower contacts are drawn where silty or shaly layers break up the dominance of sandstone.

The Pound Sandstone is interpreted to be a barrier deposit or an exceptionally well-sorted delta-front sheet sandstone, produced by waves winnowing finer sediments.

The Pound and Briery Gap Members are nearly parallel with each other (Fig. 6), nearly parallel with the base of the Foreknobs Formation, and essentially parallel with "redbed" marker zones traceable within the Mallow and Blizzard Members of the Foreknobs Formation. Consequently, the *Cyrtospirifer*-dominated Briery Gap and Pound sandstones are thought to be approximately isochronous along the Allegheny Front, each probably corresponding to an event of rapid transgression of the sea over intertidal sandy mudflats. Subsequent winnowing of the mudflat sediments produced fairly well-sorted, massive, crossbedded sandstones of a barrier deposit. The apparently isochronous top of the Pound Sandstone was arbitrarily chosen as a horizontal reference line in preparing the stratigraphic cross section of Figures 5–8. The Allegheny Front deviates only slightly from the general north-south trend of Late Devonian depositional strike, so the Briery Gap and Pound Members appear as almost blanket sandstones along the Allegheny Front. At right angles to depositional strike, the Pound and Briery Gap Sandstones change facies abruptly, so these sandstones have not been identified in the next outcrop belt 10–20 mi (16–32 km) northwest along the east limb of the Deerpark or Elkins Valley anticline.

*Unnamed member.*—This member consists of mostly light olive-gray marine sandstone and siltstone, some brownish gray "redbeds," and rare conglomeratic layers. Its top becomes younger southwestward, where it is a facies equivalent of the basal Hampshire Formation. Along the Allegheny Front in the area studied, no fossils in the uppermost member of the Foreknobs Formation appear definitely younger than Cohocton age, although at Route 250 the top of the Foreknobs Formation may extend upward into post-Cohocton.

It is preferred, at this time, to assign no formal name to the member.

### Hampshire Formation

In those sections in Figure 5 where the Pocono is preserved, the Hampshire Formation is 1,000–2,070 ft (305–631 m) thick. The subaerially deposited Hampshire strata thin southwestward chiefly by facies change into shallow marine beds of the upper unnamed member of the Foreknobs Formation. Southwest of Hopeville Gap the top of the Hampshire may have been beveled locally by pre-Pocono erosion, representing the eastern margin of an Early Mississippian island centered near Beverly, Randolph County, where Pocono strata are absent (Reger, 1931; Flowers, 1956). Present erosion has removed post-Hampshire strata from the south end of the Stony River syncline, but from Dry Run (Fig. 4) northeastward the Hampshire is overlain by the Pocono Formation.

The Hampshire Formation lacks marine fossils (because its base is placed at the top of marine strata in the delta complex); it is almost totally grayish-red in color, and composed principally of siltstone and very fine, micaceous sandstone. Generally the basal Hampshire strata are almost entirely reddish-gray nonmarine beds, but in the Dry Run and Route 250 sections a massive grayish-red sandstone 65–173 ft (20–53 m) thick, with wood debris scattered throughout, is present near the base of the formation.

### Pocono Formation

From Scherr southwestward yellowish-gray sandstone and siltstone of the Pocono contain scattered brachiopods and other marine fossils, but at La Vale it contains coaly streaks and appears totally to represent subaerial delta deposits. In the area of Figure 5 the Pocono clastics came from the same eastern source direction (Pelletier, 1958) as the Hampshire Formation, but the environment changed markedly as the effects of the Acadian orogeny waned and sediments became better winnowed. The Pocono of West Virginia was studied in detail by Dally (1956). He suggested that the scattered redbeds in the basal Pocono are tongues of the Hampshire Formation, so that to a slight degree the Pocono-Hampshire contact is diachronous, thereby precluding its being the exact boundary of the Devonian and Mississippian Systems over all the area of presumed intertonguing. However, neither the Hampshire nor Pocono contains marine fossils near the formation contact, nor have their plant fossils been studied sufficiently to document the full story.

The Pocono Formation represents the high-destructive delta sedimentation of Scott and Fisher (1969, p. 24).

### Resolution of Time and Environmental Factors

Interpretation of depositional history requires a thorough analysis of physical stratigraphic and sedimentologic traits, and then, recognition of time surfaces within the stratigraphic framework. For correlations across long distances with wide spacing of stratigraphic sections, paleontologic comparisons are most efficient. Environmental differences produce complications of facies fossils in contemporaneous strata and similar faunas in diachronous lithosomes. Increasingly, stratigraphers realize that physical correlation of closely spaced sections provides the most reliable resolution of time. Fossils were useful for general time recognition in Figures 5 and 6, but the most precise time resolution was obtained by tracing two types of physical stratigraphic indicators which mark time events: clastic pulses from source areas and eustatic sea-level changes.

Figure 7 shows the same formation boundaries as Figure 6, but instead of members and lithologic details, 13 time lines are indicated. The time lines are solid where their position is confident within 50–100 stratigraphic ft (15–30 m), and dashed where their position is known only approximately. The lithologic zones used to construct the time lines are evident from close comparison of the two figures. The cross section of Figure 7 is nearly parallel with depositional strike, and convergence of time lines would be much more apparent in a northwest-southeast section.

*Source area tectonism.*—Certain lithologic surfaces considered as time lines probably represent tectonic events affecting the acidic-crust terrane of the eastern source area of Appalachia. These events are indicated in Figure 8 by clastic supply curves along each stratigraphic column. The clastic supply curve corresponds to postulated rate of addition of volume of clastics from the southeast, not coarsening because of winnowing as the basin filled.

The base of the Brallier Formation is approximately a time line (T-0, Fig. 8) caused by the introduction of turbidite clastics into this part of the basin as the eastern source area was uplifted. The Brallier base almost exactly parallels, on the scale of this drawing, the Tully Limestone time marker, which can be traced from Mouth of Seneca to Corriganville (Hasson, 1966). Between Mouth of Seneca and Route 250 the time-marker turbidite siltstone pulse of T-1 also parallels the base of the Brallier.

Time T-2 marks the abrupt introduction of Scherr coarse clastics into the basin as the Appalachia source area was uplifted. These distinctly coarser clastic materials are thought to represent a tectonic pulse which occurred at the beginning of the Cohocton. The base of this stage is characterized by the appearance of

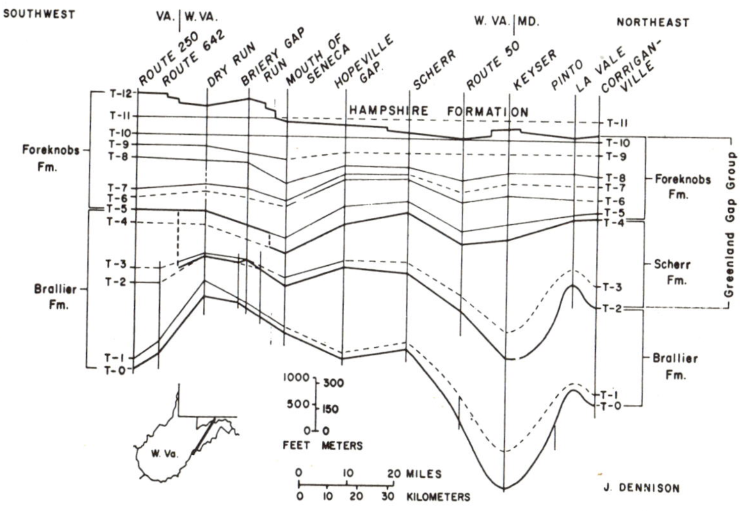


FIG. 7.—Time lines in Devonian delta strata along Allegheny Front in region indicated in Figure 4.

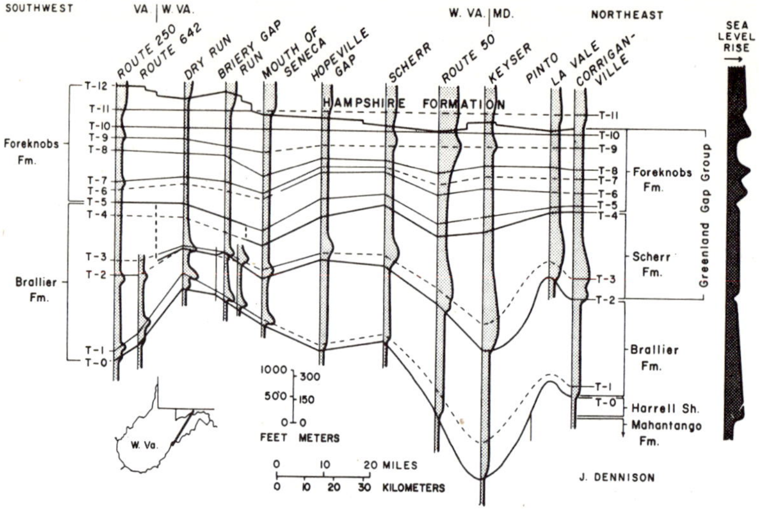


**Fig. 8.**—Sea-level changes and clastic supply from southeastern source for sections along Allegheny Front in region indicated in Figure 4.

*Cyrtospirifer* in appropriate coarse-grained, marine lithologies. The abrupt introduction of coarser sediment is evident in all sections, even though the Scherr Formation does become generally finer toward the southwest, where it changes facies into the upper Brallier Formation. A noticeable siltstone coarsening occurs at this time position in the Brallier of Virginia, where the *Cyrtospirifer* fauna is absent in the siltstone facies.

There was probably also an accompanying sea-level lowering at time T-2, as will be explained.

Time T-3 is the approximate peak of detritus from this early Cohocton pulse of uplift in Appalachia.

The coarsening at the base of the Foreknobs Formation is believed to result chiefly from infilling the basin to somewhat shallower depths, with little or no change in rate of supply of clastic material from Appalachia (Fig. 8). The base of the formation is slightly diachronous because the first influx of coarse sand arrived a little later in the southwest than the northeast, using T-5 as a more reliable time marker (a proposed sea-level fluctuation minimum) than T-4 (either a tectonic pulse or the beginning of a drop in sea level).

*Sea level fluctuation.*—Several lithologic markers are thought to have originated from eustatic fluctuations—*i.e.*, sea-level changes which affected the entire east side of the Appalachian basin (or perhaps the entire basin). Figure 8 shows the same time-lines on the column as Figure 7, but postulated eustatic sea-level changes are separated at the far right from the curves of postulated clastic supply shown alongside each stratigraphic column.

Brownish-gray "redbeds" at time T-2 probably represent a fall of sea level, so that mud flats were oxidized farther west than normal. These reddish muds were then swept westward, and locally were reduced in deeper water of Grant Bay from Dry Run to Pinto; some of the reddish color was preserved near the Augusta and Fulta delta lobes, where the shoreline lay closer to the Allegheny Front sections. The base of the Scherr Formation is definitely the time of a positive pulse of the Appalachia source area, but the pronounced coarsening of 200 ft or less of strata at this time in all sections may have resulted in part from a drop in sea level with a shoreline shift toward the west. Perhaps, however, the eastern land and adjacent sea bottom both rose to produce an apparent sea-level drop.

The "redbed" marker zones at times T-5 and T-7 are also interpreted as a drop of sea level.

At time T-9 sea level again fell, producing a brownish-gray "redbed" marker traced from Route 250 to Mouth of Seneca. At Hopeville Gap, La Vale, and Corriganville the entire Blizzard Member contains abundant brownish-gray "redbeds," so the specific marker was not distinguished. However, from Scherr to Keyser a distinct non-red zone occurs in the lower Blizzard. Sea-level lowering at time T-9 was sufficient to produce at Route 50 grayish-red coloration characteristic of Hampshire subaerial delta plain or mud flat deposits in a small distributary lobe.

Time T-11 dates a sea-level fall near the end of the Cohocton Stage. A "redbed" marker zone can be traced from Route 250 to Briery Gap Run, but this sea-level fluctuation cannot be detected in the contemporaneous delta plain deposits of the Hampshire Formation.

Times of eustatic rise leave a markedly different sedimentary imprint from the "redbed" marker zones that formed when sea level fell. Time lines T-6, T-8, and T-10 seem to correspond to sea-level rise.

Mallow Member strata were deposited farther offshore than the upper Foreknobs, thus turbidity currents left abundant markings in the Mallow. In general, Mallow strata represent a gradual shift upward stratigraphically, from well below wave base at the lowermost Mallow to within the zone of wave agitation in the upper Mallow. The sea-level drops at times T-5 and T-7 caused reddish sediment to be swept far enough westward to reach the Allegheny Front. Sea-level rise at time T-6 shifted the strandline eastward far enough that the reddish coloration of the strata was reduced, so that time T-6 is characterized by a zone with marked absence of "redbeds."

Eustatic sea-level rise at time T-8 was in water shallow enough that wave-winnowing produced coarser and better sorted marine sands (the Briery Gap Sandstone) than anywhere below this position in the sections. The sandstone extends along the entire line of sections, but has not been identified in the next belt of outcrops on the west, so it is probably a barrier-bar deposit, although it also might be simply a delta-front sheet sand (Scott and Fisher, 1969, p. 17). Its origin is not fully understood, but it is thought that, because it is so well sorted and because the middle of the sandstone body (T-8) parallels other eustatic time lines, the Briery Gap originated by sea-level change and winnowing. Well-sorted sands produced as sea level rose are more easily preserved in the sedimentary record than those deposited accompanying sea-level fall.

It is easier to document a sea-level rise at time T-10 than at T-8. This event is so distinct lithologically that the top of the Pound Sandstone (line T-10) is used as a horizontal reference line for all the stratigraphic cross sections of Figures 5–8. Widespread brownish-gray marine "redbeds" at T-9 record a time of sea-level lowering, sufficient to produce a small subaerial delta plain at Route 50, where grayish-red redbed lithology is identical with that of the Hampshire Formation. Sea-level rise after time T-9 produced a zone with a striking absence of brownish-gray marine "redbeds" above T-9 from Dry Run to Mouth of Seneca and probably to Hopeville Gap. The sea spread over the minor subaerial delta at Route 50, where *Cyrtospirifer* was collected from the Pound Sandstone. Sea-level rise at time T-10 permitted marine transgression with winnowing, sorting, and coarsening, so that the Pound environment was a barrier bar or possibly a very shallow delta-front sheet-sand deposit.

Perhaps the most significant aspect of this paper relative to petroleum accumulation is the interpretation of a eustatic sea-level rise as responsible for the origin of the better sorted Foreknobs sandstone strata which form the Briery Gap and Pound Members. They clearly have the most promising reservoir-rock lithology of all the strata studied on outcrop in the Devonian Catskill delta complex. The implications for petroleum exploration will be pointed out in the final part of this paper.

*Distributary lobes.*—In the line of sections extending along the Allegheny Front for 96 mi in Figure 6, the persistent influence of the Fulton and Augusta lobes can be discerned at the ends of the cross section, whereas an intervening Grant Bay had deeper water. This pattern is most evident at first glance in the Dry Run section because of the nearly complete absence there in the Foreknobs Formation of nearshore deposits with brownish-gray "redbeds." The Fulton lobe was a more conspicuous clastic source (island or principal river system draining Apppalachia) than the Augusta lobe. Consequently "redbed" tongues and clastic pulses tend to extend farther into Grant Bay from the northeast than from the Augusta lobe, producing the asymmetry shown in Figure 6. However, the Augusta lobe was very important at times, as evidenced by a conspicuous siltstone zone in the lower Brallier, which can be traced northeast as far as Mouth of Seneca. Beds from this group of Brallier turbidite deposits are thinner and finer northeast, away from the Augusta lobe. The southwestern extent of the Augusta lobe has not been thoroughly worked out, but certainly the lobe was not influential southwest of Roanoke.

The subaerial accumulations at Route 50 during middle and upper Blizzard deposition represent a minor and temporary subdelta on the flank of the Fulton lobe.

Late Devonian Catskill delta lobes exhibit a markedly different pattern from the Lower Mississippian strata, in which a subaerial delta lobe is conspicuous in southwest Virginia (coal deposits of Price Formation), as well as a Fulton lobe influence with Pocono coal in the eastern panhandle of West Virginia.

### Correlation Across Basin

It should be simple in principle to correlate times of specific sea-level shifts around the basin edge in shallow-water sites from Virginia through the subsurface to western New York, following approximate depositional strike. It is more difficult to resolve sea-level changes in sections perpendicular to shore, but by using Israelsky's principle (1949) it should be possible to extend time-surfaces representing sea-level maxima and minima in three dimensions throughout much of the Appalachian basin. Sea-level rise should produce landward shift of the lower delta plain in the subaerial part of the delta complex, but I have been unable to identify such a possible aid to correlations within the Hampshire Formation nonmarine strata. Direct physical correlations of Devonian delta strata should be possible from Virginia to New York by tracing strata produced by specific sea-level fluctuations. If correlations are possible by the eustatic sea-level method, the resulting time pulses should provide more sensitive time resolution from state to state than is possible by fossils alone.

In Figure 9, I propose tentative time correlations between the time lines in the strata of Figure 8 and the New York Devonian correlation chart (Rickard, 1964), using eustatic fluctuations to mark time. Fractional time interpretation can be designated by decimal fractions of whole time units in Virginia, for instance, time T-3.5. Intervals between all whole units probably are not equal amounts of time. Evaluation by numerous workers in different areas will be required to test adequately these tentatively proposed eustatic fluctuation correlations. Especially needed is careful examination of well data between the Allegheny Front and western New York.

### Implications for Petroleum Exploration

Outcrop stratigraphy along Allegheny Front yields ideas leading to a better understanding of the occurrence of petroleum sandstones in the southern Appalachian basin subsurface and to prediction of other sandstone bodies which have reservoir potential.

Shoreline prograding action was dependent on subsidence rate of the Appalachian basin, rate of uplift of the Appalachia clastic source area on the east, and the geographic location of the Fulton and Augusta lobes of the sediment distributary streams. Stratigraphic resolution on outcrop generally is not sufficient to locate specific small deltas. Eustatic sea-level rises caused eastward shifts of the shoreline by a few miles, creating permeable sand bodies by wave winnowing of original nearshore sandy muds and delta-front sands. These operating factors

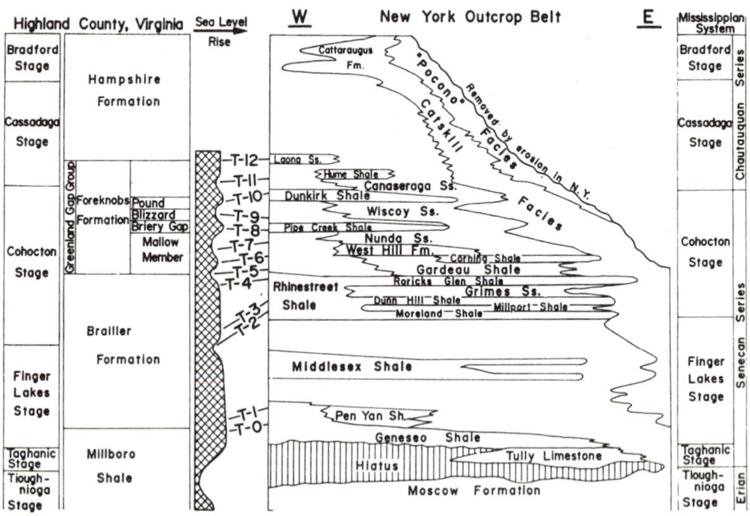


Fig. 9.—Tentative eustatic correlation from Virginia to New York. Time numbers refer to Figure 8.

can be identified in the prodelta and delta-front outcrops along the Allegheny Front.

Relatively coarse-grained and well-sorted sandstone bodies are present only in the nonred strata of the Devonian delta complex along Allegheny Front. Some rather coarse nonmarine sandstone is present in the basal part of the Hampshire Formation, at Dry Run and Route 250, at a position which appears to be just inland from the shoreline. Coarse-grained marine sandstones are present only in the Foreknobs Formation, and the only well-sorted, permeable sandstones with persistent distribution are the Pound and Briery Gap Members. Both are interpreted as sands winnowed by the sea at times of eustatically rising sea level, when mud supply was low offshore and delta sites shifted east. The Briery Gap and Pound both extend nearly parallel with depositional strike across at least four counties along the Allegheny Front, but are not recognized in outcrops 10–12 mi (16–19 km) west of the Allegheny Front.

### Producing Zones

The principal West Virginia producing sandstones in the Catskill delta are all Chautauquan Epoch sands (Fig. 2) deposited a few miles west of the limit of Catskill redbeds at that time. Figure 3 shows the areas in which the Gordon, Fourth, and the Riley-Benson sandstones have been found sufficiently permeable to yield commercial petroleum if the structural setting is also favorable. These sandstones extend across several counties in somewhat elongate areas, generally parallel with the north-south trend of the Late Devonian shoreline. Production from younger sandstones tends to be obtained somewhat farther west than from the older sandstones.

The same factors as affected shoreline progradation probably continued to operate in the latest Devonian, as the Catskill delta complex prograded westward. It is suggested that eustatic sea-level changes caused lateral shifting of the shoreline, and the resulting winnowing of muddy sands accompanying periodic rises of sea level produced the north-south trending reservoir rocks of the Benson, Fourth, Gordon, and perhaps other sandstones. Position of delta lobes probably was a major factor in determining exactly where along the coast the water depth and sediment influx were optimum for winnowing of reservoir beds.

From this depositional model, it would be predicted that the sea-level shifts that produced reservoir beds in West Virginia should also have produced favorable winnowing and reservoir rocks locally farther north in Pennsylvania and New York. Old correlations by drillers extended some of these sandstone names north-south across several states. The proposed depositional framework suggests a mechanism by which some of these long-range time correlations may indeed be generally correct, so that basin-wide eustatic sea-level shifts could have created contemporaneous sandstones trending along depositional strike, even though specific sandstone bodies may not be continuous across the basin because of configuration of the delta lobes.

### Future Petroleum Prospects

Whenever clastic pulses filled the Appalachian basin to near wave-base, and especially where eustatic sea-level rise permitted extensive winnowing to form delta-front sheet sands or barriers, potential reservoir sands should have been deposited. Because the delta front trended north-south, favorable sandstones observable in outcrop should extend generally northward into the subsurface part of the basin of West Virginia, Maryland, and Pennsylvania.

Some potential reservoir sandstones that deserve further study are indicated in parentheses in Figure 2, and their outcrop region is shown in Figure 3.

The Valley Head Sandstone (Reger, 1931, p. 371–373) may correlate with the Benson sandstone of the subsurface, or it may be a slightly older winnowed sandstone developed farther east. The Valley Head deserves outcrop study and search for it in the subsurface.

The Elkins Sandstone (Reger, 1931, p. 374–381) does not seem as well sorted as the Valley Head, but it could extend northward toward Monongalia County, West Virginia, or Garrett County, Maryland, and perhaps it may be better sorted in the north.

The Pound and Briery Gap Sandstone Members almost certainly extend to Pennsylvania and should be sought in wells in Garrett County, Maryland, and adjacent Pennsylvania. Hydrocarbons have leaked from their outcrops along the Allegheny Front, but favorable structural settings may occur northward in the subsurface.

The Parkhead sandstone (Swartz, 1913, p. 415–417) is a poorly sorted turbidite deposit in the lower "Chemung" formation over much of the area east of the line of dots shown in Figure 3. Near its type locality, in the eastern part of its occurrence, the Parkhead appears somewhat better sorted by winnowing. Hydrocarbons have leaked from it in the eastern panhandle of West Virginia, but the Parkhead should be examined for gas in the Purcell field region of Pennsylvania or in possible extensions of the same sandstone north across the fold belt and just west of Allegheny Front in Pennsylvania.

The Clearville Siltstone Member (Cate, 1963, p. 236) near the top of the Mahantango Formation in places is winnowed enough to be moderately sorted. Ellison (1965, Plate 3, p. 25) indicated a siltstone to sandstone at the Clearville position in his Frame Shale Member of the Mahantango in a composite section for Bedford and Fulton Counties, Pennsylvania. He does not refer to this coarse zone within the Frame Member by a specific stratigraphic name. In his Plate 2, he does label the Clearville siltstone. It locally forms a substratum for coral patch reefs. The Clearville probably has leaked all its hydrocarbons in most of the fold belt east of the west limit of the Clearville shown on Figure 3. The unit should be continuous in the subsurface across central Pennsylvania along the edge of the Fulton lobe and perhaps into east-central New York. Sands or silts winnowed during a sea-level drop during Clearville deposition could yield gas in the plateau of northern Pennsylvania and New York.

## Conclusions

Outcrops of Devonian delta strata along the Allegheny Front show effects of tectonic uplift of Appalachia, subsidence of the Appalachian basin, distributary lobes which persisted for tens of millions of years, and eustatic sea-level fluctuations. The most permeable and extensive sandstone bodies observed in outcrops were formed by winnowing of muddy sands as sea level rose, so that the shoreline temporarily shifted eastward. Many of the classic productive Devonian delta sandstones show a comparable geometry and probably formed by sea-level rise which reworked delta-front and lower delta-plain muddy sediments during marine advances over the Catskill delta complex. A sandstone which shows evidence of winnowing should be followed along the general north-south depositional strike, or slightly northeastward, to see if it and other contemporaneous sandstone bodies are in a structural position that might allow them to contain undiscovered hydrocarbon reserves.

## References Cited


Austin, N. C., 1969, Subsurface Tully Limestone, New York and Pennsylvania: Geol. Soc. America Abstracts with Programs for 1969, pt. 1, p. 2.

Bretsky, P. W., Jr., 1969, Evolution of Paleozoic benthic marine invertebrate communities: Palaeogeography, Palaeoclimatology, and Palaeoecology, v. 6, p. 45–59.

Burtner, R. L., 1963, Sediment dispersal patterns within the Catskill facies of southeastern New York and northeastern Pennsylvania: Pennsylvania Geol. Survey Bull. (General Geology Rept.), 4th ser., G39, p. 7–23.

Calvert, W. L., 1964, Oil and gas fields of Ohio: Ohio Div. Geol. Survey, scale 1 in. = 6 mi.

Cardwell, D. H., D. G. Patchen, and L. W. Woodfork, 1970, Oil and gas fields of West Virginia: West Virginia Geol. and Econ. Survey, 31 p.

Cate, A. S., 1963, Lithostratigraphy of some Middle and Upper Devonian rocks in the subsurface of southwestern Pennsylvania: Pennsylvania Geol. Survey Bull. (General Geology Rept.), 4th ser., G39, p. 229–240.

Dally, J. L., 1956, Stratigraphy and paleontology of the Pocono Formation in West Virginia: Ph.D. dissert., Columbia Univ., 241 p.

Dennison, J. M., 1961, Stratigraphy of Onesquethaw Stage of Devonian in West Virginia and bordering states: West Virginia Geol. and Econ. Survey Bull. 22, 87 p.

——— 1963, Stratigraphy between Tioga metabentonite and Pocono Formation along Allegheny Front in Maryland and West Virginia: Pennsylvania Geol. Survey Bull. (General Geology Rept.), 4th ser., G39, p. 213–227.

——— 1970a, Silurian stratigraphy and sedimentary tectonics of southern West Virginia and adjacent Virginia, *in* Silurian stratigraphy, central Appalachian basin: Appalachian Geol. Soc., p. 2–23.

——— 1970b, Stratigraphic divisions of Upper Devonian Greenland Gap Group ("Chemung Formation") along Allegheny Front in West Virginia, Maryland, and Highland County, Virginia: Southeastern Geology, v. 12, no. 1, p. 53–82.

——— and O. D. Naegele, 1963, Structure of Devonian strata along Allegheny Front from Corriganville, Maryland, to Spruce Knob, West Virginia: West Virginia Geol. and Econ. Survey Bull. 24, 42 p.

——— J. W. Travis, and J. H. Ferguson, 1967, Kittlelick thrust—a new fault in Mineral and Grant Counties, West Virginia: West Virginia Acad. Sci. Proc., v. 38, p. 177–185.

Dickey, P. A., 1941, Oil geology of the Titusville quadrangle, Pennsylvania: Pennsylvania Geol. Survey Bull. M22, 4th ser., 87 p.

——— R. E. Sherrill, and L. S. Matteson, 1943, Oil and gas geology of the Oil City quadrangle, Pennsylvania: Pennsylvania Geol. Survey Bull. G48, 4th ser., 201 p.

Ellison, R. L., 1965, Stratigraphy and paleontology of the Mahantango Formation in south-central Pennsylvania: Pennsylvania Geol. Survey Bull. G48, 4th ser., 298 p.

Flowers, R. R., 1956, A subsurface study of the Greenbrier Limestone in West Virginia: West Virginia Geol. Survey Rept. Inv. 15, 17 p.

Friedman, G. M., and K. G. Johnson, 1966, The Devonian Catskill deltaic complex of New York, type example of a "tectonic delta complex," *in* Deltas in their geologic framework: Houston Geol. Soc., p. 172–188.

Graham, J. A., 1952, Mineral resources of New York State: New York State Sci. Service.

Hasson, K. O., 1966, Lithostratigraphy and paleontol-

ogy of the Devonian Harrell Shale along the Allegheny Front in West Virginia and adjacent states: M. S. thesis, Tenn. Univ., 89 p.
——— and J. M. Dennison, 1968, Relation of type Millboro Shale in Virginia to Romney Group (Devonian) of West Virginia and Maryland: Geol. Soc. America Spec. Paper 115, p. 476–477.
Haught, O. L., 1965, Oil and gas report and map of Barbour and Upsher Counties, West Virginia: West Virginia Geol. and Econ. Survey Bull. 31, 15 p.
——— 1966, Structure contour map on the Benson sand in Barbour, Upsher, Lewis, and Harrison Counties, West Virginia: West Virginia Geol. and Econ. Survey.
Heckel, P. H., 1966, Stratigraphy, petrography, and depositional environment of the Tully Limestone (Devonian) in New York state and adjacent region: Ph.D. dissert., Rice Univ., 448 p.
——— 1969, Devonian Tully Limestone in Pennsylvania and comparison to type limestone in New York: Pennsylvania Geol. Survey Inf. Circ. 60, 33 p.
Hoover, K. V., 1960, Devonian-Mississippian shale sequence in Ohio: Ohio Div. Geol. Survey Inf. Circ. 27, 154 p.
Hoyt, J. H., 1969, Chenier versus barrier, genetic and stratigraphic distinction: Am. Assoc. Petroleum Geologists Bull., v. 53, p. 299–306.
Ingham, A. I., E. M. Tignor, and W. M. Nabors, 1949, McDonald and adjacent oil fields, Allegheny and Washington Counties, Pennsylvania: Pennsylvania Geol. Survey Bull. M29, 4th Ser., 113 p.
Israelsky, M. C., 1949, Oscillation chart: Am. Assoc. Petroleum Geologists Bull., v. 33, p. 92–98.
Johnson, J. G., 1970, Taghanic onlap and the end of North American Devonian provinciality: Geol. Soc. America Bull., v. 81, p. 2077–2106.
Johnson, J. H., J. M. Stapleton, and P. McGrain, 1962, Mineral resources and mineral industries of Kentucky: Kentucky Geol. Survey.
Jones, M. L., and J. M. Dennison, 1970, Oriented fossils as paleocurrent indicators in Paleozoic lutites of southern Appalachians: Jour. Sed. Petrology, v. 40, p. 642–649.
Lytle, W. S., and L. A. Heeren, 1964, Maps of the oil and gas fields, natural gas pipelines, and oil pipelines of Pennsylvania: Pennsylvania Geol. Survey.
McIver, N. L., 1970, Appalachian turbidites, *in* Studies of Appalachian geology: central and southern: New York, John Wiley and Sons, p. 69–81.
Oliver, W. A., Jr., W. de Witt, J. M. Dennison, D. H. Hoskins, and J. W. Huddle, 1967, Devonian of the Appalachian basin, United States, *in* International symposium on the Devonian System: Alberta Soc. Petroleum Geologists, v. 1, p. 1001–1040.
——— 1969, Correlation of Devonian rock units in the Appalachian basin: U. S. Geol. Survey Oil and Gas Inv. Chart OC–64.
Pelletier, B. R., 1958, Pocono paleocurrents in Pennsylvania and Maryland: Geol. Soc. America Bull., v. 69, p. 1033–1064.
Pepper, J. F., W. de Witt, Jr., and D. F. Demarest, 1954, Geology of the Bedford Shale and Berea Sandstone in the Appalachian basin: U. S. Geol. Survey Prof. Paper 259, 111 p.
Pharr, R. F., and J. L. Calver, 1963, Mineral industries and resources of Virginia: Virginia Div. Mineral Resources.
Reger, D. B., 1931, Randolph County: West Virginia Geol. Survey, 989 p.
Rickard, L. V., 1964, Correlation of the Devonian rocks in New York State: New York State Mus. and Sci. Service Geol. Survey Map and Chart Ser. no. 4.
Scott, A. J., and W. L. Fisher, 1969, Delta systems and deltaic deposition, *in* Delta systems in the exploration for oil and gas: Texas Bur. Econ. Geology Spec. Pub., p. 10–29.
Sutton, R. G., Z. P. Bowen, and A. L. McAlester, 1970, Marine shelf environments of the Upper Devonian Sonyea Group of New York: Geol. Soc. America Bull., v. 81, p. 2975–2992.
Swartz, C. K., 1913, Correlation of the Upper Devonian, *in* Middle and Upper Devonian: Maryland Geol. Survey, p. 410–444.
Tilton, J. L., W. F. Prouty, and P. H. Price, 1927, Pendelton County: West Virginia Geol. and Econ. Survey, 384 p.
Wheeler, H. E., 1963, Post-Sauk and pre-Absaroka Paleozoic stratigraphic patterns in North America: Am. Assoc. Petroleum Geologists Bull., v. 47, p. 1497–1526.
Willard, B., F. M. Swartz, and A. B. Cleaves, 1939, The Devonian of Pennsylvania: Pennsylvania Geol. Survey Bull. G19, 4th ser., 481 p.
Woodrow, D. L., and F. W. Fletcher, 1967, Late Devonian paleogeography in southeastern New York and northeastern Pennsylvania, *in* International symposium on the Devonian System: Alberta Soc. Petroleum Geologists, v. 2, p. 1327–1334.

**Reprinted from:**
BULLETIN OF THE AMERICAN ASSOCIATION OF PETROLEUM GEOLOGISTS
VOL. 55, NO. 8 (AUGUST, 1971), PP. 1206-1230, 15 FIGS., 1 TABLE

# Pennsylvanian Delta Patterns and Petroleum Occurrences in Eastern Oklahoma[1]

GLENN S. VISHER,[2] SANDRO SAITTA B.,[3] and RODERICK S. PHARES[4]

Tulsa, Oklahoma 74104, Maracaibo, Venezuela, and Tripoli, Libya

**Abstract** Data from modern deltas have made it possible to interpret the origin of the Pennsylvanian sandstone in northeastern Oklahoma. Objective criteria from cores and outcrops used to define specific areal and vertical depositional patterns include: (1) vertical patterns of sedimentary structures, bedding, and grain size; (2) clay mineralogy and detrital clasts; (3) trace fossils; and (4) detailed analysis of textures.

The historical development of a delta provides the insight for interpreting the depositional patterns observed in ancient deltaic sandstone bodies. The processes of progradation and maturation are used to develop a four-dimensional deltaic model. Six subdivisions are distinguishable: (1) lower alluvial plain; (2) upper deltaic plain; (3) lower deltaic plain; (4) subaqueous sand sheet; (5) marginal basin; and (6) marginal depositional plain. The lower alluvial plain is characterized by stream meandering, point bars, and confined channel flow. Stream gradients are reduced in the upper deltaic plain, distributary bifurcation occurs, and marsh, natural levee, and floodplain deposition is predominant. Development of the lower deltaic plain is controlled by crevassing of natural levees, tides, and floods. The subaqueous sand sheet is characterized by shallow water currents. It is modified by progradation and commonly replaced by deltaic-plain environmental units. The marginal basin and depositional plain are produced by longshore drift and reflect a balance between subsidence, sediment supply, and wave energy. These environmental units may be modified or replaced by other deltaic elements.

This model has been applied to the Bluejacket-Bartlesville sandstone of eastern Oklahoma. This stratigraphic interval was selected because of its importance as an oil reservoir, the large amount of available subsurface data, and the simplicity of the stratigraphic and structural framework.

Lower and Middle Pennsylvanian sediments were deposited during an overall transgression, but the transgression is marked by extensive regressions. These are widespread sandstone sheets generally underlain and overlain by marine shales or limestones. The sandstones are distributed over thousands of square miles, but locally they are lenticular. Each regression is in response to the outbuilding of sediment under static sea level conditions. The supply of sediment is genetically related to a single river system, and deltaic patterns related to this system can be traced through Morrowan, Atokan, and Desmoinesian strata.

## Introduction

The interpretation of the genesis of oil-bearing sandstone in the Mid-Continent has been of interest for many decades. It has been only within the past few years that sufficient data from Holocene environments have become available to provide the basis for the interpretation of ancient sandstone units. For many years interpretations were based on gross geometry, or the presence of particular fossil assemblages, but with new information on the physical processes of sedimentation much more precise environmental criteria have become available. Of even more importance is the recognition of specific vertical profiles of lithologic units, sedimentary structures, textures, and the many other specific physical and biologic aspects of a single genetic time-rock unit.

The process-response model has become the "touchstone" of sedimentologists. The studying of the fourth dimension—time—has provided insight into the interpretation of particular areal and vertical sedimentary patterns. By comparing sequences of depositional units from the Holocene to those present in ancient sediments it is possible to determine the basic physical response to a series of events; thus sedimentary patterns can be recognized in ancient stratigraphic units.

## Stratigraphic Framework

One of the most useful generalizations concerning stratigraphy is the recognition that periods of regression are followed by major transgressions. These periods account for many of the regional sedimentary patterns in North America and throughout the world. The "sequence" concept (Sloss, 1963) has allowed an

[1] Manuscript received, July 14, 1970; accepted, November 30, 1970. Presented before the Association at Calgary, Alberta, June 24, 1970.

[2] Department of Earth Sciences, University of Tulsa.

[3] Texas Petroleum Company.

[4] Mobil Oil Corporation.

The writers acknowledge the support of the National Science Foundation through grant GA 997. Enrique Vasquez, a graduate student at the University of Tulsa, was very helpful in preparing some regional data used in this study. We thank Robert J. Weimer and G. deV. Klein for their technical comments on the manuscript, clarifying ideas presented, and aid in the organization of the paper. J. M. Coleman and S. M. Gagliano have provided indirectly and directly many ideas used in developing the general concepts of deltaic sedimentation, and we are indebted to them for their published research work and for the opportunity to discuss some of the details of delta progradation and maturation.

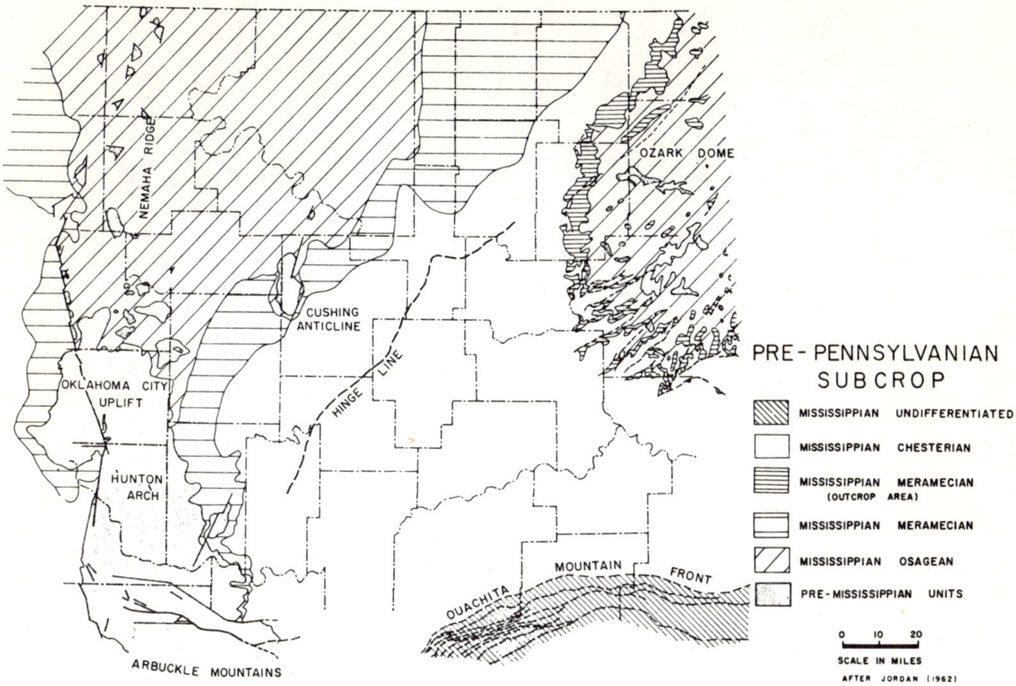


FIG. 1.—Subcrop map below pre-Pennsylvanian unconformity. Map shows Nemaha ridge and uplifts south of Oklahoma City. Ozark dome probably reflects post-Pennsylvanian truncation. Ouachita Mountain area shows no pre-Pennsylvanian truncation. After Jordan (1962).

insight into the fundamental pattern of sedimentation within a basin and into the interpretation of stratigraphic groups. The interval discussed in this paper includes rocks ranging in age from Morrowan through Desmoinesian, and the events recounted are merely a sample of those that have occurred in other places during the same period of deposition, or that occurred in a similar position within another sequence. Major unconformities separate large sections of the stratigraphic column, and the sedimentary units relating to the transgression produce a pattern that can be recognized and interpreted.

### Shelf-Geosyncline Relations

At the end of the deposition of the Kaskaskia sequence, the seas retreated into the centers of basins and to the margins of the continental shelves. When the seas left the continental interior the large subaerially exposed landmass provided sediment. The Chesteran Series of the Illinois basin and the continental margin reflect this change, and clastic sediments were deposited in channels, deltas, and shallow marine water. A pre-Pennsylvanian paleogeologic map at this time shows widespread truncation, exposed highlands, and a general progression of older to younger rocks toward the centers of basins and on the continental margins. A close approximation of this paleogeologic map in the Oklahoma area is illustrated by the subcrop below Lower and Middle Pennsylvanian strata (Fig. 1). This map illustrates the tectonic framework at the beginning of the Pennsylvanian, and indicates the presence of faulted uplifts along the Nemaha ridge, and the position of the Arbuckle Mountains, and the Oklahoma City highland areas. The edge of the continental shelf is well defined by the successively younger stratigraphic units on the south underlying the unconformity. As the actual times of uplift and erosion cannot be shown for areas not covered by Pennsylvanian strata, it is difficult to determine the time of uplift for some of the tectonic features. The Ozark dome may have been covered by Chesteran and possibly Lower or Middle Pennsylvanian strata.

### Tectonic Framework

The subcrop map shows the nature of the truncation and the position of major structural

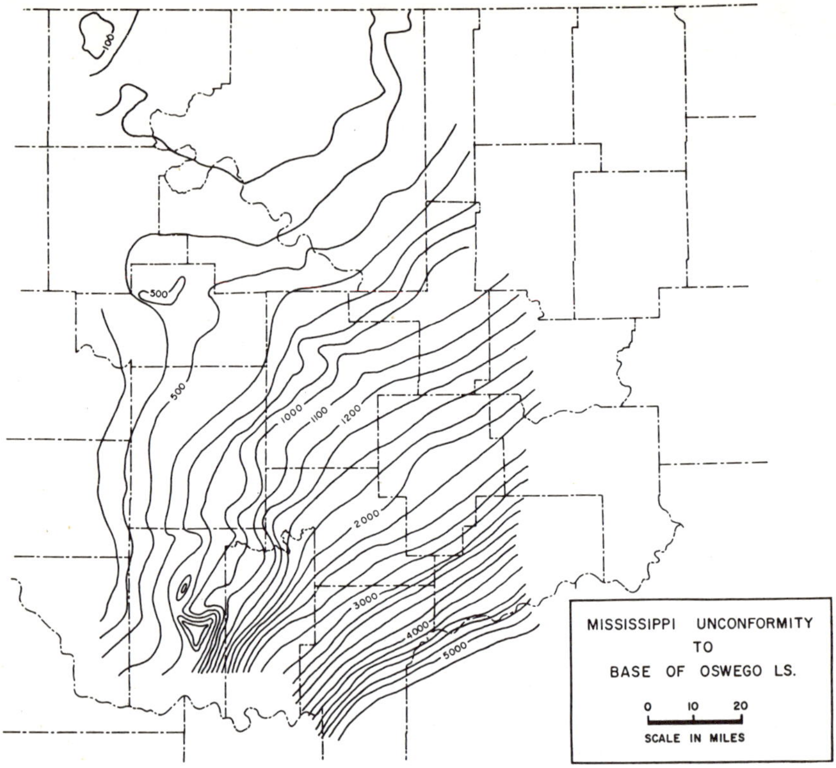


FIG. 2.—Isopach map from Mayes (Meremecian) to base of Oswego (upper Desmoinesian). This interval includes all sediments deposited during transgression from basin across Oklahoma shelf.

features. The isopach map of an interval overlying the unconformity illustrates the continuing effects of these features on deposition (Fig. 2). The interval includes the uppermost Mississippian stratigraphic units and Pennsylvanian strata including Morrowan, Atokan, and most of the Desmoinesian Series. Of major importance is the total thickness indicated by this isopach. The 5,000 ft of shallow marine strata suggests that this period was a major time of subsidence on the continental margin. It also indicates that it was a major period of onlap, as indicated by the progressive thinning toward the areas of truncated Mississippian strata. The thicknesses of sediment indicated by this isopach are minor compared with those deposited farther south and east in the Arkoma basin. In those areas more than 12,000 ft of Atokan sediment was deposited, and many thousands of additional feet of Upper Mississippian and Morrowan strata have been measured. This would indicate that a basin fill in excess of 20,000 ft occurred during this time span.

Quantities of sediment of this order of magnitude suggest that it was a time of general continental emergence and extensive erosion of older sediments from the continental interior. Only after a major period of transgression and marine onlap was the continent covered by a sea and the source of sediment eliminated.

## Transgressive Character of Lower Absaroka Sequence

The transgressive character of the lower part of the Absaroka sequence is well documented (Fig. 3). The cross section and map illustrate the successive onlap of younger time-stratigraphic sedimentary units. The onlap suggests a relative rise in sea level, and may represent in part a topographic surface that was inundated as sea level rose.

*Mississippian unconformity.*—The unconformity separating Mississippian and Pennsylvanian strata disappears into the basin. No evidence of truncation or missing stratigraphic intervals either above or below the unconformity can be demonstrated south of the "hinge line" (Fig. 3). At other places local unconformities may be identified at the base of a particular sandstone unit, but no regional erosion surface can be recognized. This suggests that warping and subsidence occurred in a progressive and regular manner, and it is not necessary to invoke tectonic events to explain the depositional pattern.

*Onlap from south to north.*—The detailed correlation of thin shelf limestone and other marker-defined units indicates the time sequence of the onlap (Fig. 3). The successive pattern of younger time-stratigraphic units extending farther onto the shelf is well illustrated. The Verdigris Limestone of late Desmoinesian age transgresses the tectonic features outlined by the subcrop map (Fig. 1). The transgression continued for possibly 15 million years and produced a sequence of sediments in excess of 5,000 ft thick. Such a transgression is similar in scope to the basal Cambrian, lower Middle Ordovician (St. Peter), or the upper Lower Devonian (Oriskany) transgressions, which represent the beginning of other sequences (Sloss, 1963).

## Depositional History

The regularity of the transgression and the presence of excellent key beds provide the basis for a detailed time-stratigraphic subdivision of the section (Table 1). Individual formats (Forgotson, 1957) can be defined, and the depositional pattern of sand, mud, and carbonate material can be determined easily. In addition, much of the subsurface work has been published on this area and the patterns of sand deposition are well known. This has provided the opportunity to document the history of a major transgressive period in detail and to interpret the depositional patterns.

### Stratigraphic Nomenclature

For many years there has been a problem in naming subsurface geologic units. The Code of Stratigraphic Nomenclature provides a basis for separating rock, time-stratigraphic, faunal zones, and time units. In the past, time-rock units have been based exclusively on biozones, but with the identification of marker beds in cores, samples, and petrophysical logs it has been suggested that another basis exists for subdividing time-stratigraphic units. The term *format* was suggested by Forgotson (1957), and the term *operational unit* was suggested by Krumbein and Sloss (1963). Busch has used a similar concept called a *Genetic Stratigraphic Sequence* (Busch, 1971).

The stratigraphic code specifically states that the rock terms, "group," "formation," "member," and "sandstone," should not carry time-stratigraphic connotations. The practice has been to use an informal term, "zone," but without stratigraphic definition. The term "format" has priority, and it is proposed that it be used as a formal time-stratigraphic subdivision of series.

For the purpose of the discussion of the depositional history of the Middle and Lower Pennsylvanian the following formats are defined: the Wewoka-Prue format from the base of the Oswego limestone to the top of the Verdigris Limestone; the upper Skinner format from the top of the Verdigris Limestone to the top of the Pink limestone; the lower Skinner-Red Fork format from the top of the Pink limestone to the top of the Inola limestone; the Bartlesville format from the top of the Inola to the top of the upper Brown limestone; the Booch-McAlester format from the top of the Brown limestone to the base of the Hartshorne Sandstone; the Atoka-Spiro format from the base of the Hartshorne Sandstone to the top of the Wapanucka Limestone; the Wapanucka-Cromwell format from the top of the Wapanucka Limestone to the top of the Caney Shale; the Mayes format from the top of the Mayes limestone to the top of the Boone Chert. The formats herein defined are shown on the accompanying cross section (Fig. 3), and the type well used for naming the units is the Earlsboro No. 1 T. J. Davis, Sec. 11, T8N, R10E.

It is believed that these are time-stratigraphic subdivisions of the appropriate series shown in Table 1. They should not be confused with rock stratigraphic units, biozones, or the informal term "zone."

### Regressive Sand Units of the Lower Absaroka Sequence

The depositional pattern is one of major regressive progradations interrupting a continuous transgressive depositional period. Each time-stratigraphic unit represents a period of regression with a common vertical pattern of depositional environments. This is well illustrated by the pattern of sandstone development within the interval (Fig. 4). The map shows an overlapping pattern of sandstone sheets branching southward and a single narrow sandstone body on the north. Also, there is evidence that younger sandstone bodies developed farther north than older units. This would suggest generally deeper water on the south.

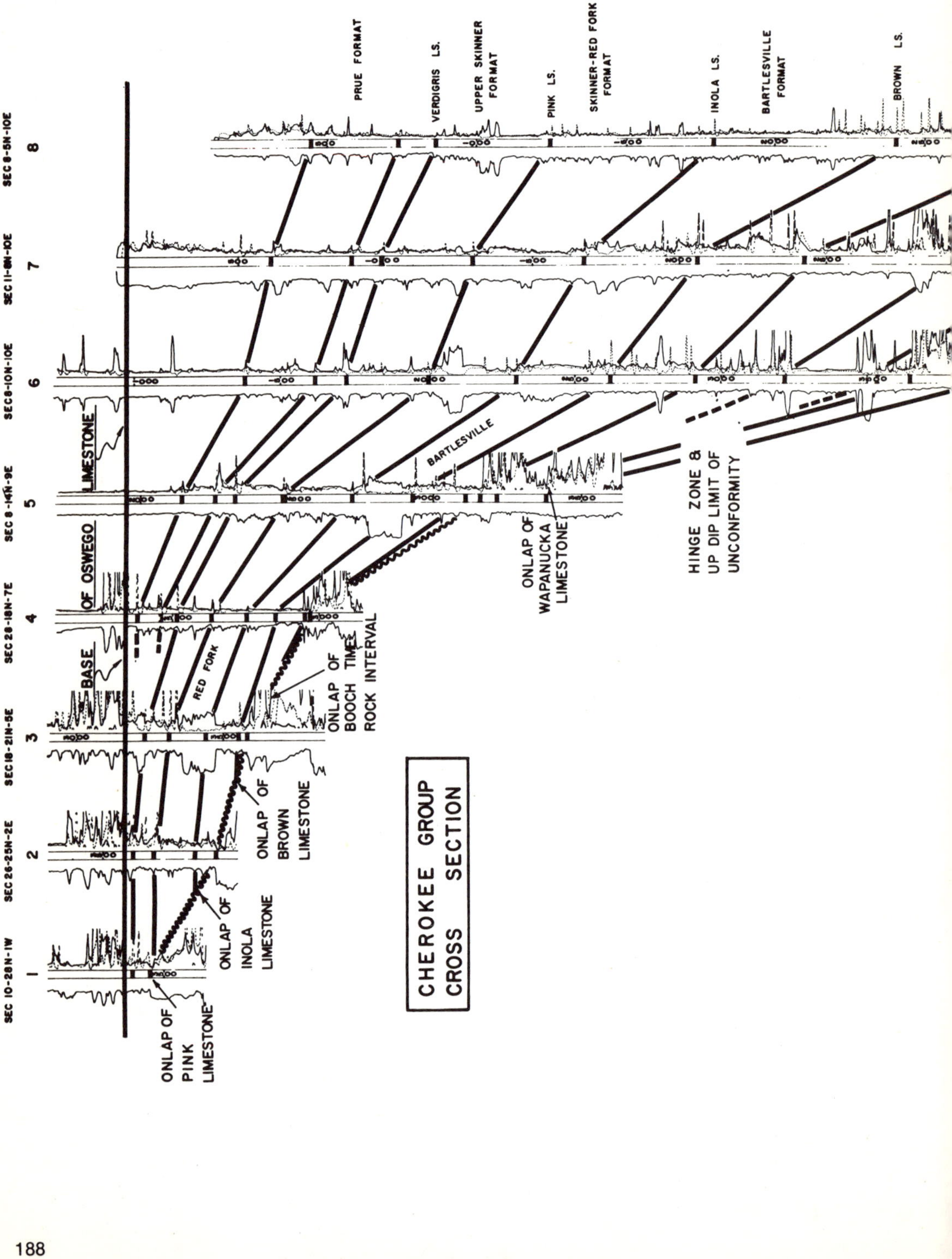
CHEROKEE GROUP CROSS SECTION
SEC 10-28N-1W
SEC 26-25N-2E
SEC 18-21N-5E
SEC 28-18N-7E
SEC 8-14N-9E
SEC 8-10N-10E
SEC 11-8N-10E
SEC 8-5N-10E
PRUE FORMAT
VERDIGRIS LS.
UPPER SKINNER FORMAT
PINK LS.
SKINNER-RED FORK FORMAT
INOLA LS.
BARTLESVILLE FORMAT
BROWN LS.
BASE OF OSWEGO LIMESTONE
RED FORK
BARTLESVILLE
ONLAP OF PINK LIMESTONE
ONLAP OF INOLA LIMESTONE
ONLAP OF BROWN LIMESTONE
ONLAP OF BOOCH TIME ROCK INTERVAL
ONLAP OF WAPANUCKA LIMESTONE
HINGE ZONE & UP DIP LIMIT OF UNCONFORMITY

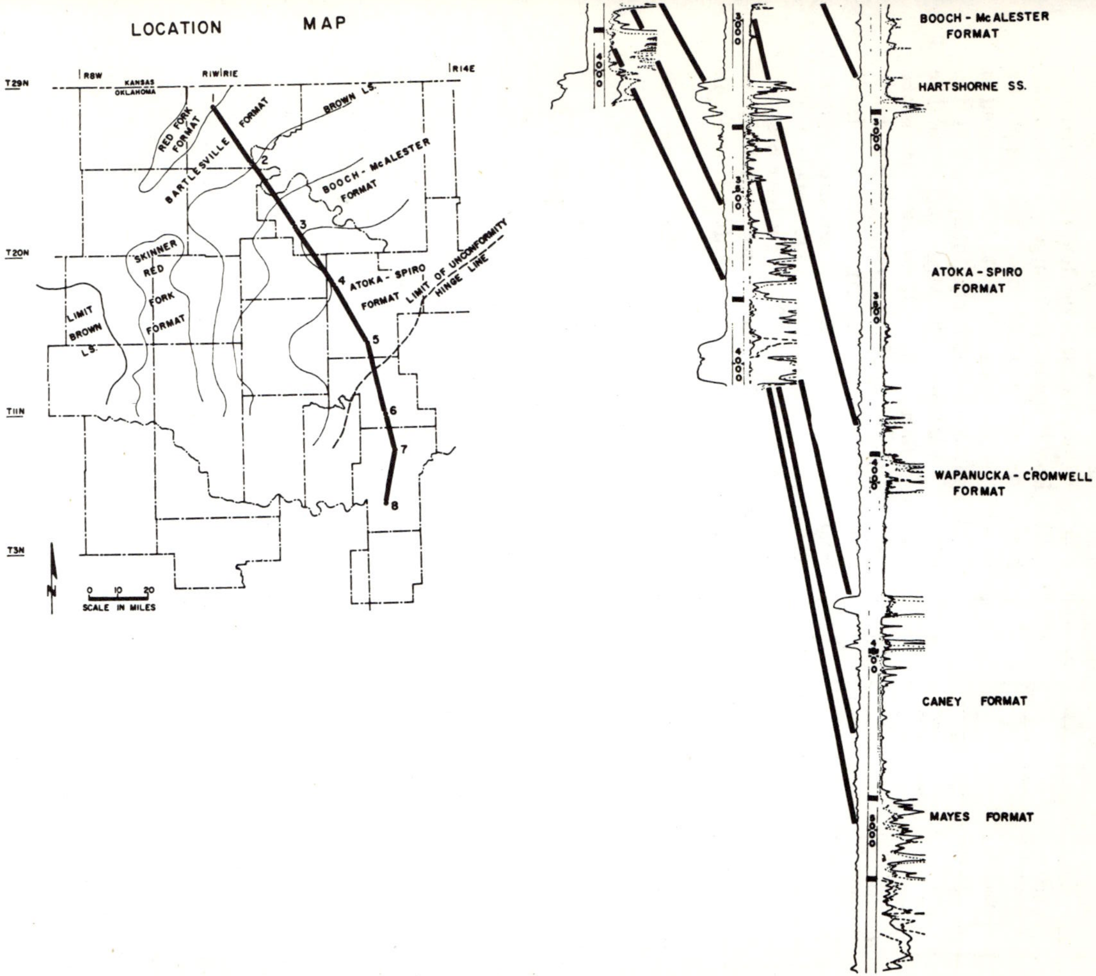


FIG. 3.—Cross section showing truncation and onlap of time-stratigraphic units. Inset shows pattern of onlap of individual units and line of section. Basinward thickening is caused by downwarp and onlap of younger depositional units.

**Table 1. Time-Rock Stratigraphic Unit***

| SYSTEM | | SERIES | FORMAT | SEQUENCE |
|---|---|---|---|---|
| PENNSYLVANIAN | Upper | Virgilian | | Absaroka Sequence |
| | | Missourian | | |
| | Middle | Desmoinesian | Wewoka-Prue | |
| | | | Upper Skinner | |
| | | | Lower Skinner-Red Fork | |
| | | | Bartlesville | |
| | | | Booch-McAlester | |
| | | Atokan | Atoka-Spiro | |
| | Lower | Morrowan | Wapanucka-Cromwell | |
| MISSISSIPPIAN | | Chesteran | Caney | Kaskaskia Sequence |
| | | Meramecian | Mayes | |
| | | Osagean | | |

* Type log Earlsboro No. 1 T. J. Davis, Sec. 11, T8N, R10E.

Each sandstone has the general shape of a large deltaic unit, and they all show a depositional pattern closely related to the isopach map (Fig. 2). Some of the sandstones have been studied in detail and environmental interpretations suggested for individual units. The similarity of the patterns and the basic depositional framework strongly suggest progradation as a result of excess sediment supply resulting in large deltaic complexes.

The continental interior was the probable source of these sands and their thickness suggests they are the product of a single large integrated drainage system. As the transgression covered more of the craton, the source was diminished, and finally, after the deposition of the Prue sandstone, the northern source did not furnish sufficient sand for continued regressive progradations.

*Depositional patterns.*—The lowermost sandstone illustrated is the Atokan (Gilcrease or Dutcher). This sandstone was interpreted as a deltaic unit (Dickey and Rohn, 1955). The pattern shows only a general distribution of thick sandstones. "Channel" sandstones have been described on the north. Another sandstone unit of some importance is the Hartshorne or McAlester (McDaniel, 1968). A deltaic origin has been postulated for this sandstone, and detailed paleocurrent measurements indicate a northeastern source direction. Sand distribution, however, has been omitted from Figure 1.

The Booch sandstone (Burgess or Savanna) has been well described in a classic paper by Busch (1959). The outline has been mapped in detail and an environmental interpretation suggested by analogy to the Mississippi delta. The overlying Bartlesville (Bluejacket) sandstone has been the focus of study during the past several years (Visher, 1968; Saitta B., 1968; Phares, 1969). Surface, paleocurrent, core, and electric-log studies have been made of this sandstone; more than 5,000 wells have been examined in detail. This pattern is well documented and little question of its deltaic origin exists.

The overlying Red Fork (Taft) Sandstone east of the Nemaha ridge has also been the subject of extensive study (Hudson, 1969). Again, the pattern of sandstone distribution is well known, and the environment of deposition documented by detailed studies of cores. The Skinner (Chelsea and Senora) sandstones also have been studied (Cole, 1969; Dogan, 1969), but the sand pattern has been omitted from Figure 4 for the sake of clarity. Essentially similar electric-log, core, and geometric patterns have been obtained. Finally, the Prue (Lagonda and Senora) sandstone is illustrated. Again the data indicate a similar depositional origin (Cole, 1969; Dogan, 1969). This sandstone is farther west and north than most of the other depositional units.

*Shelf sedimentation.*—The pattern of deposition indicated by these sandstones is one of progradation through short intervals of time. The interval from the depositional interface to the surface of the sea was filled with sediment. When the sediment-water interface reached sea level, sediment was bypassed to deeper water and the area was subsequently filled to sea level, producing a regression of the shoreline. Limestone, coal, bay sediments, and strandline sandstone deposits are associated with each of the major format units. Each is separated by a period of transgression produced by subsidence or by an equilibrium between rates of sediment supply, compaction, and the volume of the receptor area (Sloss, 1962). As the deltaic units prograded into deeper water, the rate of regression decreased and the areal extent of the deltaic unit was limited. When this happened, the relative rise of sea level and compaction

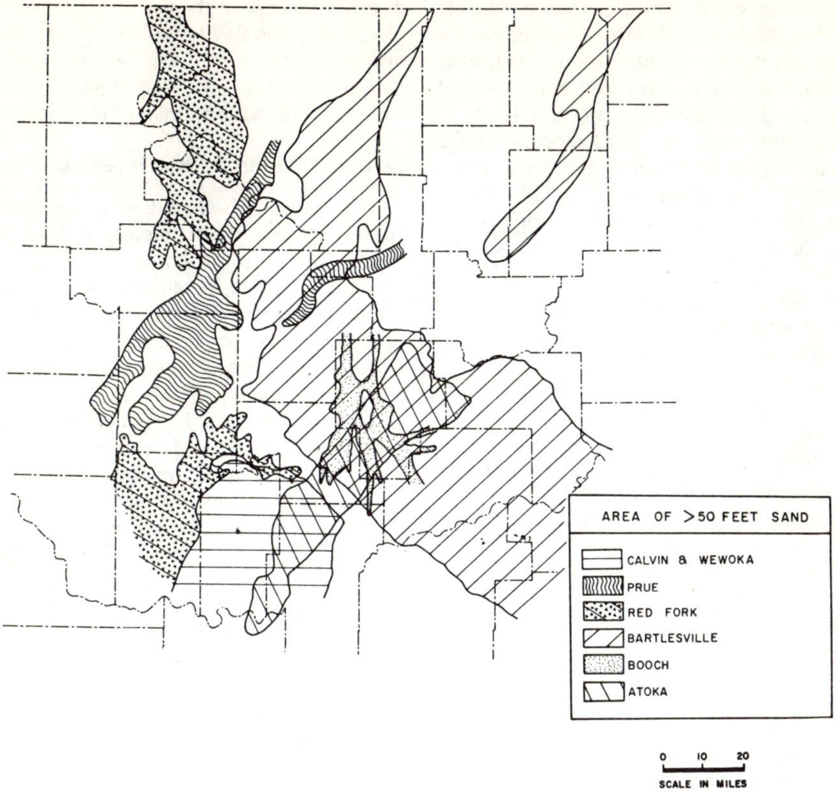


FIG. 4.—Delta patterns indicated by areas with more than 50 ft of sand deposition in individual units. Areal patterns indicate deepening water on south during transgression and compaction of previously deposited interdeltaic shale units.

could overtake the regression of the strandline and the overall transgression could begin again.

In this manner a single major drainage system could produce many individual sandstones of widely differing age, depositional environment, thickness, and geometry. Sedimentation was continuous, but the regressive period of sand deposition was discontinuous. Locally, the transgressive units of sandstone, limestone, and shale are cut out by succeeding depositional units. Generally the successive sand units do not overlie each other, but are present in areas where shale deposited during a previous regressive period would have undergone sufficient compaction to provide an embayment in the shoreline. The common channel source for these sands must have been present on the eastern Kansas shelf. Data are not available to trace these units back to this channel and to determine the time when a new distributary system developed. Transgressive marker beds were not deposited; consequently, the time-stratigraphic framework cannot be reconstructed. This area was primarily one of transportation and erosion, not of deposition.

### Tectonic Uplift and Reversal of Paleoslope

Deposition during the Early and Middle Pennsylvanian was complicated by the presence of additional source areas and of marginal positive areas. The history is one of slowly rising mountains on the southwest and a filling of the Arkoma basin by sediments derived from the east, south, southwest, and north.

*Early sediment sources.*—Evidence from unpublished studies by the writers is that the basal Atokan Spiro sandstone was derived principally from the east and northeast. Thick sands of probable deltaic origin are well developed in the southeastern part of the area. Atokan sources appear to be rather diverse (unpub. research, Sinclair Research Memorandum). Definitive evidence exists for a northern source, but also the thickness of sandstone increases toward the east in the Arkoma basin. Also, near Atoka, Oklahoma, many sandstones are present that suggest a local source from the west. A major sandstone unit, the Hartshorne, above the Atokan, thickens toward the east (McDaniel, 1968). In Arkansas this sandstone is more than 150 ft thick. The overlying Booch sandstone definitely shows a northern source (Busch, 1959), but other equivalent sandstones in the outcrop may represent a continuation of the eastern source area.

The Bartlesville sandstone shows only a northern source. Equivalents of the overlying Red Fork sandstone probably were derived from the southwest (Dogan, 1969). The sandstone geometry map (Fig. 4) also shows a separate but distinct sandstone buildup highly suggestive of a southwest source. Other evidence (Withrow, 1968) suggests that the Red Fork Sandstone west of the Nemaha ridge may have had multiple sources.

The Skinner and Prue sandstones both show the dominance of the northern sediment source. The Calvin Sandstone and sandstones in the Wewoka Formation show a pronounced southern source (Fig. 4). The Calvin sandstones are the time-stratigraphic equivalent of the upper Skinner and Prue sandstones.

*Late Desmoinesian and Missourian sediment sources.*—A new pattern of sand deposition illustrated by the Calvin Sandstone and Wewoka Formation appeared in the Middle and Late Pennsylvanian. The Prue may be the last major sandstone derived from the north during the Pennsylvanian. Later the deposition from the south became increasingly important. In the Seminole, Coffeyville, and later depositional units, large cobbles are observed in the outcrop adjacent to the Arbuckle Mountains (Taff, 1928). Evidence is that a major mountain system developed on the southwest, and coarse detritus filled the marginal basin. Evidence for the rise of the Ouachita Mountains on the southeast is not present in these Middle Pennsylvanian sedimentary units. Later erosion has removed all Upper Pennsylvanian and Permian sediments from the area of the eastern Arkoma basin, and the time of the uplift in this area cannot be determined.

The sedimentary history described indicates a reversal of the paleoslope from east-southeast to northwest during deposition of Desmoinesian sedimentary units. In addition to the change in paleoslope there subsequently occurred a change in the dip of the deposited units. The Ozark dome underwent progressive uplift and produced a primary east to west tilt. This tilting probably occurred during the uplift of the Ouachita Mountains (Huffman, 1958).

## Deltaic Origin of Bartlesville Sandstone

The history of deposition of the lower Absaroka sequence in northeastern Oklahoma has been related to progradation of large deltaic units. Deltas are developed over tens of thousands of years and are composed of a complex of depositional environments. Their interpretation necessitates bringing together a large amount of data over an area of several thousand square miles. Most studies of sedimentary rocks do not cover the range of stratigraphic information needed, or the area necessary, to determine the history of sedimentation. Details of depositional history and the determination of local environments require close control not commonly available. The Bartlesville sandstone was chosen as the basis for developing a model of deltaic sedimentation because of the availability of thousands of well logs, a nearly continuous outcrop along the eastern and northern edges of the delta, and numerous cores, and because it was one of the most prolific oil-bearing sands of the Mid-Continent.

### Sand Distribution, Paleocurrents, and Isopach Maps

The gross sandstone map was based on thickness measurements from over 5,000 well logs (Fig. 5). These were used as control points for the isopach (Fig. 6) and the gross sand maps. Additional logs were examined to determine the range of variability within a section (Phares, 1969; Saitta B., 1968). Over much of the area control density was approximately one well per square mile. This provided the detail to show the clearly developed distributary pattern that is indicated.

Paleocurrent measurements made along the outcrop support the pattern shown by the distributaries (Fig. 5). Measurements were made on trough axes of festoon crossbedded sandstones, rib-and-furrow structures formed by migrating current ripples, current lineation, aligned wood fragments, bottom scour marks and, in a very few places, by the orientation of wave ripples. Several hundred measurements were made in the field and average azimuths computed for each outcrop. In nearly every outcrop the dispersion of directional measurements rarely exceeds 90° and usually falls within a 45° spread. The type of sedimentary structures measured and the coherence of transport vectors indicated that unidirectional currents were responsible for deposition of the Bartlesville sandstone.

Study of the sand distribution patterns and the paleocurrent indicators alone provided much information on depositional environments within specific areas. A possible tidal inlet occurs on the south where paleocurrents indicate a northern transport direction. The area northwest of Muskogee shows a north-northwest transport direction which suggests crevasse deposits. The area south of Muskogee shows dispersion of the paleocurrent measurements, and examination of outcrops suggests that these may be related to bends in a distributary. The value of the sandstone-distribution and paleocurrent map is that it provides a framework for interpreting detailed information gained in specific areas.

The isopach map (Fig. 6) shows the topographic framework during the deposition of the Bartlesville. A marginal highland was present on the northwest and a basin on the southeast. Local highs occur on the Cushing structure and southwest of Tulsa. Thickness of the format increases progressively southeastward.

### Sedimentary Structures and Textures

Twenty outcrops and 10 cored sections were examined in detail. The nature of the sedimentary structures was recorded and sandstone units were sampled for textural analysis. Over 200 grain-size distributions were plotted. These data were used to develop depositional sequences to relate to electric logs and to textural plots collected from modern depositional environments.

Direct comparisons were made between the types of sedimentary structures developed within the cores and the grain-size distribution of the sand (Figs. 7–9). Representative units show festoon crossbedding, current ripples, flaser bedding, and a bioturbated sand. The shapes of the log-probability plots are similar to those described for the Mississippi delta (Visher, 1969). The types of sedimentary structures developed are the same as those described by Coleman and Gagliano (1965) for the Mississippi delta. They suggest a range in current velocities from very weak to strong. The major class of structures not common in either the cores or the outcrop are trace fossils. Some sandstones and many of the shales show the effects of bioturbation, but thick sandstone units contain only current structures.

### Vertical Sedimentary Patterns

Much of the control is based on electric logs. The vertical patterns interpreted from the electric logs are easily related to the sequence of textures and lithologies. An examination of several thousand logs indicates that six fundamental patterns are developed within the Bartlesville deltaic plain (Fig. 10). Each pattern is characteristic of a particular part or subenvironment in the delta. The environmental interpretation is based on similarity between logs from cored sections and studied outcrops. These patterns are typical and are believed to represent the sequence of depositional events that normally occur within a particular subenvironment of the delta. It is believed that they would be directly comparable to logs from other deltas of very different age and geographic distribution.

### Petrography and Mineralogy

Every sandstone sampled for textural analysis was also thin sectioned for petrographic examination. In addition, the fine-fraction from over 100 samples was used to determine the clay mineralogy of the less than 2-$\mu$ size fraction.

The detrital mineralogy was nearly constant in all thin sections examined. The sand is basically a subgraywacke. No evidence of removal of feldspar, mica, or rock fragments was observed in any area or part of the delta. Carbonaceous residues and fragments and clay chips were present in all the outcrops and cores examined. The universal occurrence of these materials was a distinguishing characteristic of the sandstones.

A wide variation in the clay minerals characterizes the fine fraction. Three clay minerals were identified: (1) kaolinite, (2) iron chlorite, and (3) illite. The proportion of these minerals appeared to be related to the environment of deposition of the sandstone (Fig. 11). Each sandstone core represented several depositional environments. Sandstone zones which showed intense bioturbation generally had the highest illite and iron chlorite content. Also the sandstone units closest to gray shales, either at the base or near the top of the current-deposited sand, contained higher ratios of illite and iron chlorite.

Sandstones with high oil saturation, no matter where they occurred, universally contained kaolinite. The suggestion provided by these data is that the original clay mineral present in these sandstones was kaolinite and that under marine conditions, or where the sands were saturated with a concentrated brine, the original clay mineral assemblage was altered to illite and iron chlorite. The scatter of the clay mineral ratios shown by Figure 11 would indicate that this relation can be altered by later migration of fluids. The formation of iron chlorite must be due to diagenetic alteration, because a divalent iron chlorite would be destroyed in the oxidizing conditions existing during sediment transport. Likewise, the illite probably represents diagenetic development.

The only other minerals that are common in these sandstones are siderite and pyrite. The siderite, which occurs as less than 2-$\mu$ rhombs, is associated with the iron chlorite and is suggestive of diagenetic development after deposition. The pyrite is typically associated with carbonaceous fragments and occurs in all depositional environments.

### Environmental Reconstruction

It is difficult to reconstruct a detailed depositional environmental pattern that spreads over an area of thousands of square miles and represents deposition during tens of thousands of years. The primary basis of the reconstruction is the vertical sequence of depositional units indicated by the electric logs (Fig. 12). The electric-log patterns suggest that there is a developmental sequence represented by the vertical profile of textures, structures, and lithologic units. By using vertical sedimentary patterns and correlating them with geometry, paleocurrents, sedimentary structures, mineralogy, textures, and petrography it is possible to reconstruct the areal pattern of particular subenvironments. Work by Coleman and Gagliano (1964) suggests that a major sixfold classification of deltaic elements can be developed for most modern deltas. These include: (1) alluvial valley; (2) upper deltaic plain; (3) lower deltaic plain; (4) subaqueous sand sheet (distributary mouth bar); (5) marginal basin; and (6) marginal depositional plain. These six units have been defined for the Bartlesville sandstone. Each area has characteristic log shapes, each may be related to a specific geomorphic area, and each is a product of specific depositional processes.

More detailed environmental interpretations have been suggested by the sedimentary structure and sandstone textures for parts of the del-

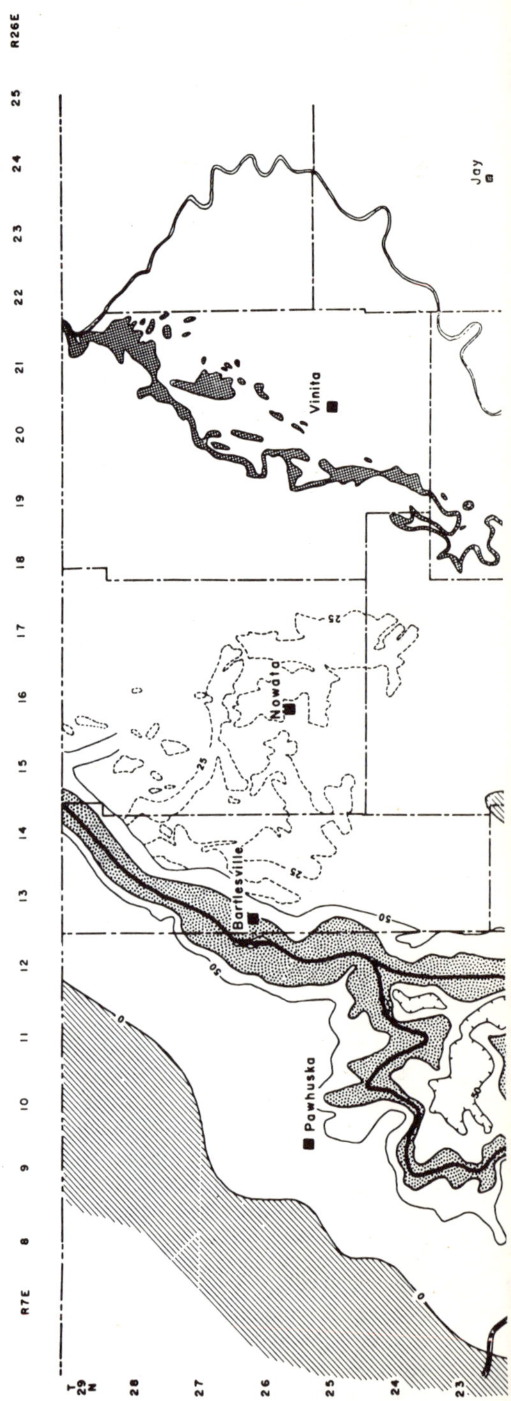

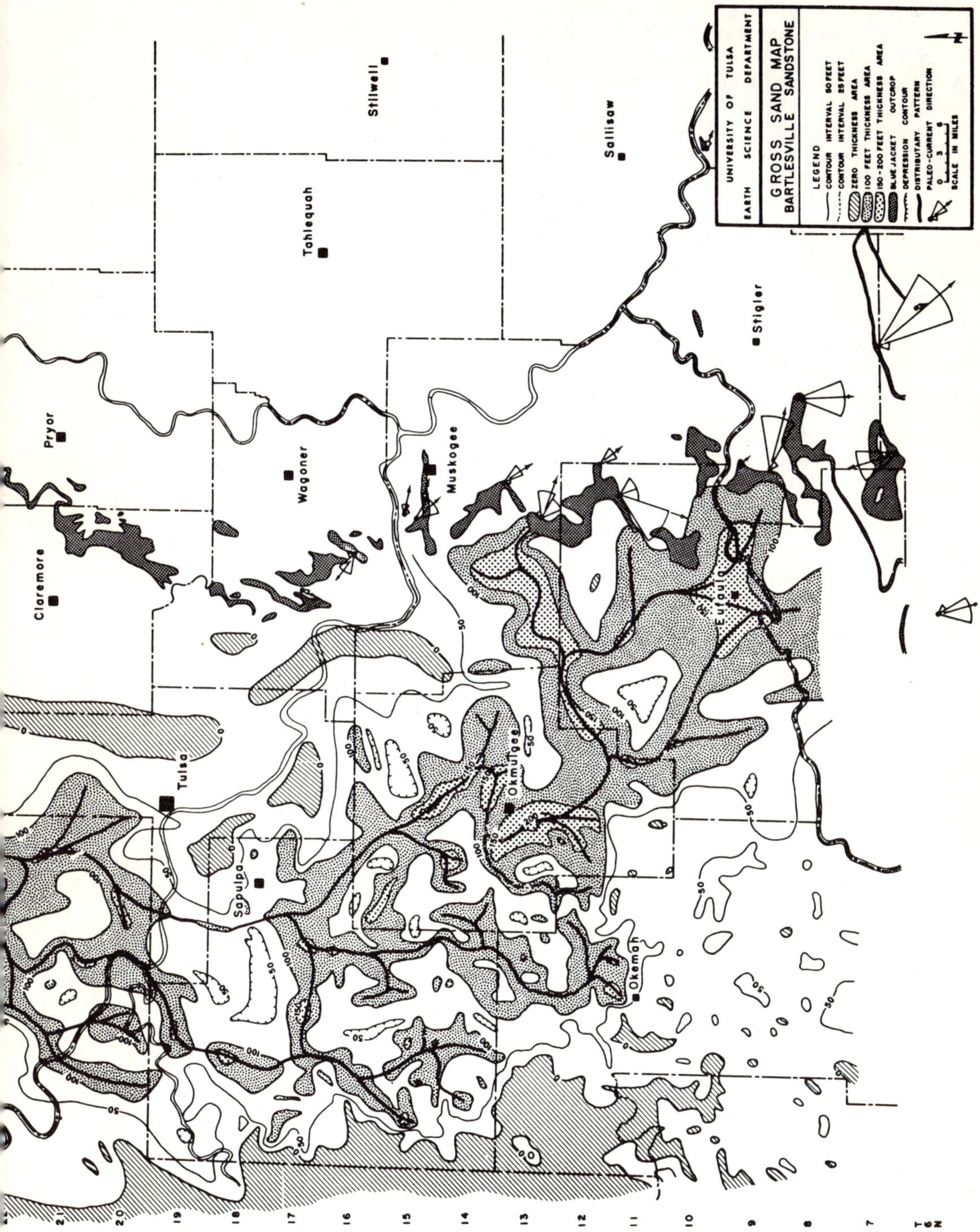


FIG. 5.—Sand distribution of Bartlesville sandstone. Map shows suggested distributary pattern and paleocurrent directions from outcrop studies.

taic complex. On the northeast a large area of sandstone has been interpreted to be a product of crevasse splay deposition. Cores from interdistributary areas show silt and fine sand deposited from suspension and flaser bedding with burrows; they suggest deposition in a bay. Certain outcrop areas are indicated by paleocurrents, structure, and textures to be tidal estuary units. On the northwest, deposition appears to be related to marine bars, possibly as a result of a later transgression. These details cannot be verified by use of electric-log shapes, but they are consistent with the pattern of deposition.

### Deltaic Model

A generalized deltaic model must incorporate several different features. It must include the elements of a delta described by Coleman and Gagliano (1964), the processes important in the building of a delta, and most important the developmental history of a deltaic sandstone. These aspects taken together can be used to explain any single depositional sequence, any particular pattern of depositional environments, and the geomorphic elements within the large progradational deltaic units developed during the deposition of the lower Absaroka sequence.

#### Concept of Delta Maturation

A delta is not deposited all at one time. It represents a history of progradation, maturation of older elements, and deposition throughout the delta. The control of these processes is dependent upon the geomorphic and tectonic framework, energy patterns, rates of sediment discharge, and other physical conditions.

The early stage of progradation of the Bartlesville sandstone was within a narrow channel only a few miles wide (Fig. 5). During this period there must have been a rapid extension of a channel from eastern Kansas. The isopach map and other data indicate that this area was a very shallow shelf, possibly restricted by depth into a complex of marshes and bays. The large river quickly traversed this area into deeper water, where progradation slowed.

The area north of Tulsa is one of many interfingering channels. Just south of Tulsa only two major channels were developed, but farther south and east these channels formed a complex of distributaries and radiating channels. The history is one of delta progradation and then diversion to form a new delta lobe farther south and east. As the delta prograded, the older elements were modified. The vertical pattern of texture in logs north of Tulsa indicates a

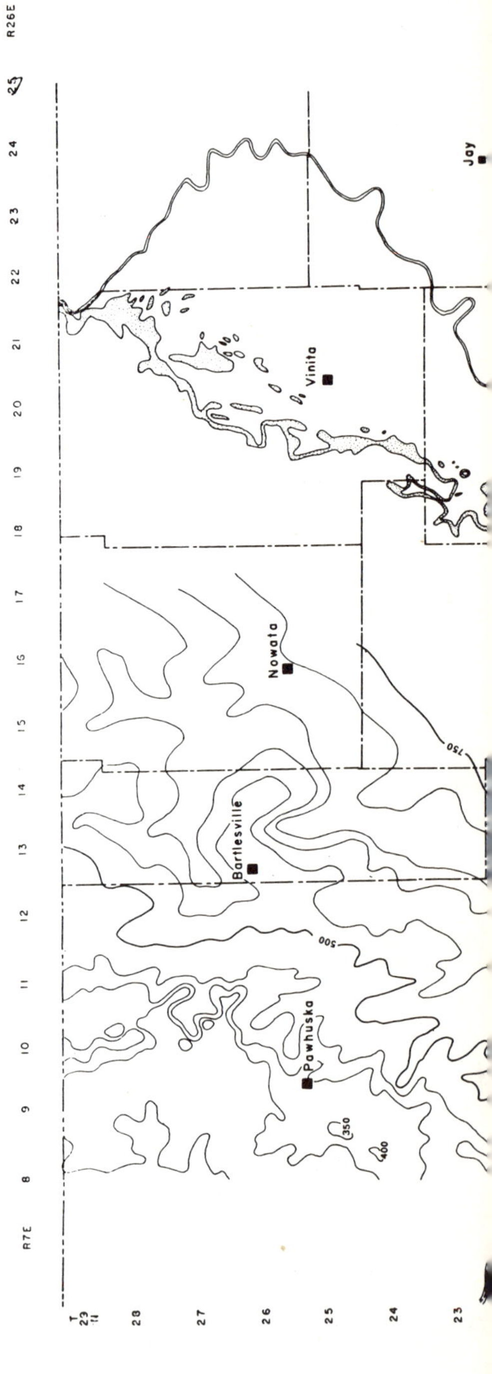

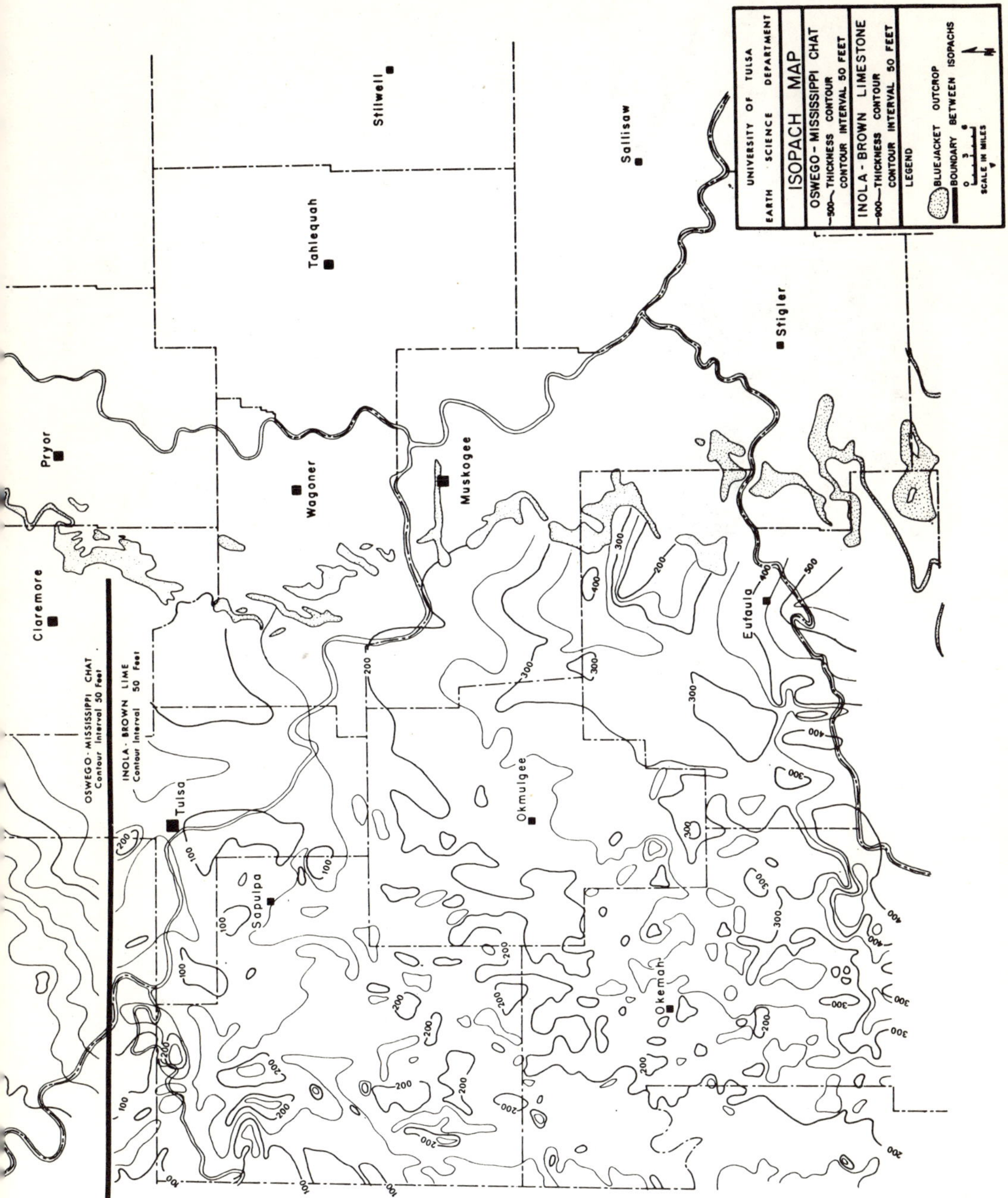


FIG. 6.—Detailed isopach map of interval including Bartlesville sandstone.

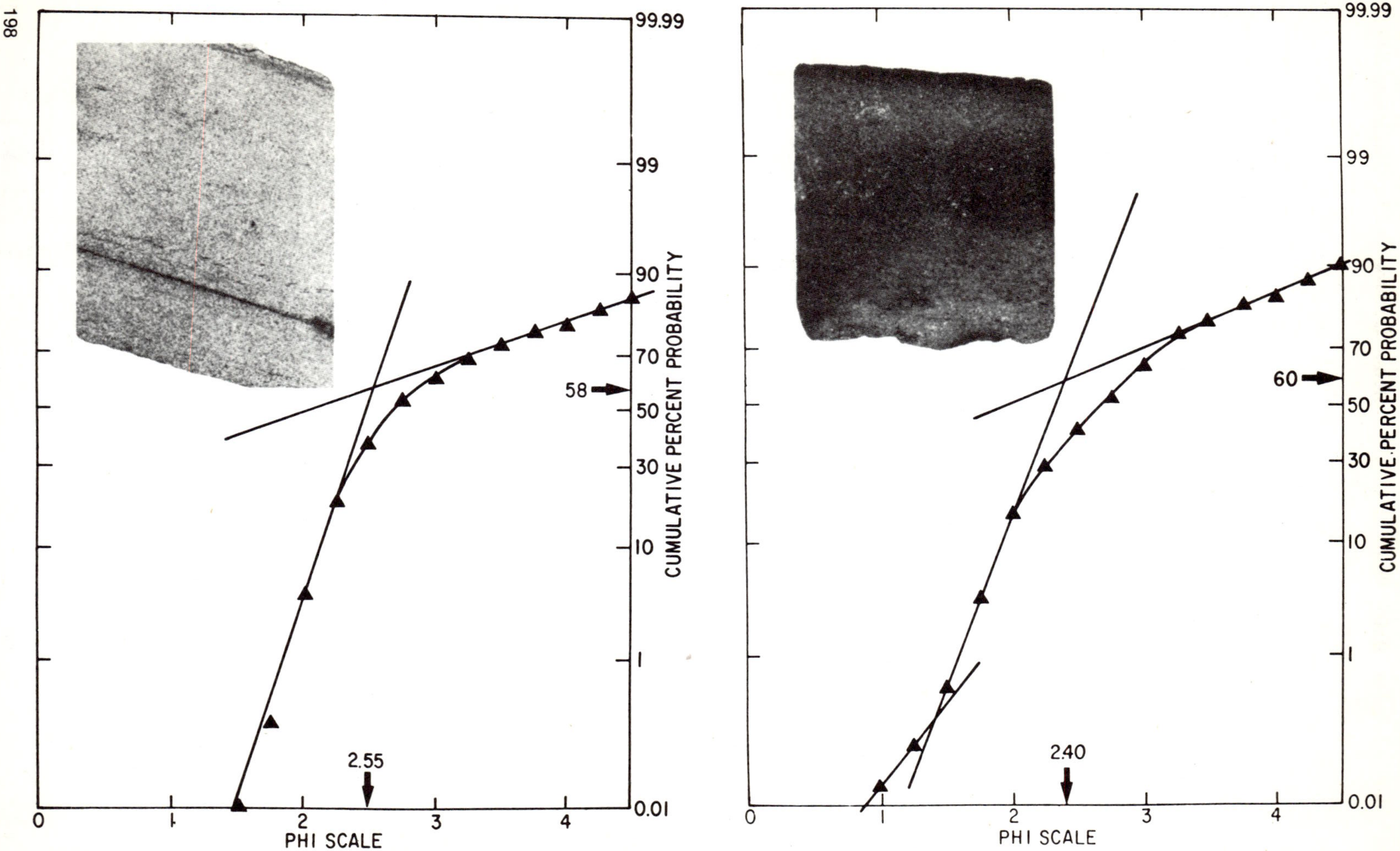


FIG. 7.—Log-probability–grain-size distributions and sedimentary structures of distributary channel deposits.

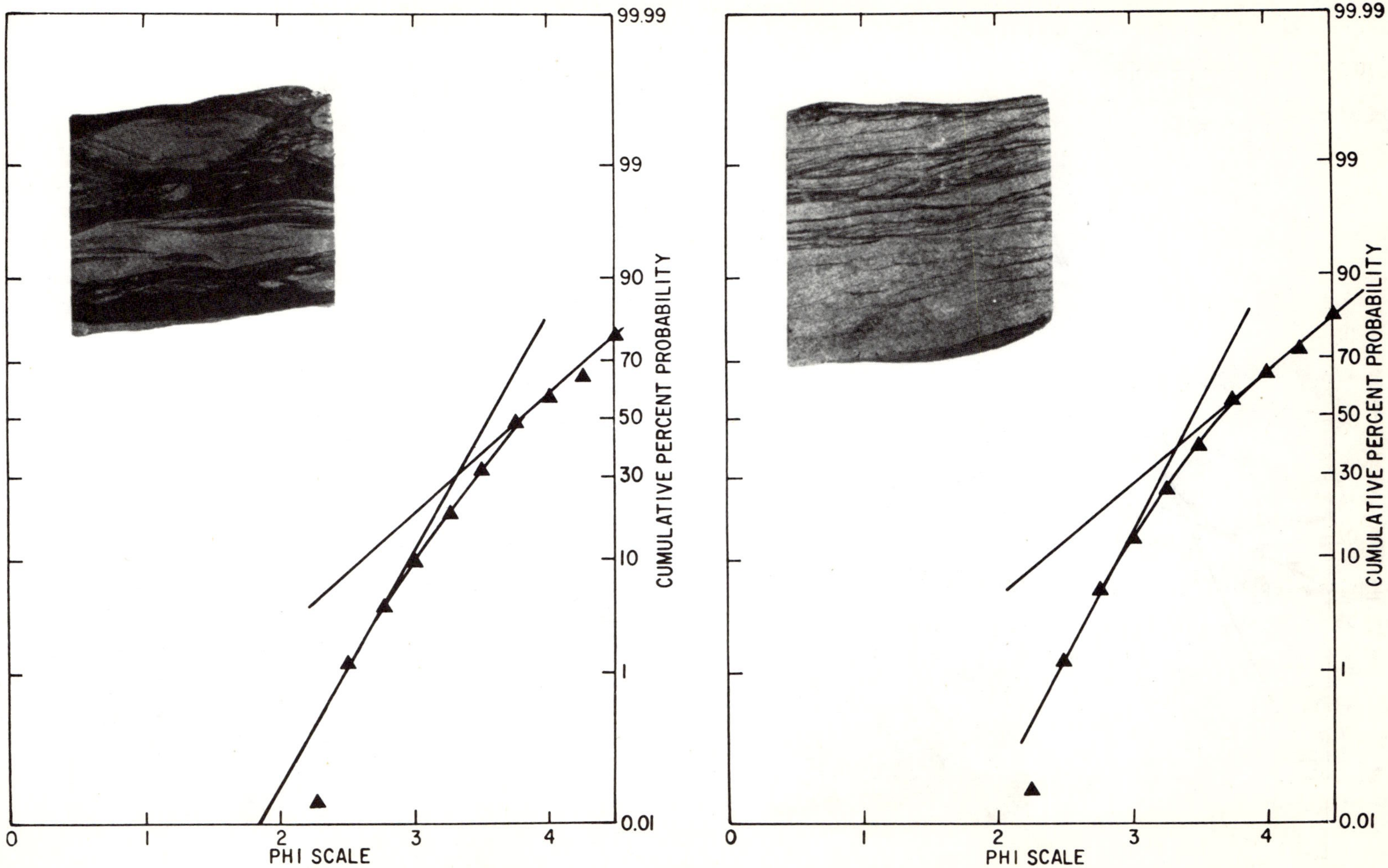


FIG. 8.—Log-probability–grain-size distributions and sedimentary structures from low energy crevasse and **channel deposits.**

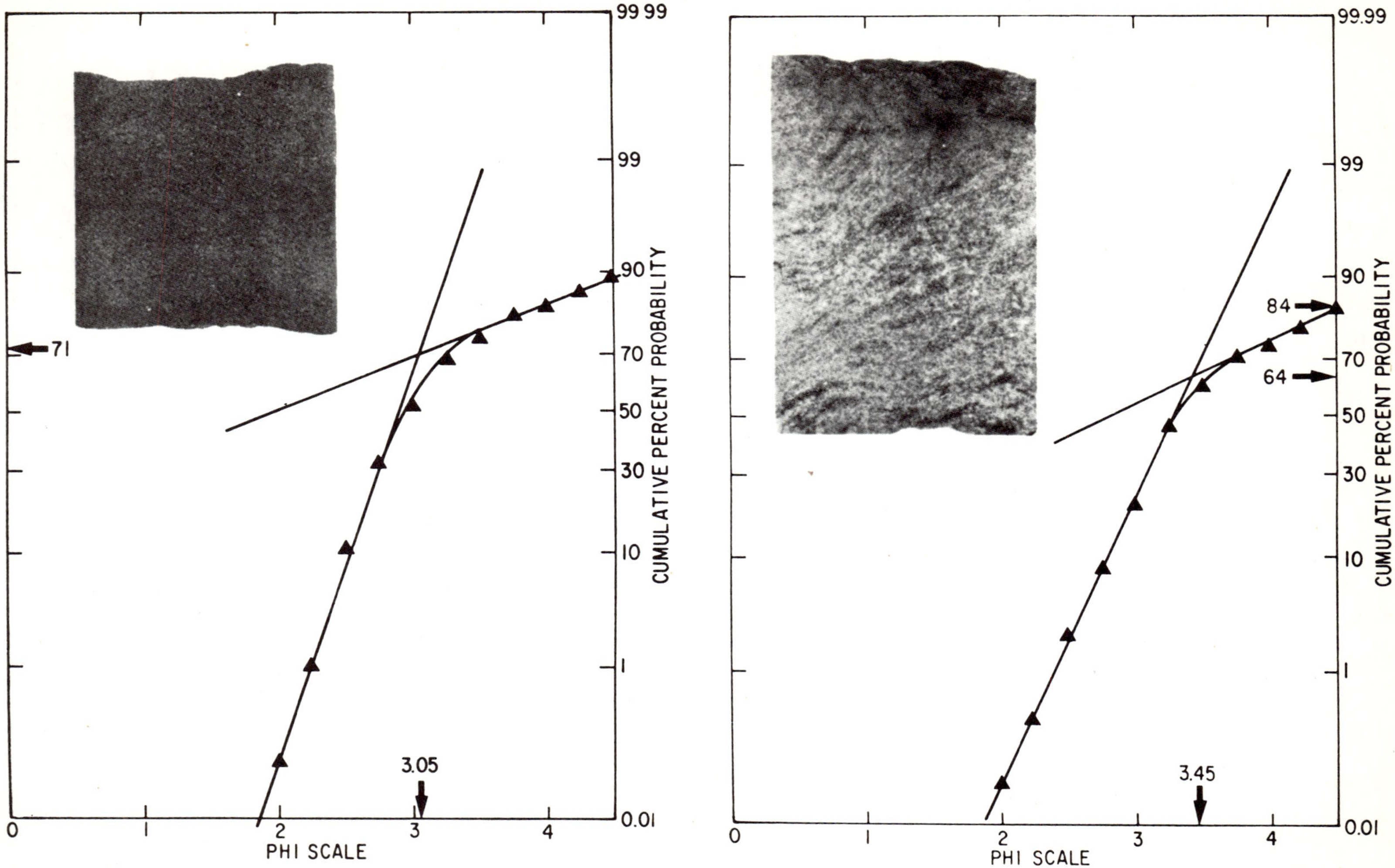


FIG. 9.—Log-probability–grain-size distributions of natural levee and transgressive marine sandstone units.

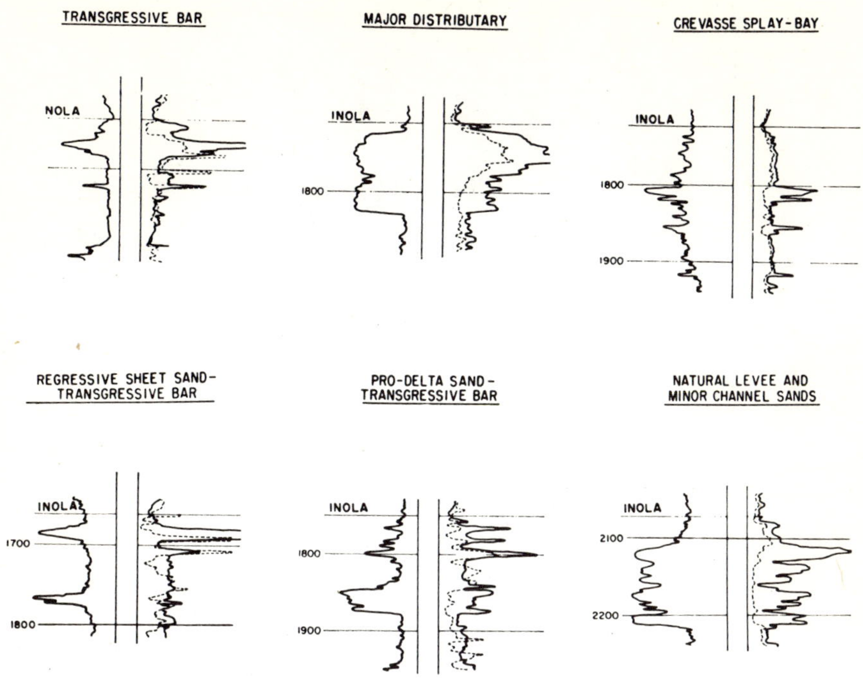


FIG. 10.—Fundamental types of SP log patterns. Individual environmental sequences are illustrated. Sequences based on lithologic and textural comparisons to cored and outcrop sections.

progressive upward decrease in grain size characteristic of meandering streams. The complex of interfingering channels may be due to lateral migration of stream channels by meandering during the formation of the lower part of the delta.

Throughout the Bartlesville delta the basal channel sandstone is underlain by marine shale. Delta-front sandstone stringers or marine sandstone units are not present, and in some localities channeling can be demonstrated. In most places in the upper and central parts of the delta, the Brown limestone closely underlies the base of the channel. Farther south and east in the outcrop, prodelta sands were deposited and preserved. This indicates that deltaic progradation over a shallow shelf can result in the development of a sandstone sheet without the underlying marine sandstone units.

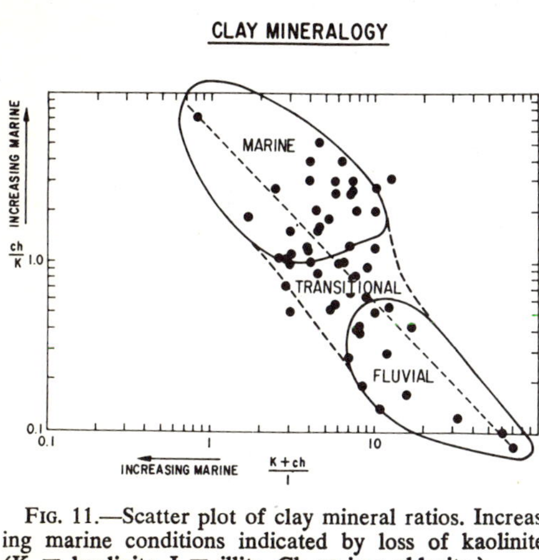


FIG. 11.—Scatter plot of clay mineral ratios. Increasing marine conditions indicated by loss of kaolinite. **(K = kaolinite, I = illite, Ch = iron chlorite.)**

When the eastern and southern parts of the delta were deposited, a true marginal marine sandstone sheet could form. This is shown by the blanket pattern of sandstone in the area of Eufaula, south of the hinge line. Only in deeper water will these marine units be developed and preserved. The marginal depositional plain also is present only on the south and west.

During the deposition of these lower units, crevasse deposits were well developed on the margins of major distributaries. These occur on the northeast, in the central area near Muskogee, and produce the thinner sandstone units within the upper and lower deltaic plains.

### Regressive and Transgressive Patterns

Progradation does not end the formation of a deltaic complex. In the upper parts of a delta, deposition continues with the filling of

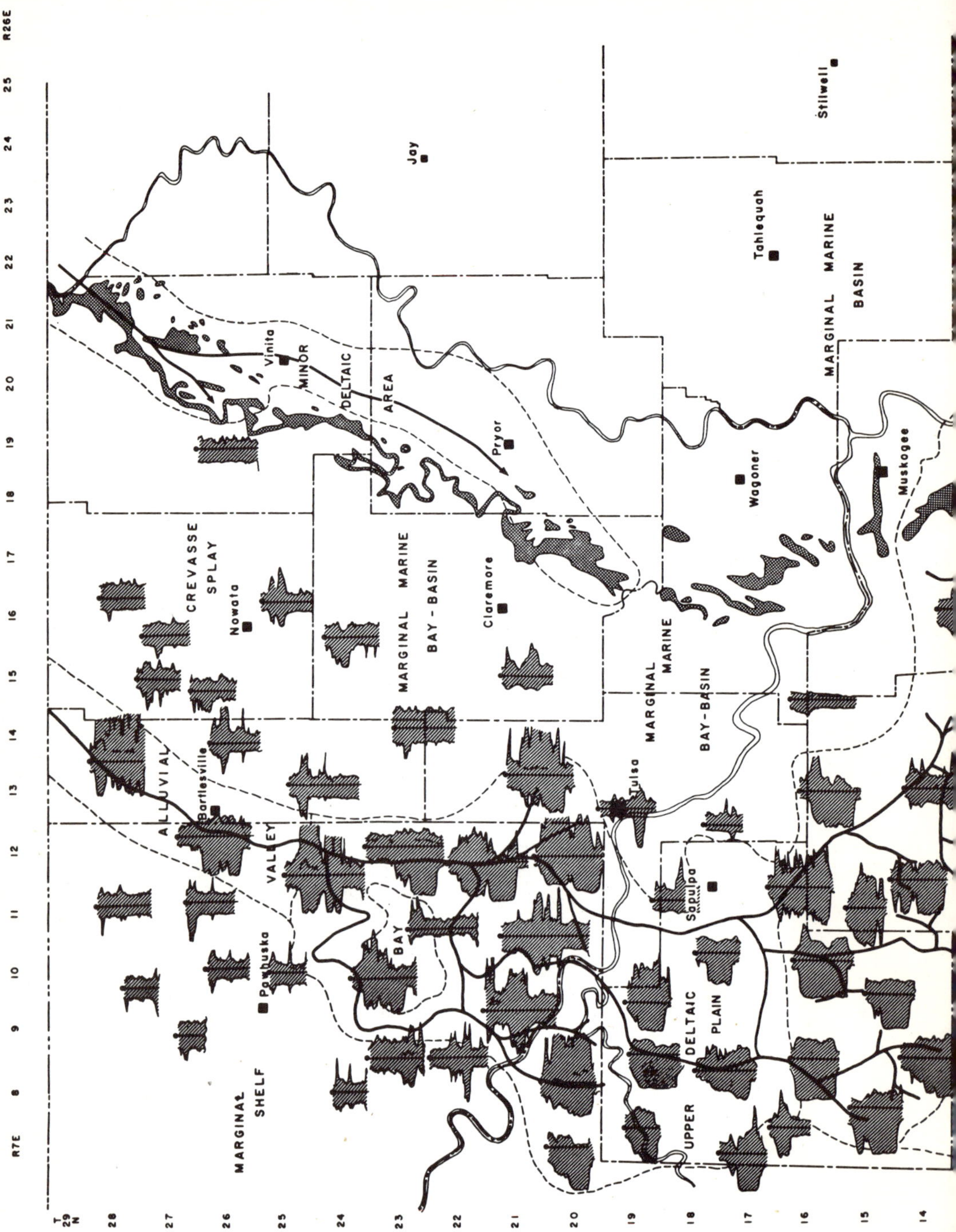


FIG. 12.—E-log patterns and environmental reconstruction of deltaic elements. Environmental interpretations based on sandstone geometry, vertical sequences, sedimentary structures, textures, and clay mineralogy.

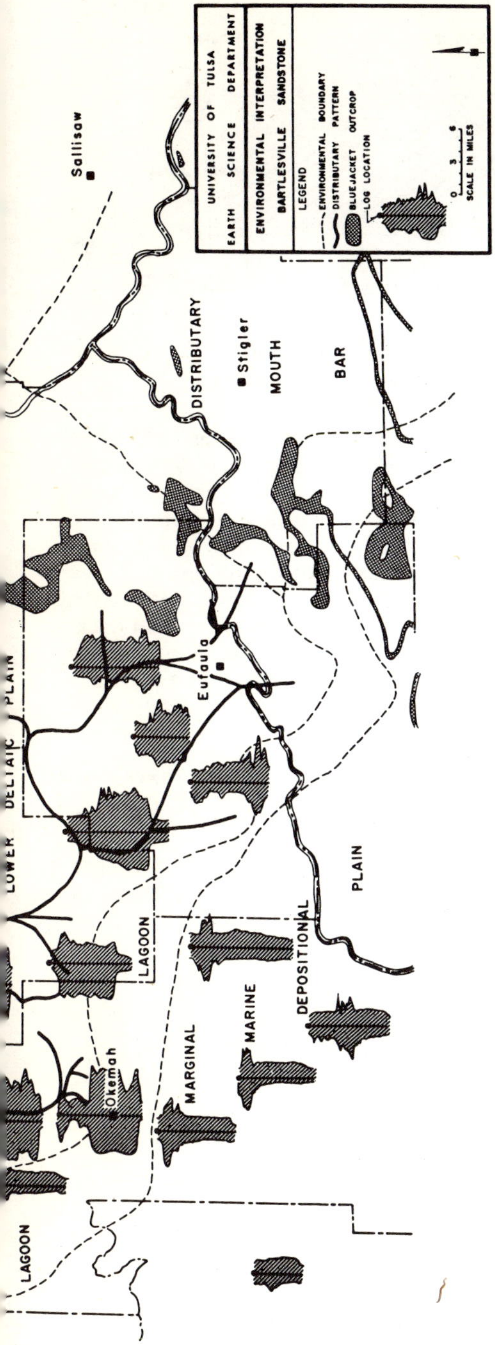


bays and marshes with sediment, and finally natural levees and overbank deposits fill the area to sea level or slightly above. The pattern can be shown by a cross section from the upper part of the delta (Fig. 13). There, crevasse deposits are illustrated on the northeast, and bay, marsh, and restricted marine deposits on the southwest.

Also illustrated is the presence of a sandstone unit across the top of the section, overlain by a limestone. This represents the closing episode of the abandonment of a delta and transgression of the sea across the delta complex. The same upper sandstone unit can be seen in logs illustrated in the south and southwest of the mapped area (Fig. 12). This transgressive sandstone covers the entire delta complex, but locally it formed a thicker marine bar sandstone separated by shale from the lower more massive sandstone unit. In other places the sandstone was produced by reworking of the natural levee and is capped by limestone.

The vertical sequence of lithologies, textures, and sedimentary structures depends on the position within the delta, the relation to waves, and the amount of compaction that may have taken place.

### Deltaic Sequences

Three deltaic sequences can be used to characterize most sections within the Bartlesville sandstone (Fig. 14). These are the products of the historical development of the delta and should be similar to other lower Absaroka deltaic units.

The interdistributary sequence is similar to the marginal depositional basin element of the delta. It is gradational into the distributary type sequence. Variations in the thickness of the sandstones are related to proximity to a distributary. The basal marginal-marine sandstone is related to early deposition on the floor of the bay when the distributary is discharging nearby. The areal distribution of this unit may be very widespread and depends upon depth of water, marine currents, and wave action. The amount of bioturbation and the type of sedimentary structures in the overlying thin sandstone depend on how open the bay is to marine waters and processes. The upper units also may be variable, but a transgressive unit of sandstone and limestone is nearly universal.

The distributary sequence depends on the position within the deltaic complex. If it has matured into an alluvial-plain type channel, it may show a marked upward decrease in grain

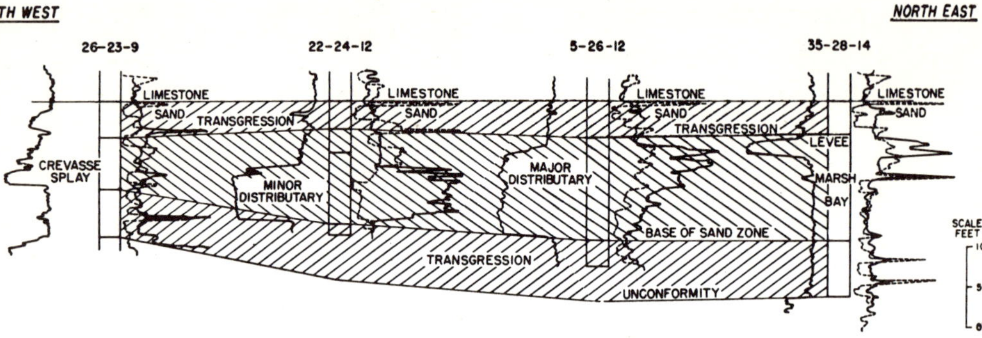


FIG. 13.—E-log section indicating development sequence within upper part of Bartlesville delta. Regressive and transgressive intervals are illustrated. Lateral variation in depositional units can be seen.

size. Also if it is a major channel near the lower part of the delta, it may be interbedded with thin siltstones and shales, have some minor bioturbation, and also show considerable scour into the underlying prodelta type sandstone and shale. In unusually thick sandstone units the transition into marine sandstone and limestone may be very abrupt without the interruption of coal and marsh units.

The prodelta sequence grades laterally into the distributary sequence and also into the marginal depositional plain. It may show either gradual or abrupt transition from marine sedimentary units into current-deposited clastics. The sequence represents a slow shallowing of water during deposition and decreasing marine influence. This entire sequence may be removed locally or replaced by the distributary sequence. Bioturbation may be associated with current features at the base of the sandstone.

Each of these three units can be recognized in the cores and outcrop sections, and commonly are suggested by the electric-log shapes (Fig. 10). The Bluejacket delta probably is an intermediate type between highly elongate and lobate delta types (Fisher, 1969). It is primarily a sandstone sheet, but the marine units are not important except in the lower part of the delta. This type of delta appears to be typical of the Pennsylvanian shelf. Probably only in the distal parts are water depth and wave energy sufficient to modify strongly the depositional pattern.

## OIL OCCURRENCE IN DELTAIC SANDSTONES

Sandstones of the lower Absaroka sequence are extremely important oil reservoirs. More than 2 billion bbl of oil has been produced from these sandstones. Most of the oil is in stratigraphic traps. It is of particular economic importance to understand the reasons for these occurrences.

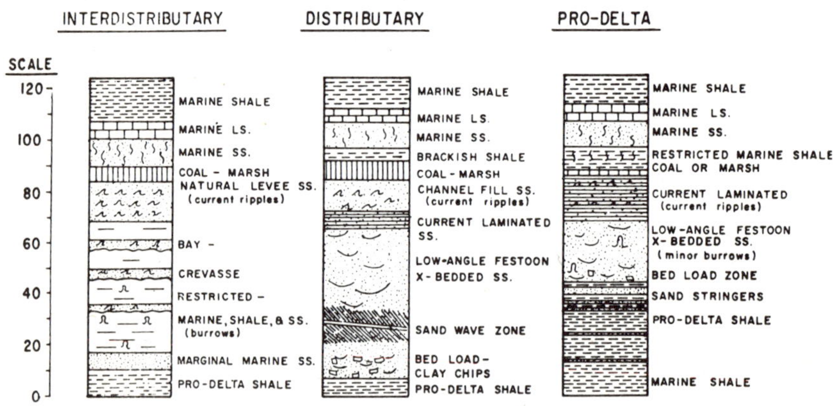


FIG. 14.—Genetic sequences of lithologic units. These patterns characterize process-response sequences within individual areas of delta.

### Regional Stratigraphic Control

Rapid sedimentary infilling of a basin during continuous transgression produces a transgressive shale or caprock at the top of each regressive cycle. Oil that is generated will be trapped in the regressive sandstone units by the overlying impermeable shale. Also, hydrocarbons migrating up the permeable reservoirs will be trapped by the onlap of the next higher regressive-transgressive cycle. During periods of major regression of the sea, the continental landmass is positive. Subaerial exposure produces high rates of runoff and erosion. This provides for rapid rates of sedimentation, increased likelihood of lateral variations in lithologic units and, because of the increased weight of sediments, rapid downwarping within basins and on the continental margins. Deep burial causes the generation of hydrocarbons. The pattern of sedimentation is such that early deposited units are buried quickly by an ever-increasing load of younger sediments.

### Environmental Control

The deltaic environment is the habitat of oil for many reasons. Rivers provide nutrients to the sea, and life grows in abundance in the waters surrounding deltas. Deltaic areas are commonly covered with a dense cover of vegetation, and organic material is one of the most characteristic detrital constituents of deltaic sandstones and shales. In addition, the rapid rates of sedimentation related to deltas provide for rapid burial of organic materials, thus preventing oxidation and loss of carbon compounds to the sea or atmosphere. Deltaic areas contain a varied complex of depositional environments. They are characterized by abrupt vertical and lateral variations in sedimentary facies, which provide excellent stratigraphic traps. Also, the association of channel and wave processes produces winnowing and the formation of relatively permeable sandstone reservoirs. Another aspect favorable for oil accumulation in deltas is the tectonic and structural instability that commonly is associated with high depositional rates. Growth faults, mud diapirs, and other early sedimentary adjustments may provide traps. A final factor may be associated with clay deposition. Poorly crystallized alumino-silicates are eroded off the surface of the continent, and these "clays" are transported to the delta. Their degraded character makes them scavengers for organic colloids and other charged particles. These "clays" are deposited with the sands in the marine environment. Shortly after burial the "clays" recrystallize, increase in grain size, and form clay minerals. They do not contain as much water or organic complexes as they did when deposited. Water loss from the shales may provide the mechanism for transporting "organic complexes" into the reservoirs. These source beds adjacent to reservoirs provide the hydrocarbon to fill the porous and permeable deltaic sandstones.

### Tectonic Reasons

The reversal of paleoslopes and the later tilting of the lower Absaroka strata provided the final fortuitous event to cause the preservation of the oil generated. Oil that was already trapped in small reservoirs and permeability pinch-outs remigrated into porous and permeable sandstone beds along the eastern flanks of the distributary channels. For this reason oil fields such as Burbank and Glenn were highly productive. Large growing structures such as the Cushing anticline also received their share of the oil.

### Bartlesville Sandstone Oil Fields

The production map from the Bartlesville sandstone explains why it has become so well known (Fig. 15). Probably 1.5 billion bbl of oil has been removed from this single sandstone reservoir (Weirich, 1968). Prospecting for oil is still proceeding in some areas, and commercial reserves are being discovered. The pattern of petroleum accumulation has been dictated by the permeability trends and later structural tilting of the area. If some idea of the tectonic and depositional patterns had been known when oil was first discovered in this sandstone, exploration could have proceeded in a systematic manner. The drilling of thousands of dry holes, the development on 5-acre spacing, and other wasteful methods could have been prevented.

## Conclusions

It has been shown that the history of the Early and Middle Pennsylvanian of eastern Oklahoma was one of delta progradation. The theme is one of transgression coupled with alternating regressive periods. Once the framework and environmental control were set, the pattern was repeated again and again. The depositional cycles of coal, thin limestone, thick sandstone, and marine shale were repeated in a systematic manner.

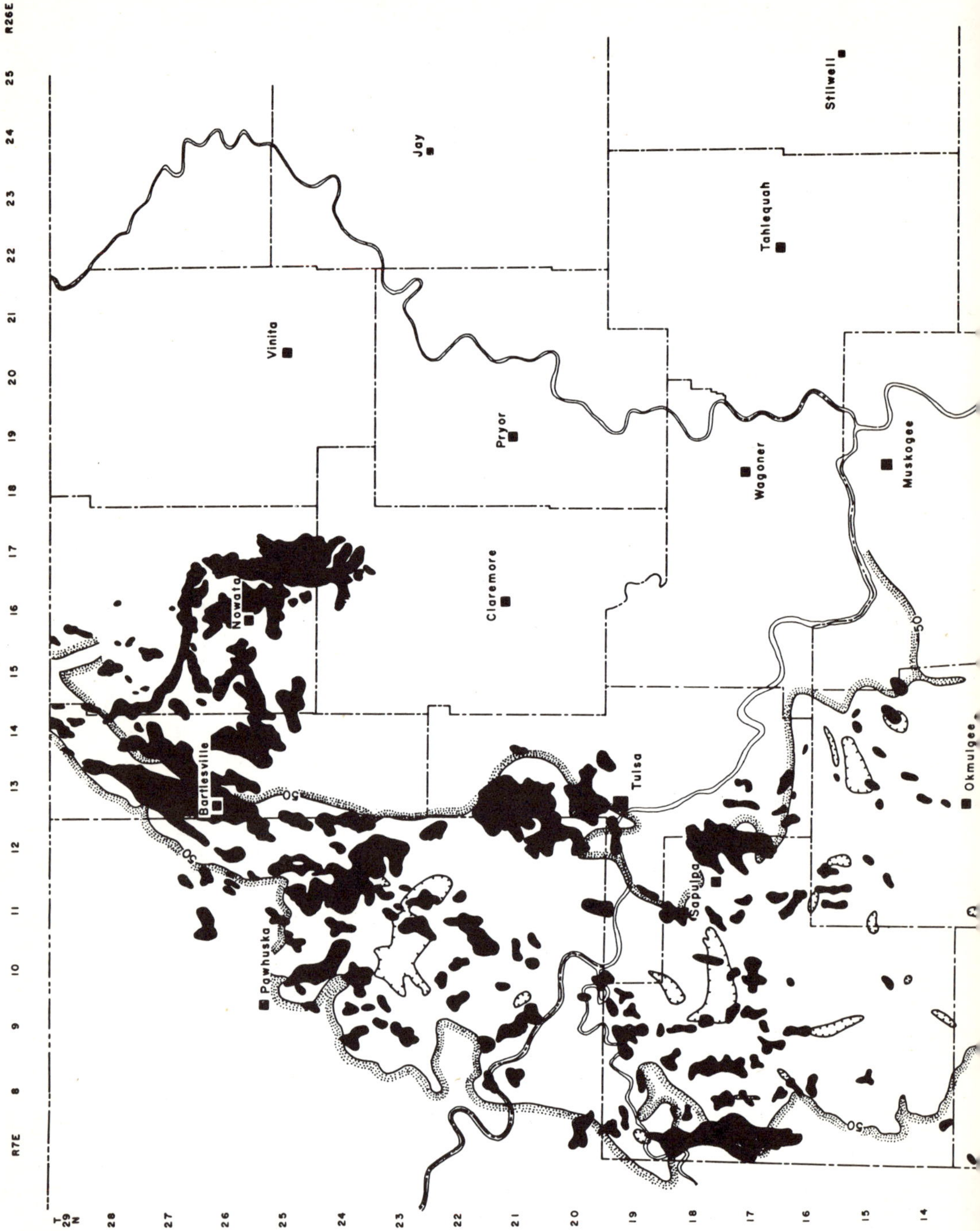


FIG. 15.—Distribution of oil fields within Bartlesville sandstone. Pattern shows concentration of oil on updip pinchout of sandstone. Combination of structural dip and sandstone distribution is basis of individual accumulations.

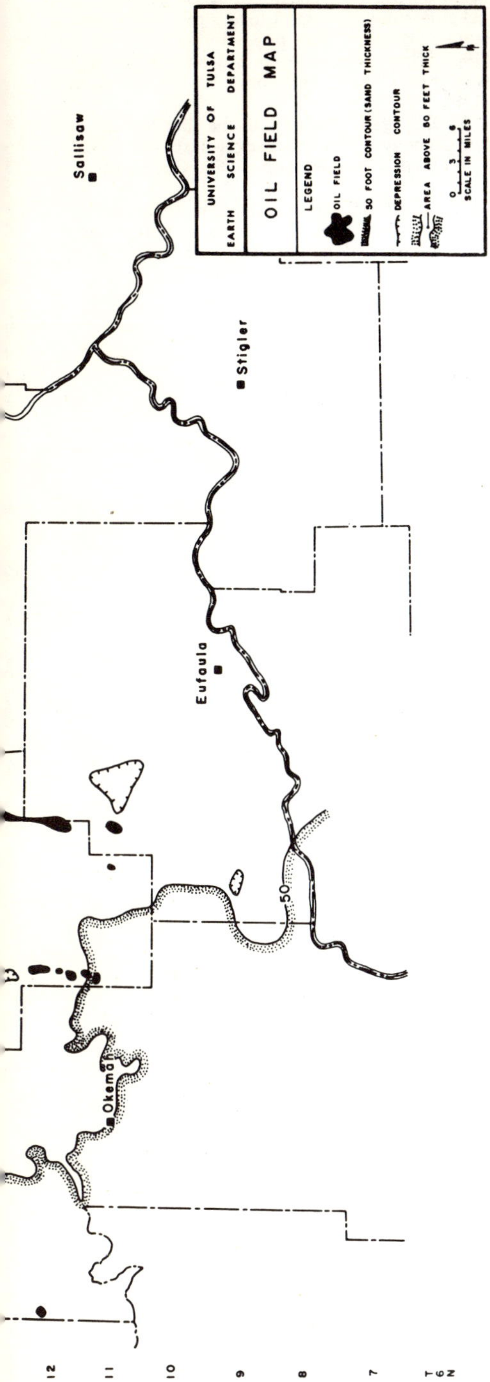


Sedimentation was nearly continuous, but depositional rates varied widely depending on time and position on the shelf. Individual limestones may represent as much time as the progradation of a delta lobe. Scour surfaces and unconformities were local and did not affect the overall sedimentation history.

It is possible to determine "ideal" cycles or patterns within the deltaic sequence. These may provide the basis for interpreting the origin of individual stratigraphic units.

## References Cited

Busch, D. A., 1959, Prospecting for stratigraphic traps: Am. Assoc. Petroleum Geologists Bull., v. 43, no. 12, p. 2829–2843.

——— 1971, Genetic units in delta prospecting: Am. Assoc. Petroleum Geologists Bull., v. 55, no. 8, p. 1137–1154.

Cole, J. G., 1969, Stratigraphic study of the Cherokee and Marmaton sequences, Pennsylvanian (Desmoinesian), east flank of the Nemaha ridge, north-central Oklahoma: Shale Shaker, v. 19, p. 134–160.

Coleman, J. M., and S. M. Gagliano, 1964, Cyclic sedimentation in the Mississippi River deltaic plain: Gulf Coast Assoc. Geol. Socs. Trans., v. 14, p. 67–80.

——— and ——— 1965, Sedimentary structures—Mississippi River deltaic plain, *in* G. V. Middleton, ed., Primary sedimentary structures and their hydrodynamic interpretation: Soc. Econ. Paleontologists and Mineralogists Spec. Pub. 12, p. 133–148.

Dickey, P. A., and R. E. Rohn, 1955, Facies control of oil occurrence: Am. Assoc. Petroleum Geologists Bull., v. 39, no. 11, p. 2306–2320.

Dogan, N., 1969, A subsurface study of Middle Pennsylvanian rocks (from the Brown limestone to the Checkerboard Limestone) in east-central Oklahoma: M.S. thesis, Tulsa Univ., 67 p.

Fisher, W. L., 1969, Facies characterization of Gulf Coast basin delta systems, with some Holocene analogues, *in* Geology of the American Mediterranean: Gulf Coast Assoc. Geol. Socs. Trans., v. 19, p. 239–261.

Forgotson, J. M., Jr., 1957, Stratigraphy of Comanchean Cretaceous Trinity Group: Am. Assoc. Petroleum Geologists Bull., v. 41, no. 10, p. 2328–2363.

Hudson, A. S., 1969, Depositional environment of the Red Fork and equivalent sandstones east of the Nemaha ridge, Kansas and Oklahoma: M.S. thesis, Tulsa Univ., 80 p.

Huffman, G. G., 1958, Geology of the flanks of the Ozark uplift: Oklahoma Geol. Survey Bull., no. 77, 115 p.

Jordan, L., 1962, Geological map and section of pre-Pennsylvanian rocks in Oklahoma: Oklahoma Geol. Survey Map GM-5.

Krumbein, W. C., and L. L. Sloss, 1963, Stratigraphy and sedimentation: San Francisco, W. H. Freeman, 660 p.

McDaniel, G. A., 1968, Application of sedimentary directional features and scalar properties to hydrocarbon exploration: Am. Assoc. Petroleum Geologists Bull., v. 52, no. 9, p. 1689–1699.

Phares, R. S., 1969, Depositional framework of the Bartlesville sandstone in northeastern Oklahoma: M.S. thesis, Tulsa Univ., 59 p.

Saitta B., S., 1968, Bluejacket formation—a subsurface study in northeastern Oklahoma: M.S. thesis, Tulsa Univ., 142 p.

Sloss, L. L., 1962, Stratigraphic models in exploration: Am. Assoc. Petroleum Geologists Bull., v. 46, no. 7, p. 1050–1057.

——— 1963, Sequences in the cratonic interior of North America: Geol. Soc. America Bull., v. 74, p. 93–114.

Taff, J. A., 1928, Preliminary report on the geology of the Arbuckle and Wichita Mountains in Indian Territory and Oklahoma: Oklahoma Geol. Survey Bull., no. 12, 95 p.

Visher, G. S., 1968, Depositional framework of the Bluejacket-Bartlesville sandstone, *in* Geology of the Bluejacket-Bartlesville sandstone of Oklahoma: Oklahoma City Geol. Soc., p. 33–51.

——— 1969, Grain size distributions and depositional processes: Jour. Sed. Petrology, v. 39, no. 3, p. 1074–1106.

Weirich, T. E., 1968, History of the Bartlesville oil sand, *in* Geology of the Bluejacket-Bartlesville sandstone of Oklahoma: Oklahoma City Geol. Soc., p. 69–72.

Withrow, P. C., 1968, Depositional environments of Pennsylvanian Red Fork Sandstone in northeastern Anadarko basin, Oklahoma: Am. Assoc. Petroleum Geologists Bull., v. 52, no. 9, p. 1638–1654.

**Reprinted from:**
BULLETIN OF THE AMERICAN ASSOCIATION OF PETROLEUM GEOLOGISTS
VOL. 55, NO. 8 (AUGUST, 1971), PP. 1137-1154, 24 FIGS., 1 TABLE

# Genetic Units in Delta Prospecting[1]

DANIEL A. BUSCH[2]
Tulsa, Oklahoma 74103

**Abstract** Deltas generally are formed at river mouths during stillstands of sea level under conditions of cyclic transgression or regression. Consequently, they are rarely isolated phenomena, but form in multiples in a predictable fashion. Reservoir facies consist of continuous and discontinuous, bifurcating channel sandstones and delta-front sheet sandstones. The channel sandstones generally thicken downward at the expense of the underlying prodelta clays and may replace selected parts of the delta-front sheet sandstones.

The lithologic components of a deltaic complex are interrelated and are referred to collectively as one type of Genetic Increment of Strata (GIS). The GIS is a vertical sequence of strata in which each lithologic component is related genetically to all the others. It is defined at the top by a time-lithologic marker bed (such as a thin limestone or bentonite) and at the base by either a time-lithologic marker bed, an unconformity, or a facies change from marine to nonmarine beds. It generally consists of the total of all marginal marine sediments deposited during one stillstand stage of a shoreline, or it may be a wedge of sediments deposited during a series of cyclic subsidences or emergences. An isopach map of a GIS clearly shows the bifurcating trends of the individual distributaries and the shape of the delta, regardless of the variable lithology of the channel fills.

A Genetic Sequence of Strata (GSS) consists of two or more contiguous GIS's and, when isopached, clearly defines the shelf, hinge line, and less stable part of a depositional basin. An isopach map of the McAlester Formation of the Arkoma basin is a good example of a GSS. The oil-productive Booch sandstone is a good example of a deltaic complex occurring within a GIS of this GSS. The upper Tonkawa, Endicott, and Red Fork sandstones of the Anadarko basin are identified as deltaic accumulations within different GIS's.

A hypothetical model serves as a basis for establishing the criteria for (1) recognizing successive stillstand positions of a shoreline, (2) predicting paleodrainage courses, (3) predicting positions of a series of deltaic reservoirs, (4) locating isolated channel sandstone reservoirs, and (5) tracing related beach sandstone reservoirs.

## Introduction

Within the past 15–20 years petroleum explorationists have come to recognize the economic significance of deltas. With this recognition has come a realization of the variabilities of deltas, both external and internal, as well as the constructive and destructive energy agents that are responsible for their development. For decades little more was known about ancient deltas than that published by Barrell (1912) on the classic Catskill-Chemung delta of the Appalachian geosyncline and the much smaller deltas of glacial Lake Bonneville described by Gilbert (1885). In more recent years several significant papers concerning ancient deltas have appeared in the literature. Busch (1953, 1959) described the Pennsylvanian Booch delta of the Arkoma basin; Nanz, Jr. (1954), the Oligocene Seeligson pool of the Gulf Coast; Weimer (1961), the Late Cretaceous Rawlins delta; Halbouty and Barber (1961), an Oligocene delta of the Gulf Coast; Rainwater (1963), a Gulf Coast Miocene delta; and Swann (1964), a Late Mississippian Michigan River delta. No two of these deltas are alike, hence it is apparent that a knowledge of their modern analogues is an essential prerequisite to their recognition and delineation in the subsurface.

Modern deltas have certain distinct differences, too, which are related to such variables as "(1) sediment input in terms of amount, caliber, rate, variation in rate, and suspended/bedload ratio, and these in turn on extent and nature of drainage area and climate; (2) nature of the discharging and reservoir water bodies particularly in regard to relative water densities; (3) reservoir energy (waves, currents, tides), both degree and kind, especially in relation to amount of sediment input; (4) depth of water into which delta is prograding; (5) nature of substrata of reservoir body in regard to compactional subsidence and storage of prograding sands; and (6) structural nature of the depositional basin" (Scott and Fisher, 1969).

The depositional environments and sediments of the delta of the Mississippi River have been extensively investigated. As a result, the Mississippi River deltaic plain undoubtedly is the most completely understood modern depositional environment anywhere in the world. Many of the sedimentary processes in operation in this area today have worldwide application in understanding both modern and ancient deltas. Ironically, however, the modern birdfoot delta of the Mississippi River is not typical of

[1] Manuscript received, July 14, 1970; accepted, January 15, 1971. Presented before the Association at Calgary, June 24, 1970.

[2] Consulting geologist.

the majority of modern and ancient deltas. This Holocene delta is the result of a combination of environmental variables that exist in only a few places in the world today.

The major emphasis of this discussion is the use of conventional geologic tools in the identification and delineation of deltas in subsurface exploration for hydrocarbons. The tools readily available to all geologists are samples, cores, and more than 30 kinds of mechanical logs. These tools, when used in conjunction with a background knowledge of modern deltas and principles of stratigraphic analysis, are adequate for delta exploration.

## Genetic Relations of Deltaic Facies

Deltas generally are formed at river mouths during stillstands of sea level under conditions of either cyclic transgression or regression. Consequently, they are rarely isolated phenomena but, rather, form in multiples in a predictable fashion. Reservoir facies consist of continuous and discontinuous, bifurcating channel sandstones and delta-front sheet sandstones. The channel sandstones generally thicken downward at the expense of prodelta clays and locally replace selected parts of the delta front sheet sandstones.

### Genetic Increment of Strata (GIS)

All marine deltas, with the exception of deep-water submarine fans, are deposited in the marginal-marine and directly adjacent subaerial environments. All the lithologic components of a delta are related genetically to each other and can be referred broadly to two basic genetic depositional units. The smaller of these is the Genetic Increment of Strata (GIS), which is defined as a vertical sequence of strata in which each lithologic component is related genetically to all the others. It is defined at the top by a time-lithologic marker bed (such as a thin limestone or bentonite) and at the base by a time-lithologic marker bed, an unconformity, or a facies change from marine to nonmarine beds. It generally consists of the total of all marginal marine sediments deposited during one stillstand stage of a shoreline, or it may be a wedge of sediments deposited during a series of cyclic subsidences or emergences (Fig. 1). An isopach map of a GIS involving a deltaic system clearly shows the bifurcating trends of the individual distributaries and the shape of the delta, regardless of the variable lithology of the channel fills.

### Genetic Sequence of Strata (GSS)

A Genetic Sequence of Strata consists of two or more contiguous GIS's (Fig. 2) and represents more or less continuous sedimentation; the stratigraphic interval must be devoid of angular unconformities, but disconformities of limited extent may be present. An isopach map of a GSS serves to define clearly the shelf, hinge line, and the less stable part of a depositional basin, as shown in Figures 2 and 4. The principal purpose of this type of map is the reconstruction of the paleodepositional configuration of a basin. Most sedimentary basins are so large that only a part may be of interest. This type of isopach map can be constructed with widely scattered control points and is, there-

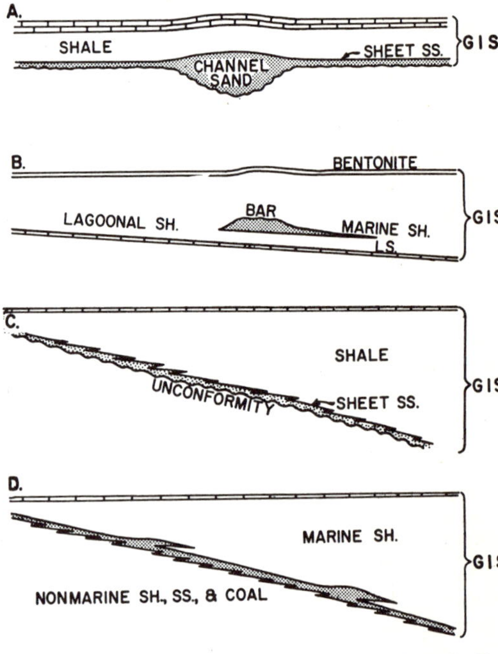


Fig. 1.—Diagrammatic illustrations of Genetic Increments of Strata (GIS). *A*. Channel and sheet sandstones with overlying shale and limestone; upper limit of GIS is defined by time-lithologic marker bed, base by unconformity. *B*. Sandstone bar bordered by lagoonal shale on one side and by marine shale on other; upper and lower limits of GIS are defined by time-lithologic marker beds (bentonite and limestone). *C*. Sheet sandstone, resulting from cyclic subsidence, overlain by basinward-diverging marine shale; upper limit of GIS is defined by time-lithologic marker bed, base by unconformity. *D*. Sheet sandstone showing two areas of locally thicker sandstone at former stillstand positions of shoreline; upper boundary is defined by time-lithologic marker bed, base by facies change from continental to marine beds.

fore, of considerable value in preliminary studies and evaluations of stratigraphic sections of parts of a depositional basin early in an exploration program. The GIS, on the other hand, is generally most applicable for detailed studies of individual sandstones.

In defining the GIS and GSS, the lithologic-time marker bed is critical because it represents contemporaneity of deposition. The most commonly used marker beds are a thin bed of limestone or bentonite. Coal, black shale, and siltstone also may serve as marker beds where it can be ascertained that they do not cross time lines. Of the five, bentonite is considered to be most reliable. The thickness of a shale sequence between two marker beds generally is fairly uniform within a local area, but normally diverges systematically in a basinward direction on a regional scale. Under starved conditions, which are somewhat exceptional, it could converge basinward. Abrupt changes in such intervals generally are due to facies changes, unconformities (at the base), or growth faulting, and if any of these occur, they must be clearly recognizable before using the GIS and GSS concepts. In essence, the GIS is a particular case of a limited interval isopach applicable to individual reservoir sandstone analysis, whereas the GSS is a case of gross interval isopach for regional stratigraphic analysis.

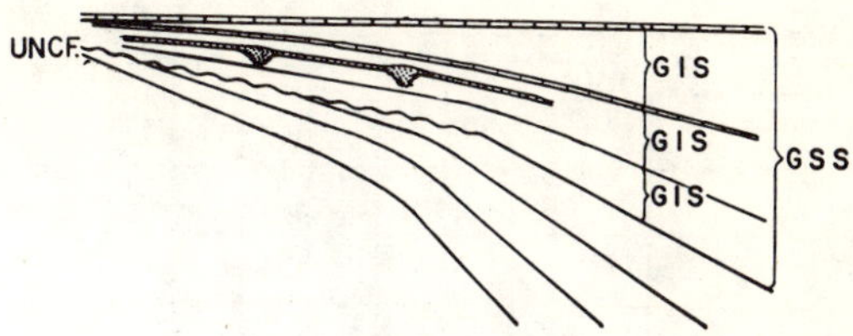


FIG. 2.—Relation of Genetic Sequence of Strata (GSS) to Genetic Increments of Strata (GIS).

## Subsurface Examples of Delta Analysis

### Booch Delta

In Early Pennsylvanian time a large delta complex was deposited in the eastern Oklahoma part of the Arkoma basin. The bulk of this Booch delta covers an area of approximately 2,000 sq mi. It was deposited largely under shoal conditions in a shelf environment. In analyzing the sandstones making up this delta, it is necessary first to consider the Genetic Sequence of Strata of which they are component parts. To do this, a series of stratigraphic profiles, such as that illustrated in Figure 3, was constructed across each of the tiers of townships shown in Figure 4. These profiles then were tied together by a series of intersecting north-south stratigraphic profiles. All wells not used in the construction of the correlation grid then were correlated with the nearest wells of the grid. In so doing, the upper and lower limits of the McAlester Formation (Genetic Sequence of Strata) were determined as a prelude to constructing an isopach map of this formation.

Figure 3 is an electric log cross section of the McAlester Formation in the Greater Seminole District of Oklahoma, illustrating the rate of thickening of this unit across the northwest shelf area of the Arkoma basin. This wedge of sediments is predominantly shale, interrupted by four members of the Booch sandstone. The middle member of the Booch sandstone increases abruptly in thickness to as much as 250 ft. In so doing, it merges vertically with the lower middle member of the Booch and even with the lower member of the Booch. The principal direction of thickening is downward, at the expense of the shale which encases the lower middle and lower Booch.

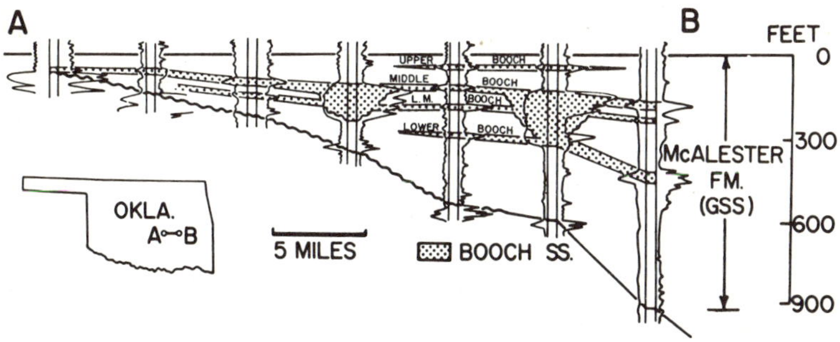


FIG. 3.—Correlation of Booch sandstone members of McAlester Formation (Pennsylvanian) of eastern Oklahoma part of Arkoma basin.

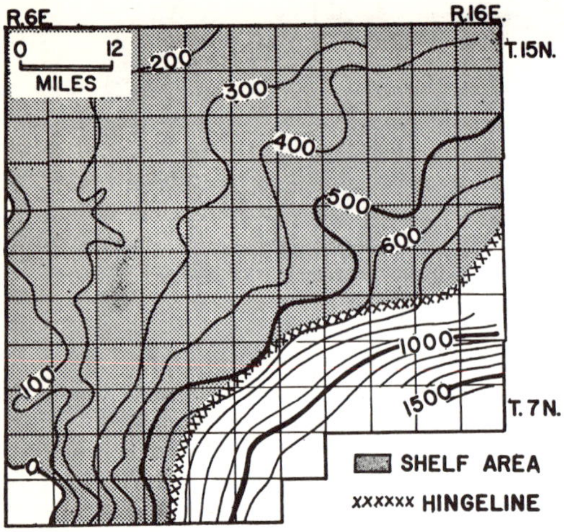


FIG. 4.—Isopach map of McAlester Formation (GSS), Greater Seminole district, eastern Oklahoma. Hinge line separates shelf environment on northwest from more rapidly subsiding basin area on southeast.

The McAlester is overlain conformably by the Savannah Formation; the shales of these two formations are lithologically indistinguishable. The contact between them has been placed arbitrarily at an easily recognized reentrant "angle" of the short normal resistivity curve of the electric logs. This "angle" is present on more than 90 percent of the 650 electric logs available for this study and occurs about 30 ft below the base of the zone known as the "Brown lime." This reentrant "angle" is the reference datum of the stratigraphic profile in Figure 3. The base of the McAlester Formation is conformable with the Atoka shale in the more basinward part of the Arkoma basin, but unconformable on much of the shelf part of this basin. It overlies unconformably successively older units on the west, in the direction of the Hunton arch. For example, from southeast to northwest it overlies Atoka shale and sandstone, on successively older formations of the Morrow Series, and finally on Mississippian shale and limestone along the western margin of the basin where post-Atoka uplift and erosion were greatest.

Electric log and sample studies of the McAlester show that basinward (southeast) this formation consists almost entirely of shale, but in the area where it is less than 600 ft thick it is characterized by an alternating series of beds of shale, sandstone, limestone, and coal. That the shelf was never covered by deep water is attested clearly by the presence of coal beds.

The isopach map of the McAlester (Fig. 4) was drawn to (1) reconstruct the shape of the northwestern part of the Arkoma depositional basin, (2) determine the trend and position of the "line of flexure" (hinge line) that existed during McAlester deposition, and (3) ascertain the principal direction of transport of McAlester sand and mud.

The McAlester Formation thickens abruptly southeast of the hinge line (Fig. 4) in the more rapidly subsiding part of the basin. In the shelf environment, northwest of the hinge line, the rate of southeast divergence is much more gradual. The greatest accumulation of sediments within the mapped area is more than 1,600 ft in T7N, R16E. The strike of the isopach contour lines changes progressively from northeast-southwest to a little west of north. This change indicates that the southeastern part of the mapped area was progressively more downwarped during McAlester deposition than the area north and northwest.

Figure 5 is a composite isopach map of the Booch sandstone units of the McAlester Formation. In the northeastern part of the area (T13–15N, R14–16E) the upper Booch sandstone is the principal unit present and is a series of delta distributary channel fills. More darkly shaded areas in the north-central part of the mapped area represent vertical merging of all four Booch sandstones. In the region west of R12E, where the Booch is less than 60 ft thick, the middle Booch is the principal sandstone present, all other members being shaly to calcareous siltstone or poorly developed sandstone. The latter (all units except the middle Booch) can be correlated only as mappable "zones" rather than as lithologic units. The lower middle and lower Booch sandstones are not present except where merged vertically with the middle Booch in the channel areas.

The most prominent development of sandstone in Figure 5 is in a thick channel, ranging from 3 to 5 mi in width, and trending southeast from T15N, R11E, to T11N, R14E. The mapped length of this channel sandstone is over 35 mi, and the maximum thickness exceeds 240 ft. It undoubtedly extends much farther southeast, but the data for mapping it in this direction were not available. The thickest parts of this channel sandstone comprise over 50 percent of the total thickness of the McAlester; the rest of the section is predominantly shale. The shale of the McAlester Formation has compacted sufficiently from the weight of the overburden, that

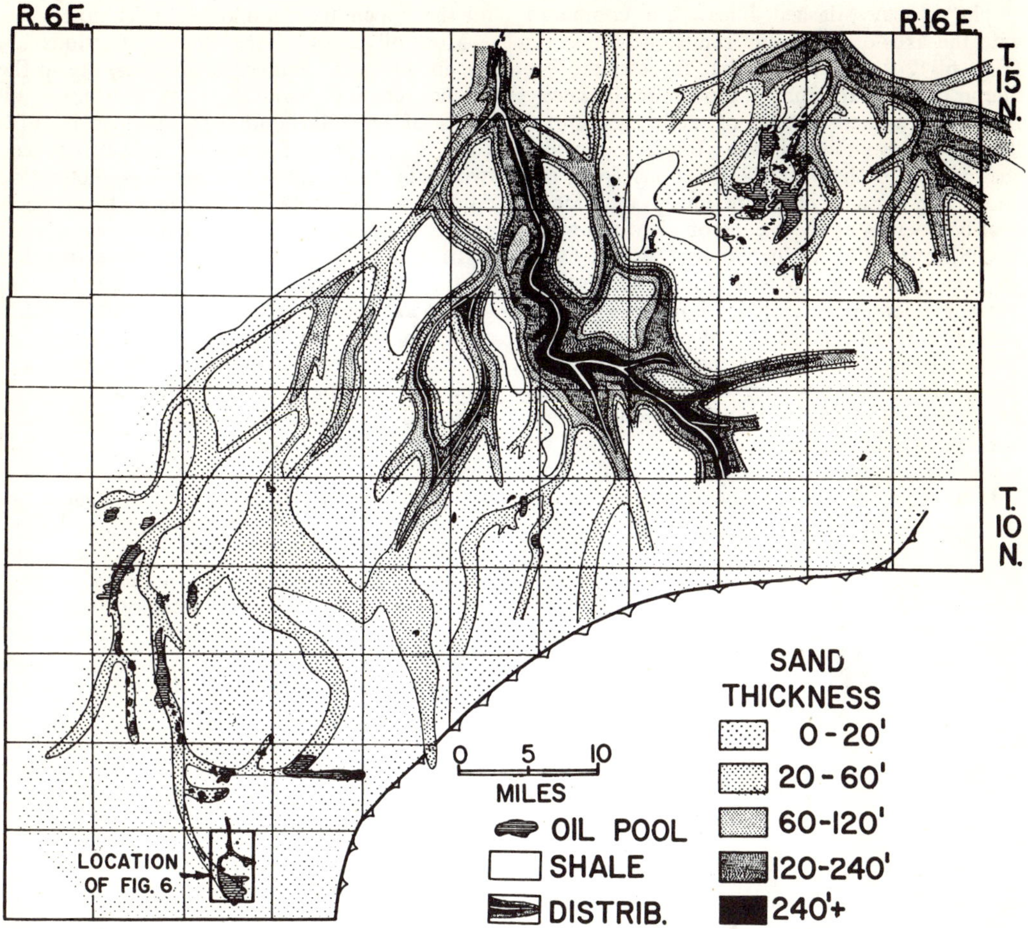


FIG. 5.—Isopach map of Booch sandstone, Greater Seminole district, eastern Oklahoma.

the location and trend of this major sandstone can be determined readily from the isopach map of the McAlester Formation (Fig. 4). As sandstone is relatively noncompactible, the northwest convexity of the 300-, 400-, and 500-ft contours is due primarily to the presence of the prominent distributary channel sandstone. This major channel sandstone and the numerous smaller channels of sandstone make up a distributary network of a large delta which had its apex on the north. It is not likely that all the distributaries shown in Figure 5 were formed simultaneously, for this illustration represents a composite of all stages of distributary development during McAlester deposition. In fact, some of the smaller distributaries close to the largest channel sandstone probably represent sand accumulation of the crevasse splay type.

The isopach map of the Booch delta, shown in Figure 5, is based on several thousand electric and drillers' logs. Such a map serves as little more than a geologic framework from which more detailed studies can be made. For example, Figure 6 is a more detailed analysis of the Booch sandstone in T6N, R8E. Production is from the middle member. The Booch here is overlain by a coal bed which consistently has a characteristic "kick" on the long normal resistivity curve of the electric logs. The structure contours of Figure 6A describe a sinuous, arcuate high over the pool proper (on the south), and a narrow, bifurcating structural ridge on the north. The maximum relief over the pool (south half) is approximately 75 ft. In the northern part of this figure the relief exceeds 120 ft. When the trends of the structural "highs" (heavy dashed lines) are compared with the axes of maximum sandstone thickness (Fig. 6B), it is apparent that the structure is due in part to variations in sandstone thickness. From logs of wells outside of, and marginal to, the pool, it is known that the Booch is lenslike and grades to shale in all directions from the pool. The base of the previously mentioned coal bed serves as the means of identifying the Booch outside the productive area. Figure 6C is an isopotential map of the Booch in which a contour interval of 25 bbl/24 hours was employed. The wells were drilled prior to the advent of hydro-frac, making the estimated daily potentials of the wells more meaningful. Trends of maximum initial potential (heavy dashed lines) coincide extremely well with trends of maximum sandstone thickness (Fig. 6B). This is to be expected inasmuch as the initial potential is partly a function of the relative permeability and amount of sandstone at any one bore-hole position. The deltaic aspect of this sandstone is much more apparent from Figure 6B and C than from Figure 6A. Postdepositional tectonism locally has modified the area; the most pronounced effect being a local structural depression in contiguous parts of Secs. 16 and 21 (Fig. 6A). Isopach and isopotential maps assist in carrying out a systematic step-out drilling development program. From the isopachs, drillsite locations most likely to extend axial trends are selected.

There are significant petrographic differences between the distributary channel sandstones and the interdistributary sheet sandstones, especially in the southwestern half of the delta shown in Figure 5. Where channel sandstones exceed 20 ft in thickness, they generally have larger geometric mean grain diameters than those of the interchannel areas, the averages being 99 and 86 $\mu$, respectively. The channel sandstones also are better sorted than those of the interchannel. In addition to these differences, the sandstones of the interchannel areas have an average clay content of 20 percent, compared with 15 percent in the channels. The higher percentage of clay and smaller grain size of the sandstone in the interchannel areas might be responsible for the relative absence of oil in these areas.

The Booch sandstones are very fine grained, and of the graywacke type; that is, they contain an abundance of fragments of preexisting rocks, and a relatively large clay matrix. The grain-size distribution and standard deviation are illustrated in Figure 7. Petrographic data for the Booch are summarized in Table 1.

Most oil pools of the Booch sandstone are not in the thick channel sandstones, but in the thinner channel sandstones. A cluster of oil pools of irregular outline is present in T13, 14N, R14E (Fig. 5), about midway between two main distributary systems. If sufficient log data were available in this area, these pools probably could be shown to occur in thinner distributary channel sandstones close to muds of the backswamp environment. Several elongate pools are in channel distributary sandstones in the southwestern part of Figure 5. All of these bear little relation to structure and probably are the result of permeability seals surrounding more porous and permeable parts of channel sandstones. Several significant oil pools (not shown on Fig. 5) occupy very selective parts of the thicker channel sandstones (middle of Figure 5). In this area the channel sandstones are medium to coarse textured and contain less interstitial clay. Such pools occur where the axes of west-plunging structural noses intersect depositional axes of major distributary channel sandstones. In this situation oil-water contacts are readily determined on the north and south sides of the pools, and abrupt loss of permeability occurs on the east and west margins where the channel sandstones merge with the silty, very fine-grained interdistributary sheet sandstones.

### Tonkawa Delta

The uppermost member of the Tonkawa (upper Missourian Series of the Pennsylvanian) is an excellent example of a deltaic gas-bearing sandstone reservoir. It is present in Beaver County in the Oklahoma Panhandle part of the Anadarko basin. This delta is described and illustrated in a recent paper by Khaiwka (1968), in which he points out that the upper member of the Tonkawa sandstone is overlain by the thin (3-ft), persistent "Haskell" limestone. Jordan (1957) noted this thin stratigraphic unit as a marker bed; it is a convenient datum of reference in a stratigraphic analysis of the directly underlying upper member of the Tonkawa. Figure 8 is an electric and microlog illustration of the three sandstone members of the Tonkawa showing their relation to the "Haskell" limestone. The downward thickening of the upper member of the Tonkawa sandstone, relative to the "Haskell" limestone reference datum, is illustrated in the lower half of Figure 8. This single stratigraphic profile is typical of many which have

A.

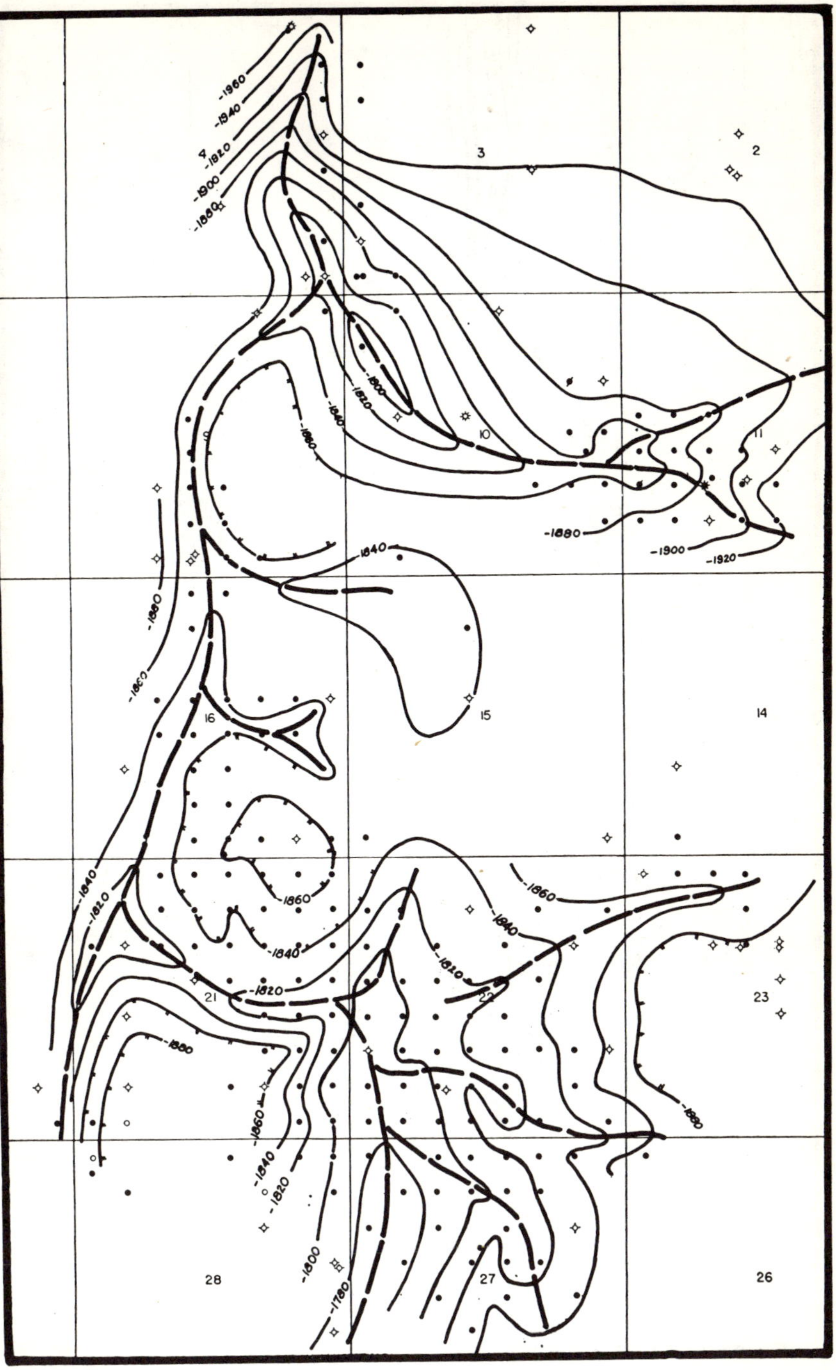


FIG. 6A.—Hawkins pool, central part of T6N, R8E, Hughes County, Oklahoma. Structure, top of Booch sandstone; C.I. = 20 ft. Figure 6B is isopach map of Booch sandstone; C.I. = 5 ft. Figure 6C is isopotential map of Booch sandstone; C.I. = 25 bbl/24 hours.

B.

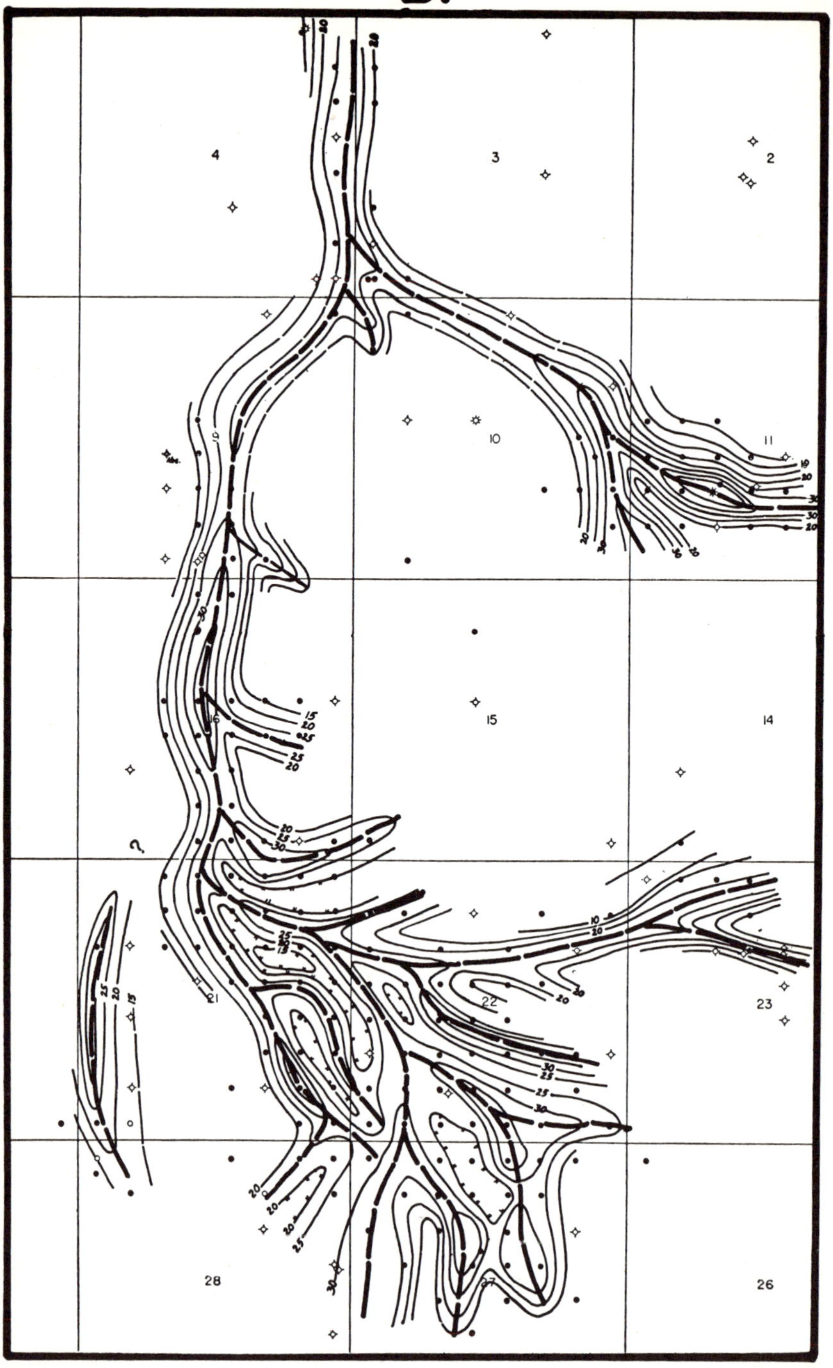


Fig. 6B.—Hawkins pool, Oklahoma. Isopach map of Booch sandstone, C.I. = 5 ft.

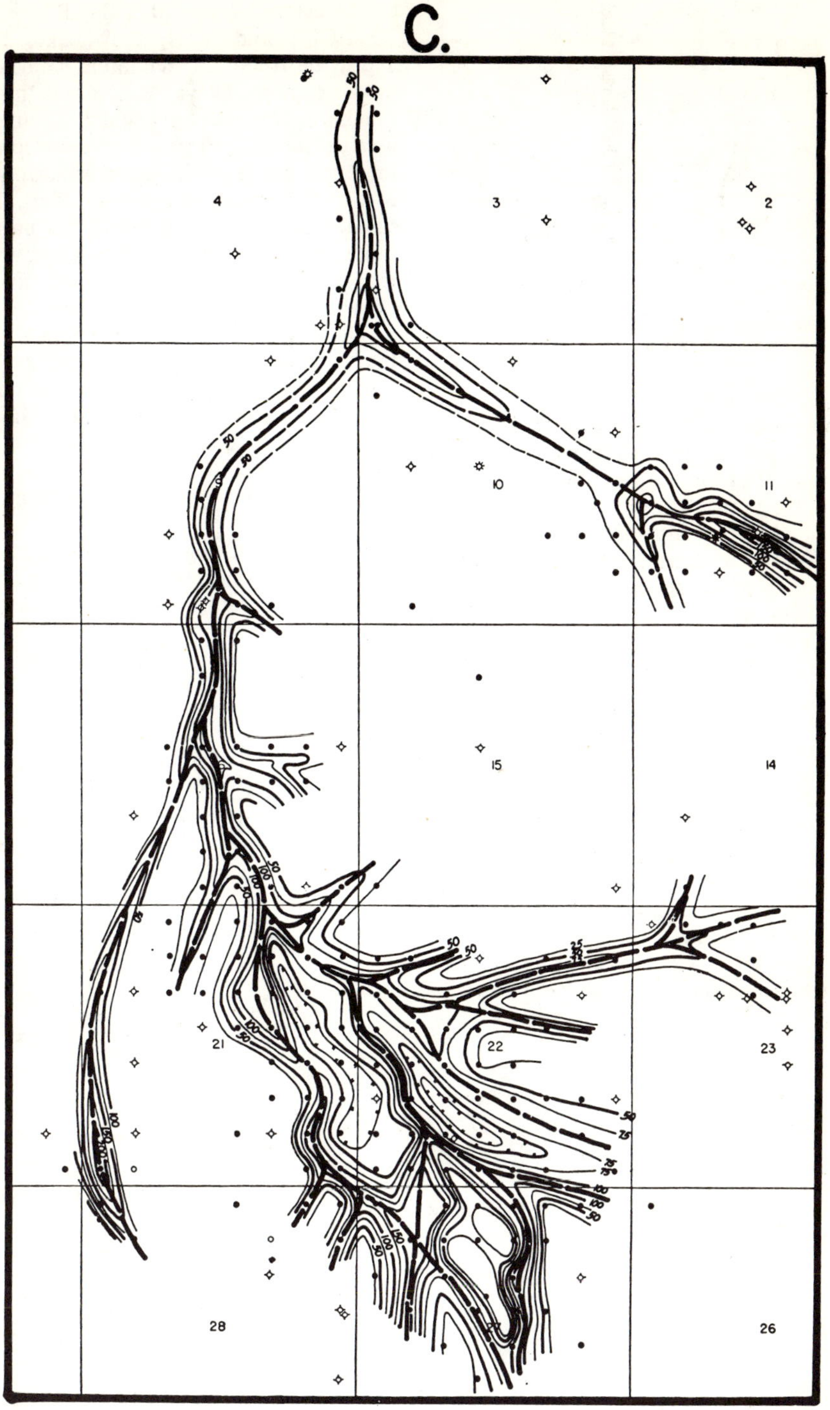


FIG. 6C.—Hawkins pool, Oklahoma. Isopotential map of Booch sandstone. C.I. = 25 bbl/24 hours.

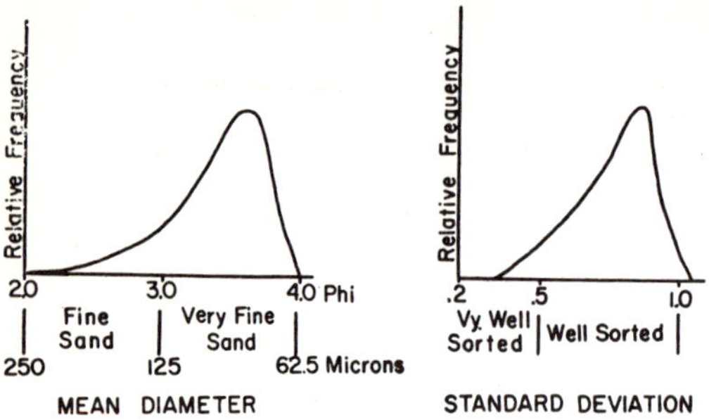


FIG. 7.—Mean diameter and standard deviation of Booch sandstone.

been drawn in the area and serves to illustrate the channel aspect of this sandstone body. Two isopach maps illustrate the distributary channel network of the upper member of the Tonkawa Sandstone. Khaiwka showed the distribution pattern of this sandstone by first constructing an isopach map of the stratigraphic interval between the "Haskell" limestone and the base of the upper member of the Tonkawa sandstone (Fig. 9). This interval is actually a genetic increment of strata. He pointed out that, "This map is essentially a submarine, paleotopographic map showing a deltaic distributary system of channels . . . it is thickest where downcutting was at a maximum; the respective valley axes are drawn along these trends of maximum down-cutting (greatest thickness)." The maximum density of well control is one well per section and, as a result, only a generalized interpretation is possible.

With Figure 9 as a guide, an interpretation of the thickness variations of the upper member of the Tonkawa sandstone was made (Fig. 10). In discussing the latter map, Khaiwka pointed out that, "Locally deltaic distributaries migrated laterally, such as in the 'four-corner area' of T's 3 and 4N, R's 24 and 25 ECM;

**Table 1. Petrographic Summary of Booch Sandstones**

| | *Range* | *Average (μ)* | *No. of Samples* |
|---|---|---|---|
| Mean grain diameter | 71–91μ | 83 | 28 |
| Stand. deviation | 0.34–1.01φ | 0.76 | 28 |
| Clastic quartz | 45–90% | 54 | 36 |
| Carbonate | 0–33% | 4 | 36 |
| Silt-size clay material | 2–40% | 16 | 36 |
| Rock fragments | 9–50% | 22 | 36 |
| Secondary silica | 9–50% | 22 | 36 |
| Other minerals | 0–3 % | 2 | 36 |

southwestern corner of T4N, R26ECM; T3N, R27ECM; and T2N, R's24 and 25 ECM." Thus, the trends of maximum sandstone thickness do not always coincide with trends of maximum downcutting which Khaiwka attributed to lateral migration of distributaries. The thickness range of the upper sandstone member of the Tonkawa is from 0 to 35 ft. The upper surface is relatively flat and the basal surface quite irregular. Interdistributary areas commonly are blanketed with thinner sandstone, all of which has reservoir capabilities.

The gas distribution in the upper member of the Tonkawa sandstone is controlled by a combination of sandstone distribution and structure, as illustrated in Figure 11. Gas migrated north-northwest until it reached the updip extent of the sandstone. A gas-water contact may be noted at the approximate position of −3,270 ft.

This type of regional stratigraphic analysis would be of considerable benefit to the exploration geologist during the developmental stages of this pool. Preferential locations with thickest sandstone development above the gas-water contact could be predicted once the depositional nature of the sandstone had been ascertained.

### Endicott Delta

The Endicott formation, in the lower Virgil Series of Pennsylvanian age in Harper County, Oklahoma, is an excellent example of a subsurface deltaic accumulation of sandstone. It was deposited in a series of bifurcating distributary

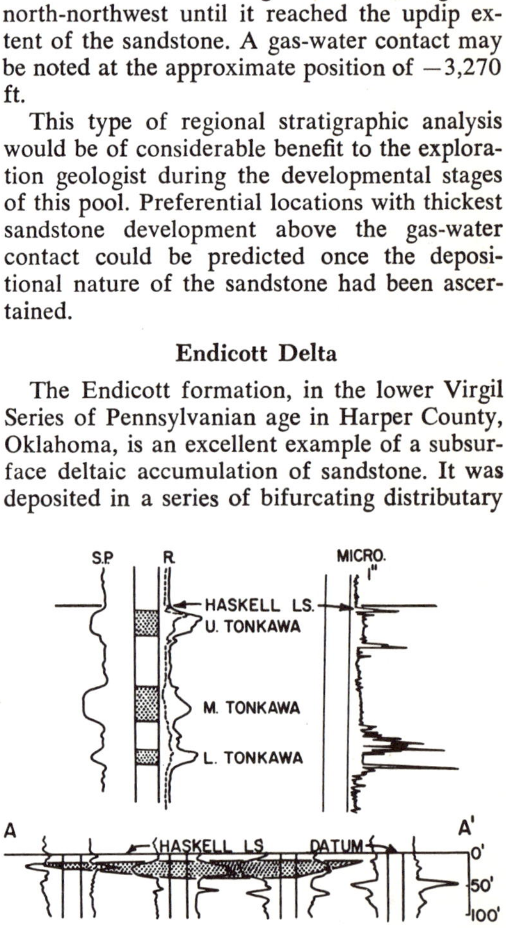


FIG. 8.—Stratigraphic cross section A-A' of Tonkawa sandstone, showing downward thickening at expense of underlying shale. Reference datum is top of "Haskell" limestone. See Figure 9 for location of cross section (modified after Khaiwka, 1968).

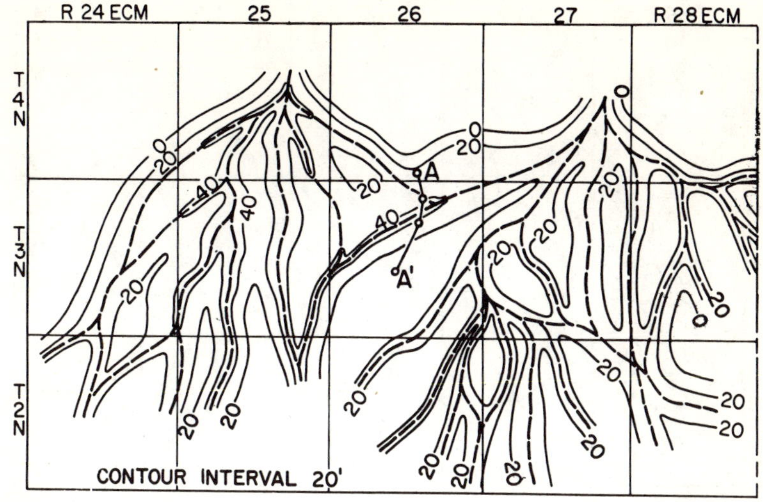


FIG. 9.—Isopach map of GIS between top of "Haskell" limestone and base of upper Tonkawa sandstone, Beaver County, Oklahoma. This map shows two laterally coalescing distributary systems (modified after Khaiwka, 1968).

channels eroded into the north flank of the Anadarko basin. The Endicott sandstone overlies unconformably either the Toronto limestone or the unnamed shale directly underlying the Toronto, as illustrated in the two profiles of Figure 12. These cross sections clearly show that the principal direction of thickening of the Endicott is downward. As the Oread Limestone serves as reference datum for these profiles, the Lovell sandstone, "Haskell" limestone, and Tonkawa sandstone, respectively, are bowed downward directly under the wells exhibiting maximum Endicott channel fill. This is attributed primarily to differential compaction of the shale section separating the Toronto limestone from the Lovell sandstone. Differential compaction of this shale was concurrent with the deposition of the Heebner shale (above the Endicott sandstone) as indicated by a tendency for the Heebner to thin slightly over areas of thick Endicott sandstone and, conversely, to be thicker in areas of thinner Endicott. This is the

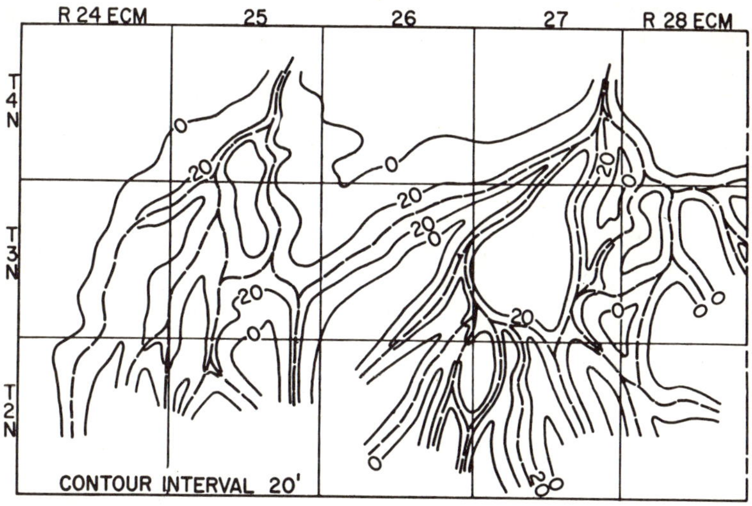


FIG. 10.—Isopach map of upper Tonkawa sandstone, Beaver County, Oklahoma (modified after Khaiwka, 1968).

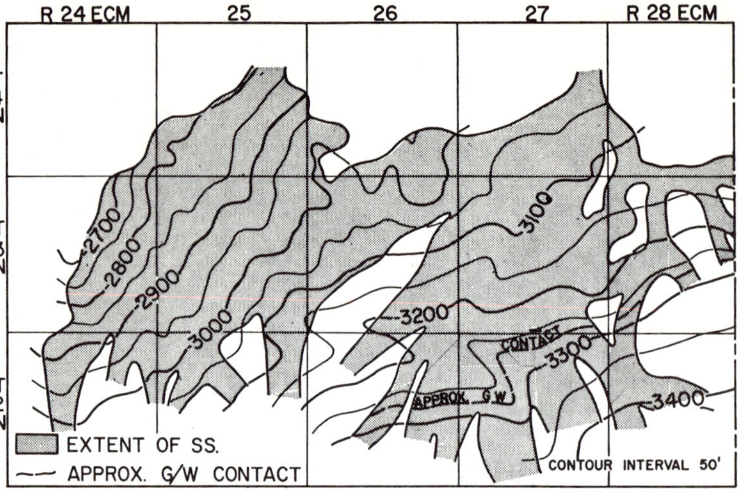


FIG. 11.—Structure map of top of upper Tonkawa sandstone, Beaver County, Oklahoma (modified after Khaiwka, 1968).

result of compensatory deposition in areas of slightly deeper water. The single apparent exception to this principle, in the easternmost well of profile A-A' of Figure 12, is explained by the fact that a remnant of the Toronto limestone is preserved below the Endicott in this well and in all probability would not compact any more than the Endicott itself.

The Oread Limestone is an excellent marker bed in this area and serves as the upper limit of a GIS that extends down to the unconformity at the base of the Endicott channel fill. An isopach map of this GIS (Fig. 13) presents a "cast" of the erosion surface that immediately preceded deposition of the Endicott sandstone. It was this irregular surface that largely controlled the positions of linear trends of thick Endicott sandstone. The axial trends of maximum downcutting in pre-Endicott time are indicated by bifurcating dashed lines. The composite picture is that of part of a deltaic complex consisting of two laterally merging distributary systems which fan out southward.

Khaiwka (1968) pointed out that "Capps (1959), and later Winter (1963), depicted the Endicott as a blanket-type sandstone." This is true if one considers only the generally smooth, rather extensive upper surface of the Endicott. The thickness of this formation ranges, however, from zero to more than 180 ft, and a thickness change of as much as 100 ft in a distance of 1 mi is not unusual. Khaiwka prepared (not published, but inspected by the writer) an isopach map of the Endicott which is almost identical with Figure 13 insofar as trends of maximum thickness are concerned. He stated

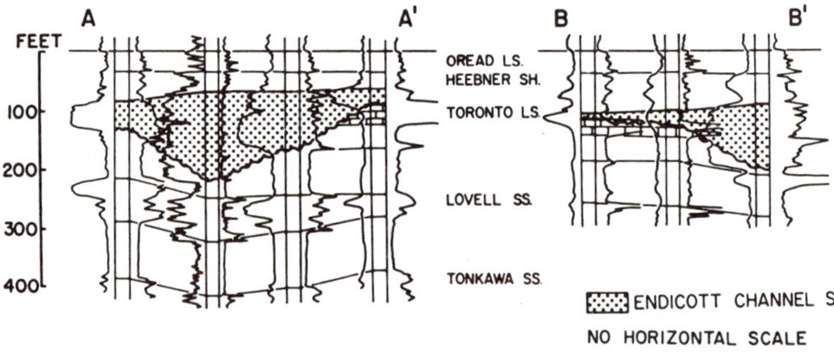


FIG. 12.—Stratigraphic profiles showing channel nature of Endicott sandstone, northwestern Oklahoma. Note that Toronto limestone was eroded away in positions of maximum sandstone thickness. See Figure 13 for locations of profiles (modified after Khaiwka, 1968).

"This similarity indicates that there was little or no migration of the deltaic distributaries. The thalwegs of the individual distributaries, with their bifurcating pattern, almost coincide on the two isopachous maps." He believes that, "The Endicott delta probably was deposited adjacent to a coastal plain of relatively high relief because sand bodies form major parts of its framework."

Limited sample and thin section studies indicate that the Endicott is a gray, fine-grained, poorly sorted, calcareous micaceous sandstone. Khaiwka (1968) has noted 85–90 percent quartz. Locally, the muscovite and chlorite compose up to 10 percent of the bulk volume. In other areas as much as 10 percent chert and 5 percent feldspar (mostly plagioclase) have been estimated from thin-section studies. The porosity ranges from fair to good, the maximum value observed being 18 percent. The sandstone is laminated with light- to dark-gray carbonaceous shale and is overlain by dark-gray and black pyritic shale with coal streaks at the base.

The structure of the Endicott is that of a south-dipping homocline devoid of any significant structural noses or closures. Although the Endicott within the area of Figure 13 has good reservoir characteristics, it apparently is devoid of the necessary trapping mechanism, such as updip permeability seal or structural closure. The only known Endicott production occurs from several wells in the Oklahoma and Texas Panhandle parts of the Anadarko basin.

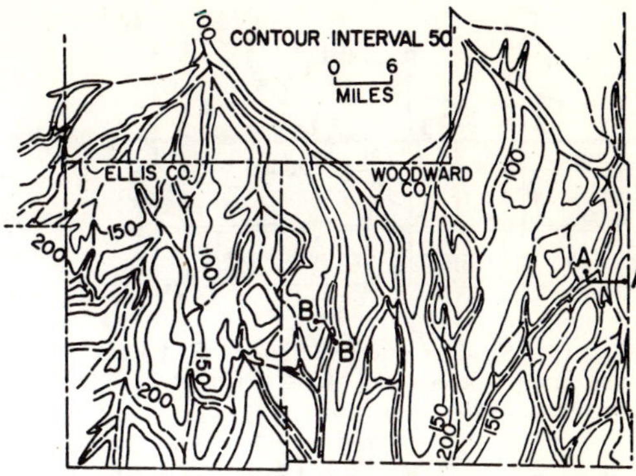


FIG. 13.—Isopach map of GIS between top of Oread Limestone and base of Endicott sandstone, northwestern Oklahoma. This is "cast" of eroded surface upon which Endicott sandstone was deposited in area of two laterally coalescing deltaic distributary systems (modified after Khaiwka, 1968).

## Red Fork Delta

The nature of the Red Fork Sandstone of Noble County, Oklahoma, has been debated by geologists for many years. It has been variously interpreted as offshore bar, isolated channel sandstone, and delta. Figure 14 is a structure map of the base of the Oswego limestone, which is a readily mappable unit about 175 ft above the Red Fork. The general structure is that of an irregular monoclinal slope toward the west. A south-southwest-plunging syncline interrupts this slope in the west and is related to the conspicuous Tonkawa fault in the north-

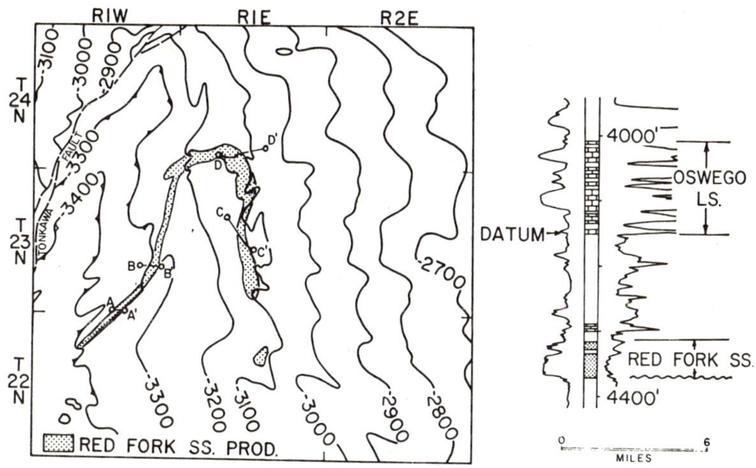


FIG. 14.—Structure of base of Oswego limestone, Noble County, Oklahoma, showing location of four stratigraphic profiles (modified after Scott, 1970).

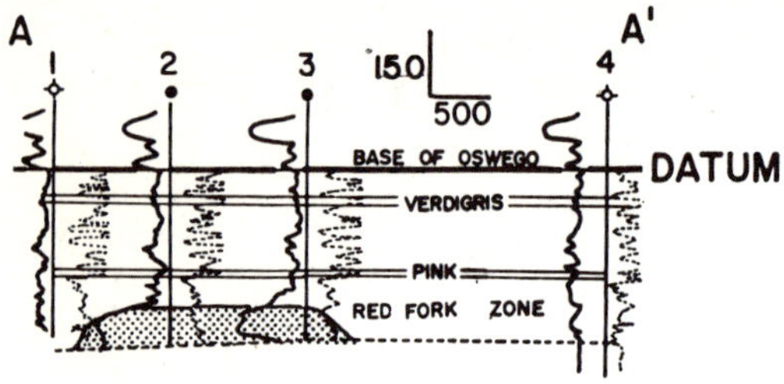


FIG. 15.—Sandbar interpretation of Red Fork Sandstone, South Ceres pool, Noble County, Oklahoma (after Neal, 1951).

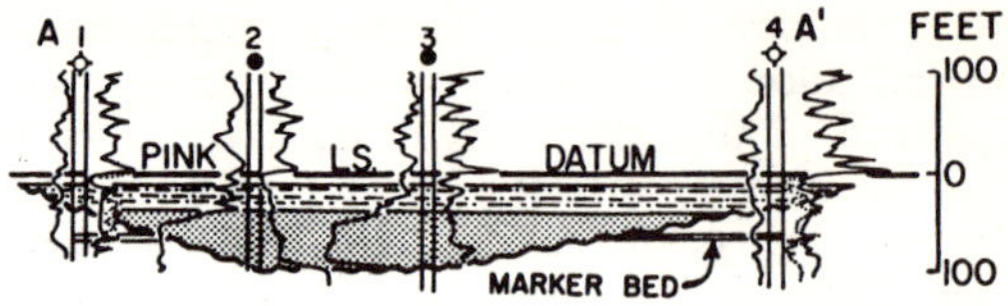


FIG. 16.—Channel interpretation of Red Fork Sandstone, South Ceres pool, Noble County, Oklahoma. See Figure 14 for location of cross section (modified after Scott, 1970).

west corner of the mapped area. The Red Fork Sandstone production describes a horseshoe-shaped area, indicated with stippling on Figure 14. The western and northern parts of this horseshoe-shaped area produce oil, whereas the eastern limb is primarily gas-productive. All production is from a very lenticular Red Fork sandstone. The influence of differential compaction (on structural configuration) of the shales laterally adjacent to the Red Fork is apparent along the west limb of the horseshoe. It is here that the −3,300-ft structural contour line of the base of the Oswego exhibits a pronounced southwestward nosing where it crosses the pool.

The locations of several stratigraphic cross sections are designated A-A′, B-B′, C-C′, and D-D′ on Figure 14. Figure 15 is a published stratigraphic profile "hung" on the base of the Oswego limestone (Neal, 1949, 1951). By this interpretation the Red Fork Sandstone increases in thickness at the expense of the shale which directly overlies it and, on this basis, has been interpreted as an offshore bar. Figure 16 (Scott, 1970) is a different interpretation of the same four electric logs used in Figure 15. The marker bed, near the base of the Red Fork Sandstone, is the key to a correct interpretation of this lenticular sandstone. It is present in wells No. 1 and 4, and eroded away in wells No. 2 and 3 where the Red Fork has its maximum thickness. By this interpretation the Red Fork is clearly a channel-type of reservoir sandstone.

Figure 17 is stratigraphic cross section B-B′ (see Fig. 14 for location) and shows an amount of downcutting similar to that of Figure 16. More than half the channel fill is silty shale. The marker bed is clearly evident in all but the No. 3 well, where it has been eroded away.

A much thicker development of channel sandstone is present along stratigraphic profile C-C′ (Fig. 18), where the maximum thickness is about 50 ft. The effects of differential compaction can be observed in the interval between the marker bed and the Pink limestone reference datum. This stratigraphic interval is thinnest in wells No. 1 and 5, where it consists primarily of compacted shale. In the other three wells noncompactible sandstone predominates and the interval is consequently thicker.

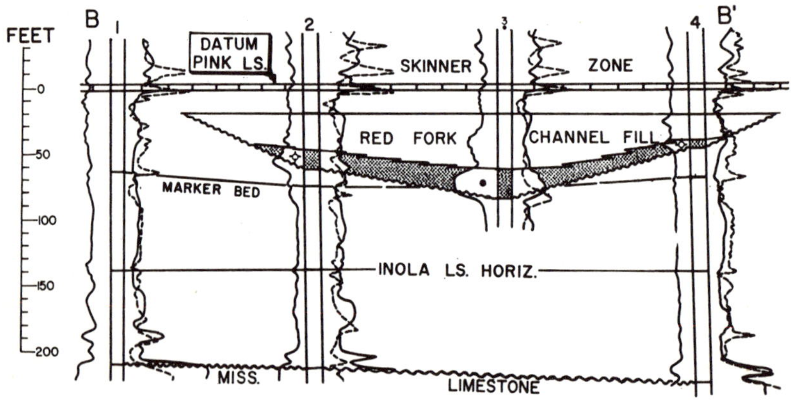


FIG. 17.—Stratigraphic profile B-B′, showing Red Fork channel-fill in which over half of fill is silty shale. See Figure 14 for location of profile (modified after Scott, 1970).

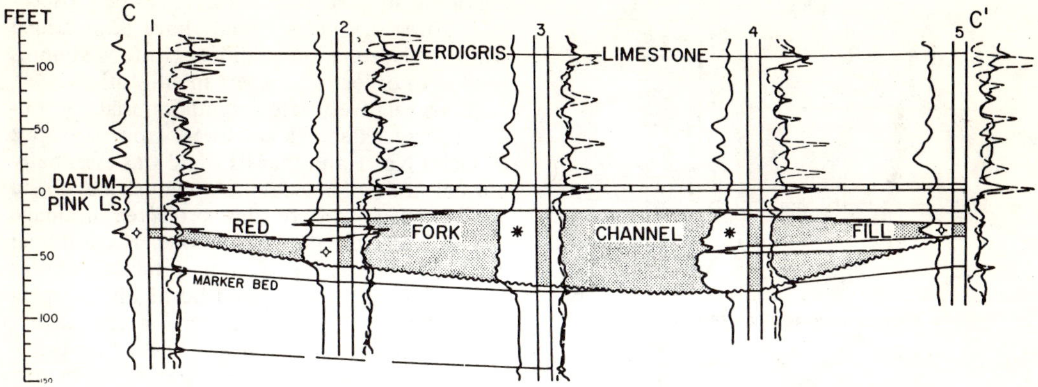


FIG. 18.—Stratigraphic profile C-C′, showing Red Fork channel-fill, predominantly sandstone. See Figure 14 for location of profile (after Scott, 1970).

Figure 19 is a stratigraphic cross section of the Red Fork channel where the valley has a maximum width. As most of the fill consists of sandstone, the effects of differential compaction are minimal. Numerous additional cross sections involving the Pink limestone marker bed interval have been constructed; all show the channel aspect of the Red Fork Sandstone, with thickening at the expense of the underlying shale.

In his study Scott (1970) considered the interval from the Pink limestone to the base of the Red Fork Sandstone as constituting a GIS and, as such, reconstructed a paleodistributary system of a delta, as shown in Figure 20. To date, oil and gas production is restricted to the two principal distributaries of this delta, and it appears that all the other distributaries are primarily crevasse splays in which the sandstones generally are thinner and more argillaceous.

### Hypothetical Model Study

Deltas, although quite varied in size, are seldom isolated phenomena along modern shorelines. Marginal marine conditions favorable to delta formation at the mouths of rivers generally extend for many miles along the shorelines. Most positive land areas, particularly in humid to semi-humid environments, are drained by a series of streams flowing to bordering embayments. Although present drainage into the Gulf of Mexico is dominated by the Mississippi and Rio Grande Rivers, there are, nevertheless, numerous smaller drainage courses emptying into the Gulf along the Texas-Louisiana coastline.

For a delta to form, the shoreline of an embayment must remain relatively stationary for a period of time. Thus, if a relatively stable shoreline condition favors development of a delta at the mouth of one river, it also should favor delta development at the mouths of most, if not all, rivers flowing into the same embayment.

If the shoreline is transgressing in response to cyclic subsidence of the basin area (or eustatic rises in sea level), conditions are ideal for the formation of rows of deltas at the successive stillstand positions of the shoreline. Like-

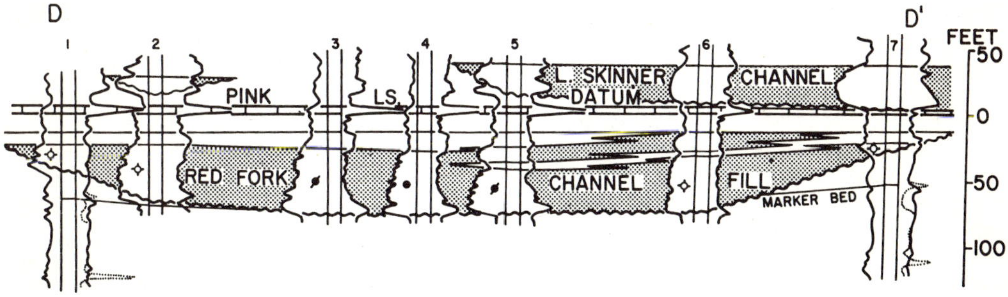


FIG. 19.—Stratigraphic profile D-D′, showing Red Fork channel-fill where channel has maximum width. See Figure 14 for location of profile (after Scott, 1970).

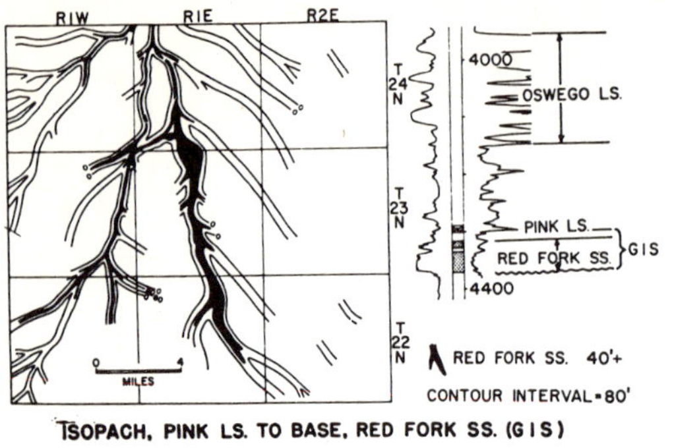


FIG. 20.—Isopach map, Pink limestone to base of Red Fork Sandstone (GIS).

wise, conditions for delta preservation are ideal when the shoreline transgresses in a cyclic manner. This can be illustrated best by reference to a hypothetical diagram (Fig. 21), which is Stage 1 of a series of three illustrations that portray a method of prospecting for ancient deltas. Figure 21 is an isopach map of a GSS that includes the oil-productive sandstone of Delta No. 1. This GSS was picked after constructing a series of stratigraphic profiles which collectively form a correlation grid of the area. The productive sandstone(s) of Delta No. 1 occurs either at, or very near, the GSS base, which is defined by an unconformity. The isopach map of this GSS shows a fairly uniform rate of southeastward divergence. Marine transgression was from southeast to northwest, and the paleodepositional strike of the shoreline was northeast.

A detailed stratigraphic analysis of Delta No. 1 reveals that the oil-productive sandstones consist of a series of southeastward diverging channel fills. All of the channel sandstones may be either at the same stratigraphic position relative to the marker bed, at the top of the GSS, or at several closely spaced intervening levels relative to the marker bed. This system of distributary channels is identified as either a single or a composite ancient deltaic complex. Its apex occurs approximately at the position of the 475-ft isopach line (dotted), and its southeastern margin occurs at about the 510-ft isopach line. Thus, the reservoir sandstones of Delta No. 1 were deposited in a strand zone shown by stippling. This strand zone is, of necessity, subparallel with the nearest contour lines of the isopach map of the GSS. As several deltas are likely to have been deposited simultaneously at the mouths of rivers which emptied into the embayment along this strand zone, several areas along the stippled zone are likely to be prospective. Three dry holes already have been drilled within this zone, so additional information is needed in order to prospect intelligently.

It is a relatively safe assumption that a drainage course extended northwest from the apex of Delta No. 1 and that its trend was essentially normal to the paleodepositional strike. Delta No. 1 occurs where this projected drainage course intersects the strand zone. If additional drainage courses could be ascertained, similar areas of intersection would be highly prospective. A review of all existing log and testing data reveals stratigraphically higher sandstone development (with an oil show) approximately 30 mi north, in the northwest part of Figure 21. Well samples show oil stain and a drill stem test recovered oil in noncommercial quantity. This well appears to be on the edge of a pool, but which edge is not known. Four possible pool outlines are indicated with dotted circles.

At this early stage of regional analysis it is necessary to drill, or cause to be drilled, a sufficient number of step-out wells from this dry hole to find the pool and cause it to be developed.

Because Delta No. 1 produces from a distributary system deposited during one of the stillstands of a shoreline under conditions of cyclic subsidence, the sandstone of Delta No. 2 (Fig. 22) was probably deposited under similar conditions—at a later stage of cyclic subsidence. In Figure 22 the position, outline, and nature of the reservoir sandstone(s) of Delta No. 2 have

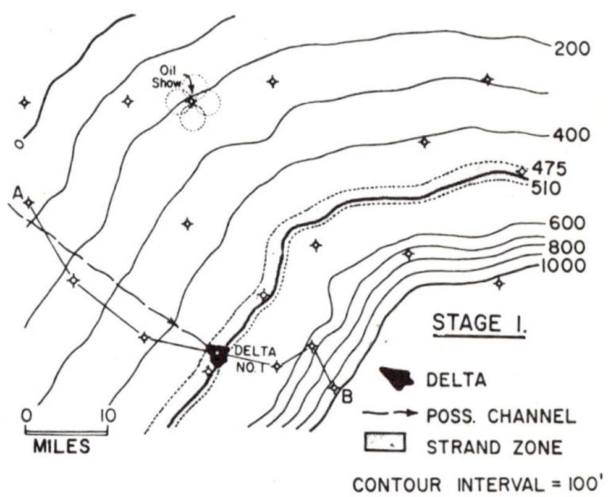


FIG. 21.—Stage 1. Hypothetical isopach map of GSS deposited under conditions of cyclic subsidence. Oil-bearing sandstone reservoir is shown where ancient drainage course discharged sediment just basinward from stillstand position of shoreline.

been determined. The drainage course of this delta is projected northwest, normal to the paleodepositional strike. Marine transgression has little or no effect on the position of a drainage course, hence it is reasonable to project the drainage through Delta No. 2 in a southeast direction. This projected drainage course intersects the strand zone on the southeast within the area labeled "Delta Prospect No. 1."

At this stage of regional analysis, Delta Prospect No. 2 assumes the same significance as Delta Prospect No. 1. It, too, lies within the area of intersection of a projected drainage course and the established second strand zone.

It is reasonable to assume that, by the time Delta Prospect Nos. 1 and 2 have been discovered and developed, numerous additional dry holes (and possibly discovery wells) will have been drilled within the area. The new data afford the means of much more detailed stratigraphic analysis and prospecting. With additional data many more pools (of various types) may be anticipated. For example, if the unconformity at the base of the GSS was fairly smooth, additional drainage channels (with deltas) should be anticipated on the east and southwest, with spacing similar to that between the two postulated channels. Additional strand zones southeast of the first strand zone, and possibly between the two established strand zones, should be anticipated.

A detailed stratigraphic profile of the GSS between Delta No. 1 and Delta Prospect No. 2 (Profile A-B, Fig. 23) will serve as a basis for determining the thickness of the shale interval

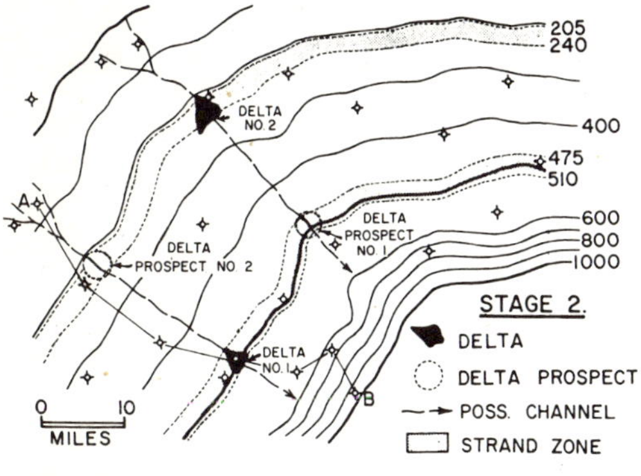


FIG. 22.—Stage 2. Hypothetical isopach map of strata deposited under conditions of cyclic subsidence. Delta No. 1 and Delta Prospect No. 1 occur where drainage channels intersect a strand zone, and Delta No. 2 and Delta Prospect No. 2 occur where headward projections of drainage channels are crossed by another strand zone.

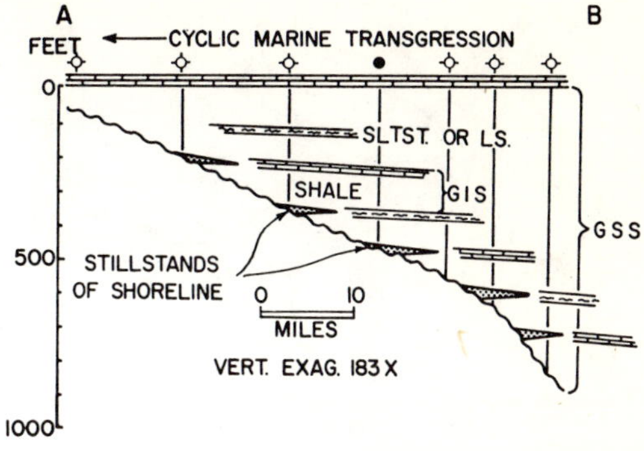


FIG. 23.—Stratigraphic profile A-B of hypothetical GSS showing sandstone development at succession of stillstand positions of shoreline.

between the pay sandstones of the two established strand zones. If this shale interval is 50 ft thick, or more, it is reasonable to postulate at least one more strand zone between them. Nonproductive sandstone developments in the dry holes between these two strand zones are tipoffs of other strand zones.

With additional well data one can determine whether the unconformity surface was fairly smooth, as well as its rate of basinward tilt at the time of cyclic marine transgression. In Figure 22 the strand zones are approximately 20 mi apart, and the amount of GSS divergence is 270 ft. A 13.5 ft/mi divergence strongly suggests the presence of one or more strand zones between the two established zones. A cyclic 25–50-ft change in sea level (or subsidence) is much more likely than 270 ft.

With the acquisition of additional drilling data, it is possible to define clearly the positions of the two drainage channels and any others present within the study area. This is accomplished by reducing the contour interval of the GSS to 50 ft or possibly 20 ft. As this GSS is, in a sense, a "cast" of the unconformity surface, the contour lines will bend upstream as illustrated in Figure 24. Selective parts of these channels may be clogged with sandstone and afford ideal conditions for oil and gas reservoirs of the simple channel type or, possibly, point bar type.

From Figure 24 it may be noted that with additional well control more strand zones have been delineated and additional deltaic reservoirs discovered and developed. The newly discovered strand zones on the southeast overlap because of the abrupt increase in the rate of divergence of the GSS on the seaward side of the hinge line. In such areas of steeper gradient

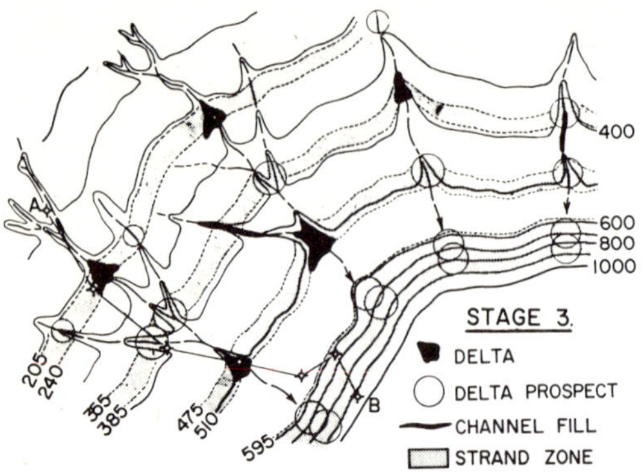


FIG. 24.—Stage 3. Hypothetical paleotopographic map, showing deltaic reservoirs and prospects in areas of intersection of strand zones and paleodrainage zones. Individual oil pools in channel sandstone also shown.

there is less lateral shift of shoreline in response to cyclic subsidence. Consequently, a part of one delta actually might overlap the apical part of an earlier delta development. In such areas the individual strand zones are likely to be narrower than on the shelf, but the sandstones, conversely, are likely to be thicker.

Where longshore currents were operative in redistributing sediments furnished at river mouths, beach sandstones may have been deposited. Such sandstones commonly are gas-bearing, for reasons not currently understood. Examples of this are the Clinton (Lower Silurian) of east-central Ohio, Atoka (Pennsylvanian) sandstones of east-central Oklahoma, Morrow (Pennsylvanian) sandstones of northwest Oklahoma and northern Texas, and the Mesa Verde (Cretaceous) sandstones of northwestern New Mexico.

The interpretation idealized in Figure 24 presents many additional prospects for exploration drilling. In construction of the figure, all postdepositional structure has been removed. With data theoretically available for the construction of Figure 24, however, it would be a simple matter to construct a structural map, preferably with the lithologic-time marker bed at the top of the GSS as datum. Thus, the modifying effect of possible structures on accumulation of hydrocarbons could be ascertained readily. In this hypothetical example the prime emphasis is one of predicting areas for favorable reservoir sandstone development, whether of deltaic distributaries, subaerial channel sandstones, or beach deposits. At this stage the exploration geologist is ready to determine the possible modifying effects of postdepositional structure on oil and gas accumulation.

## References Cited

Barrell, J., 1912, Criteria for the recognition of ancient delta deposits: Geol. Soc. America Bull., v. 23, p. 377–446.

Busch, D. A., 1953, The significance of deltas in subsurface exploration: Tulsa Geol. Soc. Digest, v. 21, p. 71–80.

——— 1959, Prospecting for stratigraphic traps: Am. Assoc. Petroleum Geologists Bull., v. 43, no. 12, p. 2829–2843.

Capps, W. H., 1959, Stratigraphic analysis of the Missourian and lower Virgilian series in northwestern Oklahoma: M.S. thesis, Oklahoma Univ.

Gilbert, G. K., 1885, The topographic features of lakeshores: U.S. Geol. Survey 5th Ann. Rept., p. 104–108.

Halbouty, M. T., and T. D. Barber, 1961, Port Acres and Port Arthur fields, Jefferson County, Texas: Gulf Coast Assoc. Geol. Socs. Trans., v. 11, p. 225–234.

Jordan, L., 1957, Subsurface stratigraphic names of Oklahoma: Oklahoma Geol. Survey Guidebook 6, 220 p.

Khaiwka, M. H., 1968, Geometry and depositional environments of Pennsylvanian reservoir sandstones, northwestern Oklahoma: Ph.D. dissert., Oklahoma Univ.

Nanz, R. N., Jr., 1954, Genesis of Oligocene sandstone reservoir, Seeligson field, Jim Wells and Kleberg Counties, Texas: Am. Assoc. Petroleum Geologists Bull., v. 38, no. 1, p. 96–117.

Neal, E. P., 1949, Southeast Ceres field, Noble County, Oklahoma: Tulsa Geol. Soc. Digest, v. 17, p. 97–99.

——— 1951, South Ceres, Oklahoma's oddest shoestring field: World Oil, v. 133, no. 7, p. 92, 94, 98.

Rainwater, E. H., 1963, The environmental control of oil and gas occurrence in terrigenous clastic rocks: Gulf Coast Assoc. Geol. Socs. Trans., v. 13, p. 79–94.

Scott, A. J., and W. L. Fisher, 1969, Delta systems and deltaic deposition, *in* Delta systems in the exploration for oil and gas, a research symposium: Texas Univ. Bur. Econ. Geology, p. 10–29.

Scott, J. D., 1970, Cherokee sandstones of a portion of Noble County, Oklahoma: M.S. thesis, Oklahoma Univ.

Swann, D. H., 1964, Late Mississippian rhythmic sediments of Mississippi Valley: Am. Assoc. Petroleum Geologists Bull., v. 48, no. 5, p. 637–658.

Weimer, R. J., 1961, Upper Cretaceous delta on tectonic foreland, northern Colorado and southern Wyoming (abs.): Am. Assoc. Petroleum Geologists Bull., v. 45, no. 3, p. 417.

Winter, J. A., 1963, Computer geology uncovers secrets of Oklahoma's Endicott sand: Oil and Gas Jour., v. 61, no. 29 (July 22), p. 96–101.